THE NEW OUTLINE
OF SCIENCE
****

THE NEW

OUTLINE OF SCIENCE

****

David Dietz

ILLUSTRATED WITH PHOTOGRAPHS AND DIAGRAMS

****

Dodd, Mead & Company
New York

GRATEFUL acknowledgment is made to the following for permission to reproduce the photographs included herein: Buffalo Museum of Science, pages 242, 417, 422, 424; California Institute of Technology, pages 413, 415; Case Western Research University, page 347; Cleveland Museum of Art, Dudley F. Allen Fund, page 351; Cleveland Museum of Natural History, page 423; Field Museum of Natural History (copyright), pages 240-41, 243, 244-45, 246, 246-247, 429, 431, 433, 436, 436-437, 438; Hale Observatories, pages 2, 9, 14, 17, 22, 50, 52, 61, 76, 77, 81, 83, 98, 113, 114, 117, 118, 119, 122, 126, 129, 133, 137, 139, 140, 144; Hawaii Department of Planning and Economic Development, pages 170, 172, 221 top; Hawaii Visitors Bureau, page 219; Manly Commercial Photographers, page 196; National Aeronautics and Space Administration, pages 21, 25, 28, 31, 32, 43; National Institute of Allergy and Infectious Diseases, page 409; National Park Service, pages 191, 192, 194, 195; Scripps Institution of Oceanography, pages 181, 182, 184; Smithsonian Institution, pages 15, 228, 229, 231, 233, 234, 236, 237, 238, 410, 427, 428; U. S. Geological Survey, pages 186, 189, 190, 198, 199, 200, 201, 202, 207, 208, 210, 211, 212, 213, 214, 216, 217, 218, 221 bottom, 223, 224, 325; University of California Lawrence Radiation Laboratory, pages 319, 326, 332, 339; Yerkes Observatory, page 45.

Copyright © 1972 by David Dietz
All rights reserved
No part of this book may be reproduced in any form without permission in writing from the publisher

ISBN: 0-396-06526-0
Library of Congress Catalog Card Number: 73-39220
Printed in the United States of America

*Dedicated to my wife, Dorothy
with grateful appreciation
for her encouragement and patience*

PREFACE

SCIENCE has made greater progress since the start of the twentieth century than in the preceding 49 centuries of written history. Scientists have probed further into the depths of the universe, the structure of the earth, the interior of the atom, and the riddle of life than ever before. Man has at last succeeded in leaving the earth and achieving the ancient dream of visiting the moon.

Science dominates modern life. The machine, product of science, sets the tempo. The findings of science influence philosophy and religion. It is impossible to understand the world of today without an understanding of science.

This book is intended to give the layman and the student a unified and organized view of modern science. With this end in view, I have divided it into four sections.

Part I is The Universe, telling what astronomers have discovered about the heavens. It takes the reader on a celestial tour, beginning with the moon, our nearest neighbor in space, and proceeding to the other members of the solar system, the stars and nebulae of our Milky Way galaxy, and finally the distant galaxies of the universe. It discusses present theories of the evolution of stars and galaxies, and the origin and nature of the expanding universe.

Part II is The Earth, the story of that tiny portion of the universe which has special interest for us because we live on it. The origin and evolution of the earth, the structure of its interior, the nature of continents and oceans, the forces that shape the crust of the earth, and the record in the rocks of the evolution of life through the geological eras are related. The newest theories of sea-floor spreading and continental drift are presented.

Part III is The Atom. It deals with the organization of matter into atoms and molecules, the structure of the atomic nucleus, and the nature of the subatomic particles which compose atoms or appear in nuclear phenomena. The nature of energy and its relation to matter, the implications of the quantum theory, and the Einstein theory of relativity are discussed.

Part IV is Life. Beginning with the chemical nature of life and the structure of the cell, it carries the story from microscopic forms of life to the climax in man. New theories of the origin of life and the new knowledge of the role of the nucleic acids, DNA and RNA, in the control of both cellular activity and heredity are discussed.

I have introduced historical data wherever they serve to portray the origin and development of modern ideas, recounting many of the great milestones in the history of science.

I am indebted to a number of good friends on the faculty of Case Western Reserve University for reading various portions of the book in manuscript and offering many valuable suggestions: Sidney W. McCuskey, Warner professor of astronomy; Francis G. Stehli, professor of geology; Robert G. Douglas, associate professor of geology; Robert S. Shankland, Ambrose Swasey professor of physics; Frank Hovorka, professor emeritus of chemistry; D. Fennell Evans, associate professor of chemistry; Robert P. Davis, associate professor of biology; and Wilford H. Wolpoff, assistant professor of anthropology. I am also indebted to Arden L. Albee, professor of geology at the California Institute of Technology, for reading part of the manuscript and offering valuable suggestions.

As science editor of the Scripps-Howard Newspapers I have attended numerous scientific meetings and spent much time at universities and scientific institutions, both here and abroad, since 1921. It was my good fortune to meet many of the great pioneers who forged the foundations of twentieth-century science, among them Albert Einstein, Lord Rutherford, Niels Bohr, Sir Oliver Lodge, Sir Arthur Eddington, Harlow Shapley, Edwin P. Hubble, Walter Baade, Robert A. Millikan, Arthur H. Compton, and Ernest O. Lawrence. Many of them became my good friends over the years.

For assistance in obtaining photographic illustrations I am

indebted to Frank H. Forrester and Irvil P. Shultz of the U. S. Geological Survey, Hugh W. Harris of the Lewis Research Center of the National Aeronautics and Space Administration, Thomas Harney of the Smithsonian Institution, Sybil E. Hamlet of the National Zoological Park, Victoria Haider and Carol Blazek of the Field Museum of Natural History, Graham Berry of the California Institute of Technology, Daniel M. Wilkes of the Lawrence Berkeley Laboratory of the University of California, Tom Wiley and Nelson Fuller of the Scripps Institution of Oceanography, Hal Marshall of the University of Arizona, Virginia L. Cummings of the Buffalo Museum of Science, and Edward J. Greaney, Jr., of the Hawaii Department of Planning and Economic Development.

The diagrams in the text are the work of Ned Glattauer.

I am also indebted to my secretary, Toni A. Jancura, for her services in preparing the final copy of the manuscript.

David Dietz

Shaker Heights, Ohio

CONTENTS

Preface   vii

PART I.   The Universe
✳✳✳✳✳✳✳✳✳✳✳✳✳✳✳✳✳✳✳✳✳✳✳✳✳

1. The Scale of the Universe   3
2. The Moon   16
3. The Planets   37
4. Asteroids, Comets, and Meteors   58
5. The Sun   71
6. The Stars   86
7. The Galaxy   106
8. The Universe   121

PART II.   The Earth
✳✳✳✳✳✳✳✳✳✳✳✳✳✳✳✳✳✳✳✳✳✳✳✳✳

9. The Rise of Geology   149
10. The Earth's Beginning   155
11. The Structure of the Earth   163
12. Continents and Oceans   171
13. The Earth's Changing Face   187
14. Earthquakes, Volcanoes, and Mountains   205
15. The Record in the Rocks   227

PART III. The Atom

16.	The Nature of Matter	251
17.	X-Rays, Radium, and the Electron	259
18.	The Structure of the Atom	268
19.	The Formation of Molecules	280
20.	The Nature of Energy	290
21.	The Quantum Theory	302
22.	The Structure of the Nucleus	314
23.	Elementary Particle Physics	331
24.	The Theory of Relativity	345

PART IV. Life

25.	From Magic to Science	359
26.	The Chemical Nature of Life	368
27.	The Cell	377
28.	Heredity and Evolution	393
29.	Bacteria and Viruses	407
30.	Plants and Animals	416
31.	Man	426
32.	Mind	440
33.	The Unity of the Universe	451

Appendix

Table One.	The Planets	461
Table Two.	The First-Magnitude Stars	462
Table Three.	The Geologic Time Scale	463
Table Four.	The Chemical Elements	464

Bibliography	469
Index	477

PART 1

****

The Universe

View by moonlight of the dome of the 200-inch telescope on Palomar Mountain. The telescope can be seen through the open shutters.

one

THE SCALE OF THE UNIVERSE

The heavens declare the glory of God; and the firmament showeth his handywork.
—PSALMS XIX: 1

IT is only within the last few decades that astronomers have realized the true grandeur of the universe. We now know that the most distant objects to be seen with today's giant telescopes are so far away that their light, traveling at the rate of 186,000 miles a second, takes more than 6 billion years to reach us.

In ancient times it was believed that the earth was the center of the universe. Today we know that our earth is but one of nine planets revolving around the sun. The sun is one of 100 billion stars that form our galaxy. Our galaxy is but one of trillions of galaxies scattered through the boundless ocean of space. No one has any idea of the total number of galaxies. It may be a million times a trillion. It may be infinite, if space is infinite.

It is interesting to compare our present picture of the universe with that which held sway in Biblical days, and to trace the steps in the evolution of the modern view.

We can reconstruct the ancient view from the opening chapters of Genesis. The earth was believed to be a great flat plain, stretching away in all directions. It was surrounded by the seas and supported upon them. The sky was a real canopy or roof over the earth. When it rained, it was because the windows in this roof were opened and the waters which were above were allowed to leak through. The stars were so many lamps or lights. The sun and moon were larger lights, "lights in the firmament of heaven to divide the day from the night," as Genesis phrases it.

This picture of the universe was derived from an earlier one which had long been held by the Babylonians. However, the Babylonians regarded the heavenly bodies as luminous gods. The sun was Shamash, the sun god. The moon was Sin, the moon god. Sin was a male deity and his wife was Ishtar, the planet Venus. The chief god was Marduk, the planet Jupiter. The stars were all minor gods or angels. They comprised "the hosts of heaven."

The Egyptian notion of the universe was very much like that of the Babylonians. To begin with, the Greeks also held similar ideas. Homer, writing in the ninth century B.C., described the earth as a great circular disk floating upon the ocean and surrounded by what he called the River Ocean. The sun was the chariot of the sun god who drove across the heavens each day.

But with the rise of the Greek philosophers in the sixth century B.C., astronomy took an important step forward. The philosophers brushed aside the notion of luminous gods and sought to construct a rational picture of the universe.

If you watch the sky for several hours, you will see that the heavens seem to be turning around the North Star. The stars do not change their positions with relation to each other, so that it is easy to imagine that the stars are attached to the interior of a solid dome. However, if you watch the sky night after night, you will note that the planets move against the background of stars. Our word "planet" comes from the Greek word for "wanderer."

The Greek philosophers developed a concept of the universe which took these facts into consideration. According to it, the earth was a sphere at the center of the universe. It was surrounded by a series of concentric transparent spheres which revolved around the earth at different rates of speed, carrying the heavenly bodies around with them. The moon was attached to the first sphere, Mercury to the second, Venus to the third, the sun to the fourth, Mars to the fifth, Jupiter to the sixth, Saturn to the seventh, and

THE SCALE OF THE UNIVERSE

the fixed stars, as they were called, to the eighth.

However, as better observations were made of the planets, it became apparent that this scheme was too simple. The planets do not always appear to move steadily eastward against the background of stars. At times the motion of a planet is westward, or retrograde, for a brief period. Today we know that this is a geometrical effect resulting from the combination of the planet's own motion with the motion of the earth in its own orbit. But it was difficult for the Greeks to explain, since they believed that the planets, being celestial objects, must exhibit what they regarded as perfect motion, namely, uniform circular motion.

A solution was offered in the fourth century B.C. by Eudoxus, a disciple of Plato. He imagined a nest of four concentric spheres for each planet. The planet was attached to the innermost sphere. Each sphere turned on an axis that fitted into the next sphere outside it. These axes were so arranged that the direction of spin

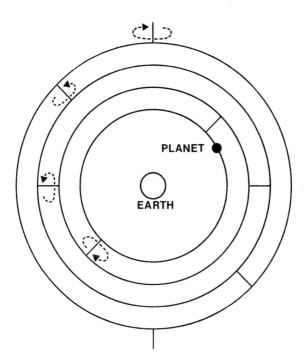

Figure 1. The spheres of Eudoxus

varied from one sphere to the next. The composite motion of the spheres approximated the motion of the planet. This is shown in Figure 1.

This system was further elaborated by Aristotle, who wound up with a nest of 56 concentric spheres kept in motion by the rotation of the outermost sphere, which he called the prime mover.

About 280 B.C. Aristarchos advanced the theory that the sun and not the earth was the center of the universe. The apparent motions of the heavens, he contended, could be explained most easily by assuming that the earth rotated on its axis once in 24 hours while both it and the planets revolved about the sun. Here, in essence, was the Copernican system 17 centuries before Copernicus.

However, his theory was rejected for the same reason that many astronomers later refused to go along with Copernicus. If the earth moved in a great circle around the sun, the stars ought to show a slight shift or change of position when viewed from opposite sides of the orbit. But no such shift, known as stellar parallax, was observed. Aristarchos insisted that this was because of the immense distances of the stars. This explanation was brushed aside as incredible. It is interesting to note that astronomers did not succeed in detecting the parallax of the nearest stars until the nineteenth century.

After 250 B.C. a theory of the universe that was much simpler than Aristotle's complicated nest of concentric spheres gradually took shape. It was the theory of epicycles, first suggested by Appolonius in about 230 B.C. It owed much of its development to Hipparchos in about 130 B.C. and was finally perfected by Claudius Ptolemy in A.D. 140.

According to this theory, the motion of a planet was the combination of two circles. The earth was at the center of a large circle known as the deferent. A point on its circumference, sometimes called the fictitious planet, revolved around the earth. This point was the center of a smaller circle called the epicycle. The real planet revolved around the circumference of the epicycle as the epicycle was carried around the deferent. This is shown in Figure 2.

For 14 long centuries after the days of Ptolemy the Ptolemaic system reigned supreme. As more precise data were gathered concerning planetary motions, the system was extended by adding

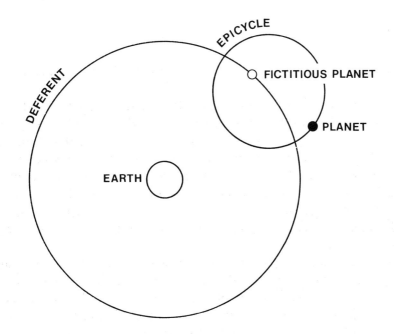

Figure 2. The theory of epicycles

epicycle to epicycle until a complicated system employing 80 deferents and epicycles was evolved. And then in 1543 the end of its long reign was heralded.

The revolution in astronomy, which put the sun, instead of the earth, at the center of the solar system, began with the publication in 1543 of the great work of the Polish astronomer Nicholas Copernicus, *De revolutionibus orbium coelestium* (Concerning the Revolutions of the Heavenly Spheres). But this was only the beginning. Establishment of the new era in astronomy required the achievements of his illustrious successors, Tycho Brahe, Johannes Kepler, and Galileo Galilei. Even so, the revolution was not completed until 150 years after his death with the formulation by Sir Isaac Newton of the law of universal gravitation.

COPERNICUS

Copernicus was a canon in the Frauenburg Cathedral in Polish East Prussia. In his university days he had studied medicine and law as well as mathematics and astronomy. His ecclesiastical duties were light and he gave some time to the gratuitous practice of medicine among the poor. His chief interest, however, was astronomy.

He spent almost 30 years writing and editing his book, but hesitated to publish it, fearing that it would arouse violent opposition. When he finally sent it to the printer, he was old and ill. The story is that a copy was placed in his hands a few hours before his death on May 24, 1543. Death spared him the knowledge that a colleague had inserted an anonymous preface saying that his theory was to be regarded only as a geometrical way of representing the heavens and that it was not meant to set forth the true structure of the universe.

Copernicus adhered to the ancient notion of deferents and epicycles. But by placing the sun at the center of the universe, he was able to reduce the number from 80 to 34. Most of his contemporaries, however, regarded his system as just another version of the Ptolemaic system with earth and sun transposed. They felt that acceptance of a moving earth was too high a price to pay for his simpler geometry.

TYCHO BRAHE

The next advance in the astronomical revolution was made by Tycho Brahe, who was born in 1546, three years after the death of Copernicus. The son of a Danish nobleman, he had an ungovernable temper. When he was 20, he lost the tip of his nose in a duel and thereafter wore an artificial nose made of a mixture of gold and silver. The King of Denmark built him an observatory on the island of Hveen. He always donned his court robes when he entered it, explaining that when he studied the heavens, he stood in the court of the King of Kings.

Tycho carried on observations of the motions of the planets for more than 30 years with greater precision than any previous astronomer. He refused, however, to accept the Copernican system because he could not observe any parallax in the stars. He proposed a compromise system, based to some extent on suggestions which had been put forward by some of the Greek philosophers. According to this, the moon and sun revolved around a stationary earth while the planets revolved around the sun.

KEPLER

In 1600 Tycho invited the German mathematician Johannes Kepler to join him as an assistant. The history of science offers no greater contrast than that of Tycho, the imperious hot-tempered nobleman, and Kepler, who suffered from ill health and poverty all his life. Kepler was born with faulty eyesight which became worse as the result of an attack of smallpox when he was four. But he was a mathematical genius. The science of astronomy leaped forward when good fortune brought about the union of Tycho's observations and Kepler's mathematical skill.

Tycho gave Kepler the task of computing the orbit of Mars from the observations of the planet's apparent motions which he had recorded over the years. Kepler boasted that he would solve the problem in 8 days. Instead it took him 8 years. But in doing so, he demolished the ancient notion that the celestial bodies moved

The 200-inch telescope of the Palomar Observatory

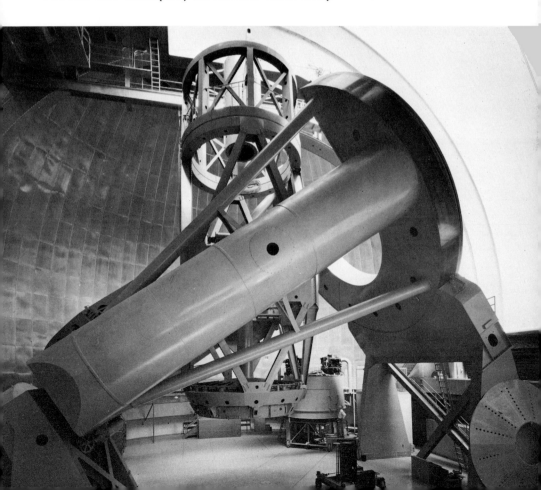

only in perfect circles and brushed aside the whole cumbersome contraption of deferents and epicycles. He proved that Mars went around the sun in an orbit which was a flattened circle, or ellipse, and that it moved fastest when closest to the sun, slowest when farthest from it. Subsequently he proved that this was true of all the planets.

Kepler formulated three laws of planetary motion, still known to astronomers as Kepler's laws. The first states that the orbit of each planet is an ellipse with the sun at one focus. The second law states that the speed of a planet in its orbit varies so that a line from the sun to the planet sweeps over equal areas in equal intervals of time. The third law, called the harmonic law, connects the periods of revolution of the planets to their distances from the sun. It states that the square of the period is proportional to the cube of the average distance.

GALILEO

Meanwhile, a brilliant young Italian professor, Galileo Galilei, was performing experiments that upset ideas about motion that had been believed since the days of ancient Greece. Aristotle had taught that heavy objects fall faster than light ones and that a body continues in motion only as long as a force acts on it. Galileo proved that both these notions were wrong. His brilliant lectures attracted students from all parts of Europe. But he had a sharp tongue and a biting sarcasm that made him bitter enemies.

In 1609 Galileo heard of the invention of the telescope and proceeded to make one for himself by fitting spectacle lenses into the ends of an organ pipe. His first telescope magnified only 3 times, but he soon built superior ones, obtaining a magnification of about 32 times.

Until Galileo turned one of his little telescopes upon the heavens no one had seen any more of the universe than the caveman of the Old Stone Age, 500,000 years ago. Now a new era began. One sensational discovery after another met Galileo's gaze. His telescope revealed that the surface of the moon was irregular and mountainous, that there were many more stars than can be seen with the unaided eye, and that the Milky Way was composed of innumerable faint stars. His most spectacular discovery was the four moons of Jupiter.

His discoveries made him famous throughout Europe. But a storm of opposition was beginning to gather against him. It came from two sources: one was the entrenched university professors who were firm disciples of Aristotle; the other was a segment of the clergy that insisted that Galileo was contradicting the Bible by his championship of the Copernican theory. Holy Writ, they pointed out, said that Joshua made the sun stand still.

In a letter to Kepler, Galileo told how the principal professor of philosophy at Padua refused to look through his telescope. "Why are you not here?" he wrote. "What shouts of laughter we should have at this glorious folly." But the passing years brought events that were no laughing matter. In his old age he was arrested by the Inquisition and brought to trial for his espousal of the Copernican system. He was found guilty, made to recant, and sentenced to spend the rest of his life under house arrest.

His end was tragic. His eyesight began to fail and during the last four years of his life he was completely blind.

"Alas," he wrote in one of his letters, "your dear friend and servant is totally blind. Henceforth this heaven, this universe, which by wonderful observations I had enlarged a hundred and a thousand times beyond the conception of former ages, is shrunk for me into the narrow space which I myself fill in it. So it pleases God; it shall therefore please me also."

NEWTON

Galileo died in 1642, the same year Isaac Newton was born. It seems prophetic that he should have been born in that year, for it was Newton who carried Galileo's work to its logical conclusion, refining Galileo's laws of motion and applying them to Kepler's laws of planetary motion. Newton showed that Kepler's laws were the natural outcome of his own law of universal gravitation, thus completing the revolution which Copernicus had begun.

In 1664, when Newton was a student at Cambridge University, the Great Plague broke out in London. Fear of it spread to Cambridge and the university closed its doors. Newton returned to the farmhouse in Woolsthorpe, Lincolnshire, where he had been born, remaining there until the university reopened in 1667.

Those two years at Woolsthorpe saw the incredible flowering of Newton's genius. He made three of the greatest discoveries in the history of science: the differential calculus, the composition of light, and the law of universal gravitation. Strangely enough, because of a curious secretiveness that characterized his entire life, he made no effort to publish his discoveries or even to tell anyone about them.

Unlike Kepler, Newton realized that the planets did not require a constant push to keep them moving. What was needed was a force to hold them in their orbits and keep them from flying off into space in a straight line. The fall of an apple which he witnessed in the orchard at Woolsthorpe set him to thinking that this same force of gravity might keep the moon in its orbit around the earth and the planets in their orbits around the sun. His calculations convinced him that this was so.

Years later, in 1687, he set forth his conclusions in his monumental work, *Philosophiae naturalis principia mathematica* (The Mathematical Principles of Natural Philosophy), since known to the whole world simply as the Principia.

In Newton's day five planets were known in addition to the earth, the same five which the ancients had known. In 1781 Sir William Herschel, who had given up music to become a stargazer, added the planet Uranus. Half a century later the planet Neptune was discovered. In 1930 astronomers at the Lowell Observatory at Flagstaff, Arizona, brought the number of planets to nine with the discovery of Pluto. But even more important than this extension of the solar system has been the way in which astronomers have pushed out the boundaries of the universe. Each time a bigger telescope was built, they saw farther out into space.

Let us look at a few statistics about the universe. It is 240,000 miles from the earth to the moon. It is 93 million miles from the earth to the sun. Pluto, the outermost planet in the solar system, is more than 3 billion miles from the sun. But the nearest star is 25 trillion miles away. See how fast the universe grows! First thousands of miles, then millions, then billions, then trillions!

A MODEL OF THE UNIVERSE

We can try to visualize this state of affairs by pretending to build a model of the universe. Let us begin with the sun and its family of planets, the solar system. We will let a globe 1 foot in diameter represent the sun. Then, on the same scale, the earth will be represented by a tiny seed approximately 1/10 of an inch in diameter. If we place the seed 100 feet from the globe, we will then have a model to scale of the sun, the earth, and the distance between. We have made a model, therefore, in which 100 feet represents 93 million miles.

Let us obtain eight other seeds to represent the eight other planets. We will place the first about 38 feet from the globe to represent Mercury. We will place the second one 72 feet from the globe to represent Venus. Next comes the seed which we have already placed 100 feet from the globe to represent the earth. The remaining six seeds will be placed at greater distances from the globe, representing, in order, Mars, Jupiter, Saturn, Uranus, Neptune, and Pluto. If we have kept our model to scale, we will find that the outermost seed, representing Pluto, is about 1 mile from the globe.

Suppose we undertake now to add the stars to our model. Our sun, as we know today, is a star. Or to put it the other way around, every star is a sun. So let us get another globe to represent the nearest star. Where shall we place it in our model? Five miles away? That would be too close. Five hundred miles away? Still too close.

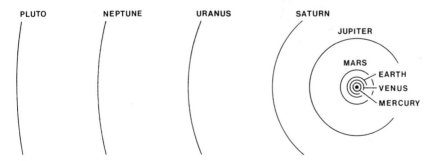

Figure 3. The solar system

14   THE UNIVERSE

To make a long story short, we should have to place the globe 5,000 miles away. And now, if we use the entire surface of the earth for our model, we will have room for only three or four stars. Since we know that there are 100 billion stars in our own galaxy, it is useless to go on with the model.

The nearest star is approximately 25 trillion miles away. But the mile is an inadequate yardstick for measuring stellar distances. Astronomers find it more convenient to use the light-year. This is the distance that a beam of light, traveling 186,000 miles a second, covers in one years. It is approximately 6 trillion miles. The nearest star is a little more than 4 light-years away. Other stars in our galaxy are 10 light-years away, still others 100 and 1000. The stars on the outer edge of our galaxy are more than 50,000 light-years away.

It is awe-inspiring to reflect upon the significance of these figures. If a star is 100 light-years away, its light takes 100 years to reach the earth. When you go outdoors on a clear night, the particular beam of light which reaches your eye from some star may have started toward you just about the time George Washington was crossing the Delaware. The light from another star may have started

The 100-inch telescope of the Mt. Wilson Observatory

A replica of one of Galileo's first telescopes

on its journey when King Tut was ruling in Egypt. And the light from a third may have started on its journey when dinosaurs roamed the face of the earth.

Stars in our galaxy are, on the average, 4 or 5 light-years apart. But this distance is dwarfed by the distances separating the galaxies in the vast oceans of space. The average distance between neighboring galaxies is about a million light-years. This means that the universe is mostly empty space. It has been calculated that there is 10 billion trillion times more empty space than stellar material in our galaxy. However, in the universe at large there is 10 million billion trillion times as much empty space as stellar material.

Let us close this chapter with one more startling thought. You are never twice in the same place. You may challenge this statement since you have your favorite chair at the breakfast table. But as you attack your bacon and eggs the earth is turning on its axis and circling the sun. Even so, you will say, you are back where you started at the end of a year. This is true with respect to the sun.

But the sun is not standing still. The whole galaxy is turning like some giant cosmic carrousel. It takes the sun about 200 million years to make one trip around the galaxy. Of course it carries the earth and the rest of the solar system along with it. Finally, you might ask if the galaxy is moving through space as well as rotating. Astronomers do not yet have a conclusive answer to that question.

So now, if you have not lost your courage, we are ready to go on a trip through the universe. We shall begin with our nearest neighbor in space, the moon.

two

THE MOON

> *How like a queen comes forth the lonely Moon*
> *From the slow opening curtains of the clouds;*
> *Walking in beauty to her midnight throne!*
> —CROLY

FEW sights rival the beauty of the moon. Poets have lavished their finest phrases upon it. Long before the dawn of written history the moon held an interest for mankind exceeded among the heavenly bodies only by the sun. With the passage of time men began to note the regularity with which the moon goes through its phases. The earliest calendars were lunar calendars. Each month in the ancient Babylonian and Hebrew calendars began with the new moon.

Fantasies of a "man in the moon," inspired by the appearance of the full moon, go back to ancient times and are found in regions as widely separated as the Scandinavian countries and the Polynesian Islands of the Pacific. Some legends, however, describe a woman in the moon, a hare, a cat, or a toad. In the Middle Ages it was supposed that the moon had a smooth crystalline surface like a mirror or a burnished shield and that the dark spots were merely reflections of the earth. Some of the ancient Greeks, however, had imagined that there were mountains and seas upon the moon. But nothing was known until Galileo turned his little telescope on the moon. It revealed mountains and valleys and dark areas that he mistook for seas.

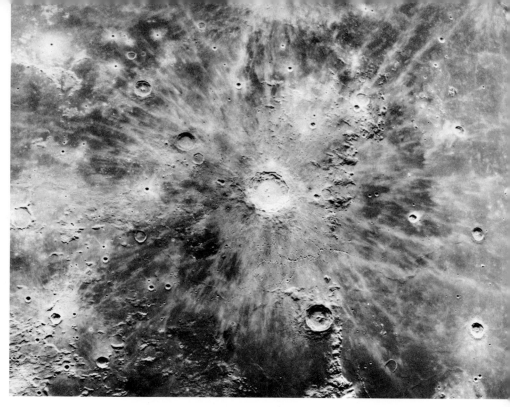

Copernicus Crater seen through the 100-inch telescope of the Mt. Wilson Observatory. It is 57 miles in diameter.

Our knowledge of the moon has been growing from that day to this. Each time a larger telescope was built, it disclosed new details of the lunar surface. The best photographs made with today's giant telescopes show the moon as it would appear to the unaided eye at a distance of 200 miles and reveal craters and other details about 1,000 feet across. However, a new era of lunar exploration began with the dawn of the Space Age. Unmanned American and Russian spacecraft have revealed a wealth of fine detail about the lunar surface, while the Apollo astronauts brought back samples of lunar rocks and soil.

The moon is the earth's nearest neighbor in the great ocean of space. It accompanies the earth on its journey around the sun, revolving around the earth as the earth revolves around the sun. The orbit of the moon is an ellipse. Its average distance from the earth is approximately 240,000 miles, more exactly, 238,856 miles. The distance varies from a minimum of 221,463 miles to a maximum of 252,710 miles.

MEASURING THE MOON'S DISTANCE

Perhaps you are wondering how astronomers arrived at these figures. The method is simple, though the execution is extremely difficult. It is essentially that used in surveying and known as triangulation. The surveyor, wishing to know the distance across a river, lays off a base line, let us say, 100 feet long. From either end of the line he points his transit at an object on the other bank, measuring the angle from the base line to the object. He then has a triangle of which he knows the base and two of the angles. By trigonometry he is then able to calculate the altitude of the triangle which is the distance across the river.

In applying this method, astronomers at two widely separated observatories turn their telescopes on the moon at the same time. The distance between the two observatories is the base line. But because this line is short compared to the distance to the moon, the astronomers do not attempt to measure the angles at the base of the triangle. Instead they calculate the angle at the apex from the shift in the position of the moon against the background of stars as seen from the two observatories. This shift is known as parallax. This is shown in Figure 4.

A simple experiment will help you visualize the phenomenon of parallax. Hold a pencil at arm's length and close your left eye. With your right eye note the position of the pencil against the background of the wall opposite you. Now close your right eye and open your left. You will be surprised to note how the pencil has shifted against the background. In this experiment the base line is the distance between your two eyes.

THE PHASES OF THE MOON

The moon is a dark body, shining only by reflected sunlight. This fact, coupled with its revolution around the earth, accounts for the phases of the moon. When the moon is between us and the sun, the side of the moon illuminated by the sun is turned away from us and the dark side is turned toward us. The moon is then entirely invisible and we have what astronomers call new moon.

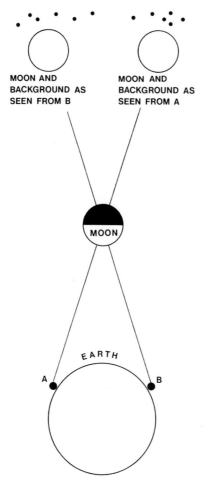

Figure 4. Measuring the moon's parallax

About two nights later the moon has moved sufficiently for us to see a little of the lighted side, and it appears as a thin crescent which is known popularly, though incorrectly, as the new moon. The crescent grows fatter each night and approximately a week after new moon half of the lighted side is visible. We now have first quarter. A week after first quarter, the moon is on the opposite side of the earth from the sun, and all of the illuminated side is toward us. It is full moon. Next we see less of the lighted side each night. A week after full moon it is last quarter. In another week we are back to new moon. This is shown in Figure 5.

20 THE UNIVERSE

The apparent motion of the moon in the sky is a little difficult to understand at first, for it is a mixture of the moon's actual motion around the earth added to an apparent motion which results from the rotation of the earth. The earth's rotation on its axis causes the moon to appear to rise in the east and set in the west in the same fashion that it makes the sun and stars appear to rise and set. But the revolution of the moon around the earth makes the moon appear to be moving eastward against the background of stars. If the moon is noted near a certain group of stars tonight, it will be found 13 degrees to the east of them tomorrow night. As a result of this eastward motion the moon rises about 50 minutes later each night.

Since the moon is on the same side of the earth as the sun at new moon, it rises at sunrise and sets at sunset. Each day thereafter it rises later and therefore is higher in the sky at sunset. When we first catch sight of the thin silvery crescent popularly known as the

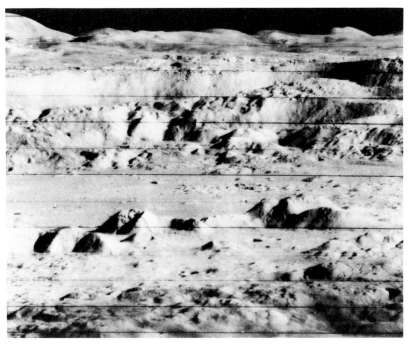

A close-up of Copernicus Crater taken by Lunar Orbiter 2 from an altitude of 28.4 miles. The mountains in the foreground are 1000 feet high.

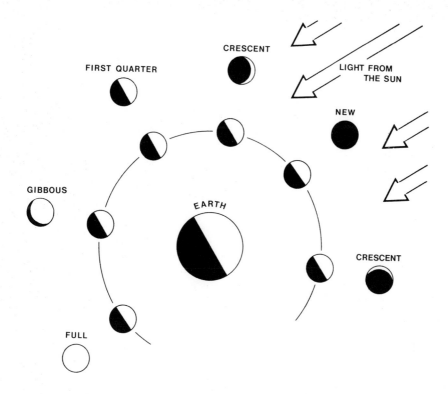

Figure 5. The phases of the moon

new moon, it is low in the west and sets soon after the sun. But by first quarter the moon is rising about noon. Consequently, it is high in the sky at sunset and does not set until about midnight. At full moon the moon rises as the sun sets and so there is moonlight all night. At last quarter the moon rises about midnight and sets about noon.

The moon takes approximately 27.3 days to go once around the earth. This period is known as the sidereal month. But because of the earth's motion around the sun, the time from new moon to new moon is longer, approximately 29.5 days. This is known as the synodic month.

Since ancient times, men have wondered why the full moon, as it rises over the eastern horizon, looks so much larger than it does when it is high in the sky. Measurements prove that this is an optical illusion. Numerous explanations have been offered for it, but none is entirely satisfactory. Some psychologists believe that it results from the way in which we judge size in relation to distance.

When the moon is near the horizon, they say, it seems more distant than when it is overhead. Consequently, we judge it to be larger.

Often, when the moon appears as a slender crescent in the western sky, a night or two after new moon, it is possible to see dimly the dark portion of the moon's disk. This is spoken of as "the old moon in the new moon's arms." It results from the fact that sunlight, reflected from the earth, is striking the dark side of the moon. The moon is being faintly illuminated by "earthlight." If an astronaut were on the moon at the time, he would see a "full earth" in the sky.

THE MOON'S MOTIONS

The moon rotates on its axis in exactly the same length of time that it revolves around the earth, namely, 27.3 days. Consequently, it always keeps the same face turned toward the earth. The other side of the moon was a complete mystery until the first photos of it were sent back by the Russian spacecraft Luna 3.

However, certain irregularities in the motions of the moon result in oscillations, known as librations, and because of these we see a little more than half of the moon's surface, about 59 per cent. These librations are of three kinds. One, due to the angle at which the axis of rotation of the moon is tipped with respect to its orbit, permits us to see a little beyond the north pole of the moon at times, and at other times a little beyond the south pole. The second is caused by the fact that the moon is sometimes a little fast in its motion around the earth and at other times a little slow. Because of this, we sometimes see a little further around the eastern edge of the moon and sometimes a little further around the western edge. The third libration is known as the diurnal libration. This results from the fact that at the time of rising and setting we see a little over the moon's upper edge, seeing more of the moon than we do when it is high in the sky.

Northern portion of moon, showing Mare Imbrium, bordered by the Lunar Alps, Carpathians, and Apennines

DIAMETER AND DENSITY

The moon has a diameter of 2,160 miles, slightly more than one-fourth that of the earth. This makes it a quite large satellite for a small planet. While some of the satellites of Jupiter and Saturn are larger than our own moon, they do not begin to approach these giant planets in size. From another planet our earth-moon system, as astronomers sometimes call it, would look like a double planet. The moon is the chief cause of the tides in the oceans of the earth, although the sun also has some tidal effect.

The density of the moon is only about 60 per cent that of the earth. Consequently, the mass of the moon is only about 1/80 that of the earth. The force of gravity on the surface of the moon is only one-sixth as strong as it is on the surface of the earth. Since weight is a measure of the pull of gravity, a person who weighs 120 pounds on earth would weigh only 20 pounds on the moon. A person who could jump 4 feet high on earth would be able to leap to a height of 24 feet on the moon.

THE LUNAR SURFACE

To our unaided eye, the moon appears as though it must be a veritable fairyland. But big telescopes and spacecraft equipped with television show that the lunar surface is a barren, rugged, lifeless place of unbelievable desolation and eternal silence, for there is neither air nor water on the moon. The forces of erosion, wind and rain and running water, which shaped the surface of the earth, are missing on the moon. Consequently, we see it as it looked billions of years ago.

An amazing amount of detail can be seen on the moon with a small telescope or even binoculars. Some impression of the lunar features can be gained even with opera glasses. You will discover, as Galileo did in 1609, that the man in the moon is a combination of dark and light areas. The man in the moon is apparent to the unaided eye at full moon. But this is not the best time to study the moon with binoculars or a telescope. The rays of the moon strike the lunar surface full face at full moon and shadows are lacking. At other times the sun's rays strike the moon at an angle and

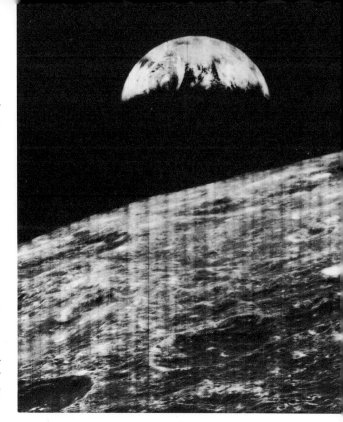

Lunar Orbiter 1 made this photograph of the crescent earth rising above the surface of the moon.

BELOW: *The lunar surface photographed by Ranger 7 from an altitude of 63.4 miles*

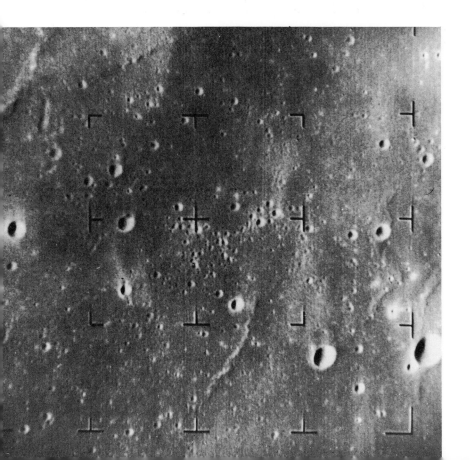

mountains and crater walls cast shadows, bringing them out in sharp relief.

THE MARIA

When Galileo first looked at the moon through his telescope, he thought the great dark areas were seas and called them *"maria,"* the Latin word for "seas." (The singular is *mare*, pronounced "mah-reh.") Although it has long been known that they are great flat plains, they are still called by the fanciful names given them in the seventeenth century—the Mare Imbrium (Sea of Showers), Mare Serenitatus (Sea of Serenity), Mare Tranquillitatis (Sea of Tranquility), Mare Frigoris (Sea of Cold), and so on.

The maria account for a little less than half of the visible hemisphere of the moon. They are hundreds of miles in diameter. Mare Imbrium has a diameter of 700 miles. Mare Serenitatis is 430 miles in diameter. The maria consist of two general types, circular and irregular. The circular maria are bordered in large part by mountain ranges. The outline of the irregular maria are ragged and in every case they are joined to circular maria through gaps in the mountain ranges bordering the circular ones. The maria are depressed below the level of the rest of the moon and give every indication of having been formed by huge pools of molten lava which first appeared in the circular maria and then spread into the irregular ones.

There are many great cracks, known as rills, and many ridges in the maria, as though the lava had cracked and wrinkled as it solidified. Some of the cracks are 2 or 3 miles wide and more than 100 miles long. In places they seem to be filled with solidified lava.

The walls of buried craters, sometimes called ghost craters, protrude to varying degrees from the irregular maria, strongly suggesting that they were submerged by the spreading lava. No ghost craters occur in the circular maria, indicating that any craters that once existed there had been destroyed by the creation of the maria. However, there is a scattering of sharply defined craters of all sizes on both types of maria. These were obviously created after the maria had taken shape.

MOUNTAIN RANGES

There are 10 major mountain ranges on the visible hemisphere of the moon. Three of them, the Lunar Apennines, the Caucasus, and the Alps, border the Mare Imbrium. The Apennines are the most magnificent, extending in a great curve for 640 miles. Their rugged peaks rise abruptly from the Mare Imbrium to heights of 18,000 feet, but slope away gently on the far side, with numerous valleys and ravines. A great irregular crack, or rill, parallels the inner edge of the Apennines as though the Mare Imbrium had subsided under the weight of the lava. But in places the crack is filled with solidified lava, converting it into a ridge.

A striking feature of the Lunar Alps is a gigantic cut through the center of the range, known as the Alpine Valley. It is 75 miles long, 6 miles wide, and has a level floor.

CRATERS

The wild profusion of craters is the dominant and most spectacular feature of the moon. Big telescopes reveal some 32,000 craters on the visible side of the moon, ranging in diameter from more than 100 miles to about 1000 feet. But the Ranger photographs disclose tiny craters less than 18 inches in diameter. There are even smaller craters, so that the total number defies counting. Craters are found everywhere on the moon. They dot the maria. They occur in the mountains. The southern area of the visible hemisphere is a confused hodgepodge of big and little craters. There are craters within craters. In numerous cases craters break through the walls of other craters. Some have smooth, flat interiors, but others have one or more central peaks.

The craters can be divided into several general types. One, known as the walled plains, or bulwark plains, includes many of the largest on the moon, ranging in diameter from 50 to more than 100 miles. They possess interiors as level as the maria and mountain walls that rise abruptly from the crater's floor. The interior plain is sometimes level with the outside terrain, sometimes lower, occasionally higher. The mountain walls are comparatively low, rising

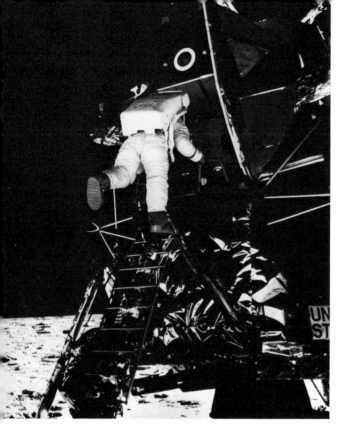

Astronaut Edwin E. Aldrin, Jr., descends to the moon's surface from the lunar module of Apollo 11 on July 20, 1969.

only a few thousand feet above the surrounding territory. However, the interior plain may be depressed as much as 2 miles. Clavius, the largest crater on the visible side of the moon, is a walled plain 146 miles in diameter.

Another type of crater, known as the ring-mountain crater, ranges up to 70 or 80 miles in diameter. It has a saucer-shaped floor. The inner side of the mountain wall rises steeply in a series of terraces. The crest of the wall shows an irregular array of ridges. The outer side slopes away gradually and is cut by numerous valleys. In many cases one or more peaks rise from the center of the crater to heights of a mile or more. However, these peaks never exceed the crater's wall in height.

Two theories were advanced in the nineteenth century for the formation of the craters. One held that the craters were volcanic in origin, the other that they were formed by the impact of meteorites. Most astronomers today favor the second theory. It is assumed that not only the craters, but the maria as well, were formed some billions of years ago when the solar system was young and space within it contained a great many meteorites of considerable size, cosmic debris that was left over when the planets and their satellites took shape. The size of a crater would depend upon the size of the meteorite which created it.

A huge meteorite, weighing some billions of tons, crashing into the moon at a speed of 20 miles or more a second, would bury itself in the lunar surface. The sudden stoppage of its motion would convert a large part of its energy into heat, vaporizing much of the meteorite and the lunar rocks under it. The expansion of this gaseous material would constitute a tremendous explosion which would create the crater and its mountain wall, at the same time hurling millions of tons of shattered rock to a distance of 10 miles or more. Calculations show that it would take a meteorite weighing 25 billion tons to create a crater 50 miles in diameter. It is thought that the meteorite which created Clavius weighed 200 billion tons.

Beyond many large craters and radiating away from them in all directions is a vast number of small craters. These craters, known as secondaries, have very low, irregular rims, in many cases none at all. It seems apparent that they were gouged out of the lunar surface by material hurled out of the big craters.

The circular maria differ from the walled plains chiefly in size. The mountain ranges bordering the Mare Imbrium show the same characteristics as do the mountain walls of the larger craters. A number of today's foremost students of the moon are convinced that the circular maria were created by meteorites larger than those responsible for the largest craters. It is calculated that the meteorite which created the Mare Imbrium was 120 miles in diameter. Its impact caused an explosion equal to a billion hydrogen bombs.

It is thought that the impact of such a meteorite opened up great cracks in the face of the moon from which vast seas of molten lava poured forth, forming the floors of the circular maria and overflowing into the irregular maria.

A number of grooves, or furrows, radiate from the Mare Imbrium for hundreds of miles, plowing their way through every feature in their path. The Alpine Valley is probably one of these grooves. It is thought that they were created by pieces of the huge meteorite when it exploded.

It is interesting to note that a comparison of lunar craters with craters on earth of known meteorite origin, such as Meteor Crater in Arizona, as well as a comparison with the craters caused by artillery shells and nuclear explosions, reveals striking similarities.

The question is often asked why the earth does not possess more craters. The answer is that the forces of erosion, wind and rain and rushing water, have completely changed the face of the earth. It is probable that when the earth took shape some 4.6 billion years ago, its surface resembled that of the moon.

While most astronomers agree today that both craters and maria resulted from meteorite impact, there are some features of the moon which appear to be the result of volcanic activity. These are the so-called domes, many of which resemble such terrestrial volcanoes as Mt. Vesuvius.

THE RAYS

We come finally to a feature of the lunar surface which puzzled astronomers for many centuries. These are the bright streaks radiating from a number of craters, known as rays. They are 5 to 10 miles wide and up to 1,500 miles in length. They run over maria, mountains, and craters without any regard to the territory traversed, overlaying every feature in their path. The rays do not cast shadows and are best seen at full moon. The craters from which they originate are known as ray craters. There are about a dozen of them. Several of the more conspicuous ones, such as Copernicus, Kepler, and Aristarchus, occur in maria, but Tycho, a very conspicuous ray crater, is situated among the hodgepodge of craters in the southern area of the lunar surface. The ray craters are the most perfectly formed of all the craters. They have bright rims which match the brightness of the rays. It appears certain that they are the youngest craters and that the rays are splash marks resulting from their formation.

Despite the silvery appearance of the moon to the unaided eye, the lunar surface is actually a neutral gray. This is the result of the bombardment of the lunar surface by streams of atomic particles from the sun. The moon has no atmosphere to ward off the particles. Laboratory experiments have established that such bombard-

The hidden side of the moon photographed by Lunar Orbiter 3. The large crater in the center of the picture is 150 miles in diameter.

ment of rocks causes them to grow darker. The present belief is that the rims of the ray craters and the rays are bright because they have not yet had time to grow as dark as the rest of the lunar surface. Recent studies have disclosed that some of the older craters have very faint, darker systems of rays. In no case, however, do these dark rays overlay the bright rays of the younger craters.

ABOVE: *Apollo 11, atop its Saturn 5 rocket, blasts off for the moon on July 16, 1969.*

TOP RIGHT: *Astronaut Aldrin beside the American flag on the lunar surface*

RIGHT: *Apollo 11, atop its Saturn 5 booster, on the way to the launching platform*

The rays, therefore, are thought to be pulverized material thrown out by the meteorite impacts which created the ray craters. Photographs taken by the Ranger spacecraft show that the rays are not solid streaks but are made up of many feather-shaped elements arrayed side by side. Tiny craters are numerous at the far ends of these little streaks.

You can try an amusing little experiment which illustrates the creation of the rays. Put a little mound of flour or talcum powder on a sheet of dark paper or cloth. Now hit the mound a sharp blow with the back of a teaspoon. The powder will splash out in all directions and form a system of rays.

THE LUNAR SPACECRAFT

Our knowledge of the lunar surface has been greatly expanded since the dawn of the Space Age by American and Russian unmanned spacecraft and by the American Apollo spaceships which put astronauts on the lunar surface.

The American program began with the Ranger lunar probes, which crashed on the lunar surface but sent back television pictures up to the moment in which they crashed. These showed many small craters never before seen, some only 18 inches in diameter. Subsequently both the United States and Russia launched unmanned spacecraft which made gentle, or soft, landings on the lunar surface, and others which orbited the moon.

Surveyor 1, the first American spacecraft to make a soft landing on the moon, revealed a firm, gritty surface of soil-like material, neutral gray in color, strewn with small pebbles and larger rocks. Some were on top of the soil, others half buried.

The third spacecraft to make a soft landing on the moon was the Russian Luna 13. It was equipped with a device which drove a steel rod about a foot into the lunar surface and measured the force in doing so. This indicated that the mechanical properties of the lunar surface are similar to those of terrestrial soil.

The American Surveyor 3, which landed on the moon on April 19, 1967, was equipped with the world's first Space Age shovel, a small steel-tipped scoop. This was mounted on a mechanical arm and controlled by ground-based radio. Television pictures from the spacecraft showed its action. It dug several trenches of various depths. It was significant that the walls of these trenches did not collapse. More force was needed as the trenches became deeper, indicating that the lunar soil became more compact with depth.

The American Apollo 11 arrived at the moon on July 19, 1969. The next day astronauts Neil Armstrong and Edwin Aldrin descended to the surface of the moon in the lunar module. They landed on the Mare Tranquillitatis, or Sea of Tranquility. They spent 2 hours and 21 minutes walking on the moon as the portable television camera which they set up enabled millions of people in all parts of the world to watch them. As the two walked about their boots left crisp footprints in the fine-grained lunar soil.

The soil was powdery and weakly cohesive. There was a profusion of rocks of all sizes, some on the surface, some partly buried. A number of rocks at the centers of small craters showed little glassy patches on their surfaces that glistened like drops of solder.

The lunar rocks brought back by the Apollo 11 astronauts range in color from light gray to medium gray. They appear to be igneous rocks such as would result from the solidification of molten material. Many of the rocks show tiny glass-lined pits, thought to be the result of melting caused by the impact of tiny meteorites moving at very high speeds.

Chemical analysis shows that the lunar rocks contain the same chemical elements as terrestrial basaltic rocks. However, they contain 10 times as much titanium and more zirconium and yttrium than terrestrial basalts. They contain no carbon compounds or water.

About 50 per cent of the lunar soil brought back by the astronauts consists of tiny glass globules and teardrops ranging in color from various shades of brown to yellow, green, and red. These tiny bits of glass are believed to have formed when the impact of meteorites vaporized lunar material which subsequently condensed into liquid droplets and then solidified. The lunar soil, most astronomers and geologists are inclined to believe, is the rubble created by the impact of meteorites. The top layer of the soil appears to be pulverized rock.

THE MOON

Luna 3, launched by the Russians on October 4, 1959, sent back the first pictures of the far side of the moon. The pictures were not very distinct, but they showed that the hidden side of the moon is very rugged and contains very few maria. The Russians promptly named the largest one the Mare Moscovium, or Sea of Moscow.

Excellent pictures of the hidden side of the moon were subsequently obtained by the five American Lunar Orbiters, making it possible to map the hidden side. Names were assigned to its craters in 1970 by a commission of the International Astronomical Union. The Orbiters also made striking photographs of the visible side of the moon. One of the most spectacular shows the Copernicus crater from an altitude of 28 miles. A historic picture taken by Lunar Orbiter 1 on August 23, 1966, shows a crescent earth in the sky above the lunar surface.

The temperature of the lunar surface varies over an extremely wide range. Because of the slow rotation of the moon, any area of the moon experiences 2 weeks of sunlight followed by 2 weeks of darkness. During the long day the temperature rises to about that of boiling water, 212° F. During the long night it drops to about —243° F. It has been suggested that this variation in temperature has caused some exfoliation, or flaking, of the lunar rocks. Flaking may also occur at a greater rate during a lunar eclipse. Measurements made in one eclipse showed that the temperature dropped in one hour from 160° F. to —110° F.

In recent years some astronomers have insisted that they have detected color changes at the bottom of some craters that indicate the presence of gases. It is possible that there may be some emission of gases from cracks in the lunar surface. All attempts, however, to detect a lunar atmosphere, indicate that if there is one, its density is less than a trillionth that of the earth's atmosphere.

ECLIPSES

Eclipses occur when the sun, earth, and moon are in a straight line. If the moon is between the sun and the earth, there is a solar eclipse. If the moon is on the opposite side of the earth, there is a lunar eclipse. It will be seen, therefore, that a solar eclipse can occur only at new moon and a lunar eclipse at full moon.

If the moon's orbit lay in the plane of the earth's orbit, there would be a solar eclipse at every new moon and a lunar eclipse at every full moon. However, the moon's orbit is inclined to the earth's orbit at an angle of 5 degrees 38 minutes. Consequently, an eclipse can occur only if the moon is at one of the so-called nodes, the points where the moon's orbit crosses that of the earth, at new moon or full moon. Otherwise the shadow of the moon at new moon is either above or below the earth, while the moon, at full moon, is either above or below the shadow of the earth.

The earth's huge cone-shaped shadow is approximately 859,000 miles long. At the distance of the moon it is about 5,700 miles in diameter. In a total lunar eclipse, the moon takes about an hour to become totally immersed in the earth's shadow and remains in the shadow for about 2 hours, taking another hour to emerge. During the period of totality the moon shines with a dull coppery color due to the refraction of sunlight from the earth's atmosphere into the shadow cone. An eclipse of the moon is visible from an entire hemisphere of the earth. It is a beautiful and impressive sight, but unlike a solar eclipse, holds little interest for professional astronomers.

A total solar eclipse is the most spectacular of all astronomical events and of great importance to astronomers since it reveals important details of the solar atmosphere (see Chapter 5).

The conical shadow of the moon is only 232,100 miles long, less than the average distance of the moon from the earth. If the shadow fails to reach the earth, the result is an annular eclipse, a bright ring of sun remaining in view around the dark disk of the moon. When the shadow reaches the earth, the eclipse is total only within the narrow track of the shadow on the earth's surface. This may range from less than a mile to 167 miles. The period of totality is short, lasting from a few seconds to a maximum of about 8 minutes.

An average year has two solar and two lunar eclipses but the number may vary from a minimum of two, both solar, to a maximum of seven.

three

THE PLANETS

*Then felt I like some watcher of the skies
When a new planet swims into his ken.*
—K<small>EATS</small>

INTERPLANETARY spacecraft, radar, and radio telescopes have opened a new chapter in the study of the solar system. In many cases the use of these new techniques has confirmed conclusions previously reached by astronomers with the aid of ground-based telescopes and auxiliary instruments. But in others it has yielded surprising findings that are not yet fully understood.

Astronomers today are aware of nine planets in the solar system. In order from the sun they are Mercury, Venus, Earth, Mars, Jupiter, Saturn, Uranus, Neptune, and Pluto. It will be seen, therefore, that two are closer to the sun than is the earth; six are farther away. The four nearest the sun, Mercury, Venus, Earth, and Mars, are known as the terrestrial planets because of their general resemblance to the earth. They are in the same size range and possess solid surfaces and relatively thin atmospheres. The next four, Jupiter, Saturn, Uranus, and Neptune, are known as the major planets. They are giants compared to the terrestrial planets, possess extensive atmospheres, and are quite different from the terrestrial planets in every respect. This leaves Pluto, the outermost planet, in a class by itself. It is believed that Pluto resembles the terrestrial planets in size and character.

During the last half of the nineteenth century a persistent search was made for a planet closer to the sun than Mercury. It was thought that its gravitational attraction would explain certain peculiarities in the orbit of Mercury. Some astronomers even thought they had caught sight of it near the sun when the sky was darkened by a total solar eclipse. The name Vulcan was proposed for it. It is now certain that no such planet exists.

Since the discovery of Pluto in 1930, it has been suggested that there may be other small planets in the outer reaches of the solar system. But to date none has been found.

The nine known planets possess a total of 32 moons. Mercury and Venus have none, the earth has one, Mars two, Jupiter 12, Saturn 10, Uranus five, and Neptune two. Pluto has no known moon.

In addition the solar system contains some thousands of little planets, or asteroids, in orbits between those of Mars and Jupiter, and an incredible number of comets and meteorites, as well as some interplanetary dust and gas. Even so the solar system is mostly empty space, and all these entities together equal only a little more than 1/1000 of the mass of the sun. The sun accounts for almost 99.9 per cent of the mass of the solar system.

All of the planets go around the sun counter-clockwise, as seen from a point far to the north of the earth's orbit, in orbits that are ellipses. These orbits lie approximately in the plane of the earth's orbit, known technically as the ecliptic.

Numerical data concerning the planets are summarized in Table I of the Appendix.

MERCURY

Most people go through life without ever catching a glimpse of Mercury. This is because the planet, as seen from the earth, is so close to the sun that it is ordinarily lost in the bright glare of the sun's light. As the planet goes around its orbit, it appears to oscillate back and forth, first on one side of the sun, then the other. When it is farthest east of the sun, it can be seen low on the western horizon after sunset, about as bright as a first-magnitude star in the dim twilight. When it is farthest west of the sun, it can be seen low in the eastern sky before sunrise. At first the Greeks did not recognize it as the same object in the eastern and western sky, calling it Apollo when a morning star and Mercury when an evening star.

Because it is closest to the sun, Mercury receives more heat and light than any other planet. It revolves around the sun at a higher rate of speed than any other. Its orbit is the most flattened. It is the smallest of the planets.

It has a diameter of 3,010 miles, about 1.5 times the diameter of the moon. Its average distance from the sun is 36 million miles. It takes 88 days to make one revolution around the sun. For many years astronomers believed that it rotated once on its axis in the same length of time, thus always keeping the same face toward the sun. This would mean that one face was always bathed in sunlight, while the other remained in perpetual darkness. However, radar observations indicate that the planet has a rotation period of 59 days.

The proximity of Mercury to the sun makes it extremely difficult for astronomers to study it. Turbulence of the earth's atmosphere prevents satisfactory observations when the planet is near the horizon. Observations are best made in the daytime with suitable screens to protect the telescope as much as possible from scattered sunlight.

The telescope reveals certain dark markings on the planet, but the most competent observers do not agree as to their exact appearance. Some astronomers think that the surface of the planet is much like that of the moon and that the dark markings are similar to the lunar maria.

Measurements of the temperature of the sunlit side of Mercury, made in 1936 by Edison Pettit and Seth B. Nicholson at the Mt. Wilson Observatory by attaching a thermocouple to the 100-inch telescope, indicate a surface temperature of about 650° F., high enough to melt lead. Part of the solar energy absorbed by the planet's surface is given off again in the form of extremely short radio waves. These indicate an even higher temperature for the planet.

It was long assumed that the dark side of the planet must

have a temperature in the neighborhood of —400° F. However, observations made with the 210-foot radio telescope in Australia in 1964 indicate that the dark side of the planet has a temperature of about 62° F. If this is correct, it can only be explained on the basis that the planet has some slight atmosphere capable of transferring heat from the sunlit side to the dark side. Observations by the French astronomer Audouin Dollfus suggest that Mercury may have an atmosphere with a density of about 1/300 that of the earth's atmosphere.

Because of the angle at which the orbit of Mercury is inclined to the orbit of the earth, the planet, as seen from the earth, normally passes in the sky above or below the sun. Occasionally it passes directly between us and the sun and is visible as a tiny dark spot slowly moving across the bright face of the sun. The phenomenon is known as a transit. The last one, visible in the United States, occurred on May 9, 1970.

VENUS

Beautiful Venus is the brightest starlike object to be seen in the heavens, shining with an enchanting blue-white light. At times it is so bright that on a moonless night it casts shadows.

As seen from the earth it behaves like Mercury, oscillating back and forth, first to the east of the sun, then to the west of it. But because its orbit is larger than that of Mercury, it climbs higher into the sky when farthest from the sun. When it is east of the sun, we see it in the western sky at sunset. It is then the evening star, the first point of light to break out in the dim twilight. As the twilight deepens, it grows brighter, a beautiful celestial jewel in the gathering darkness. When it is west of the sun, it appears in the eastern sky before sunrise. It is then the morning star, shining brightly until it is lost in the rays of the rising sun. As with Mercury, the Greeks originally thought that they were dealing with two separate planets, calling it Hesperus when an evening star and Phosphorus when a morning star.

When Galileo turned his little telescope upon Venus, he discovered that the planet went through phases like the moon. This is true also of Mercury and is the result of the fact that both Venus and Mercury are closer to the sun than is the earth.

Venus is 7,600 miles in diameter. Its average distance from the sun is 67 million miles. It takes 225 days to make one revolution around the sun.

The distance between Venus and the earth varies greatly, depending upon where the two are in their respective orbits. Venus is closest to the earth when between the earth and the sun, a configuration known as inferior conjunction. At such times Venus is approximately 26 million miles from the earth. This is closer than the approach of any other heavenly body with the exception of the moon, an occasional comet, and a half dozen asteroids.

Nevertheless, Venus is in many ways a planet of mystery. This is because it is surrounded by a thick layer of dense white clouds. When astronomers turn their telescopes on Venus, they see only the outer side of these heavy clouds. It is impossible to see the surface of the planet. Because Venus is about the size of the earth, it has been called the earth's twin sister, but it obviously is no identical twin.

Spectroscopic observations indicate that the atmosphere of Venus contains 500 times as much carbon dioxide as that of the earth. However, it appears to contain less than $1/10$ as much water vapor and less than $1/1000$ as much oxygen. It is assumed that, as in the case of the earth, the chief constituent of the atmosphere is nitrogen. However, nitrogen cannot be detected by observation since it has no lines in the visible portion of the spectrum.

The dense cloud cover, which shows only vague, changing dark markings, makes it difficult to determine the rotation of Venus on its axis. At one time it was assumed that the planet turns in the same length of time it goes around the sun, namely, 225 days. However, recent radar studies indicate that the planet has a retrograde, or clockwise, rotation with a period of about 247 days.

Thermocouple measurements give a temperature of about $-36°$ F. for the outer side of the cloud cover. However, measurement of the extremely short radio waves emanating from the planet indicate a temperature of at least $600°$ F. for the surface of the planet. The American spacecraft Mariner 2, which sailed by Venus on December 14, 1962, at a distance of 21,600 miles, showed that the surface temperature is 800° F., hot enough to melt tin and lead.

The high temperature is not surprising in view of the large amount of carbon dioxide in the atmosphere of Venus. The carbon dioxide produces a "greenhouse effect," acting like the glass roof of a greenhouse which permits the entry of sunlight but prevents the escape of the longer heat waves reflected from the ground which has been warmed by the sunlight.

Mariner 2 disclosed that the cloud layer begins at an altitude of 15 miles above the planet's surface and extends to an altitude of 60 miles. Various theories have been advanced for the nature of the clouds. One is that they consist of ice crystals; another that they are dust. A third is that they consist of hydrocarbons and that the planet possesses oceans of oil instead of water. It seems most probable that they are ice crystals.

In view of the high temperature of the surface, it seems highly improbable that life as we know it exists on Venus. However, it has been pointed out that there could be high plateaus or mountain peaks where the temperature was far lower than that of the surface in general. Here on earth there are snow-capped mountains in the temperate zones and even the torrid zone. Another suggestion, perhaps fantastic, is that there might be some form of life floating in the cloud layer.

Like Mercury, Venus sometimes passes directly between us and the sun, appearing as a dark spot moving across the bright face of the sun. However, such transits are extremely rare. The last one occurred in 1882. The next one will take place in 2004.

THE EARTH

If any of the other planets were the abode of intelligent life, their inhabitants would see our earth shining in the sky just as we see the other planets. They would also see our moon as a starlike object oscillating back and forth first to the west of the earth, then to the east. They would probably think of the earth-moon system as a double planet. When the earth is nearest Venus, it would appear six times as bright in the sky of Venus as Venus appears to us. This

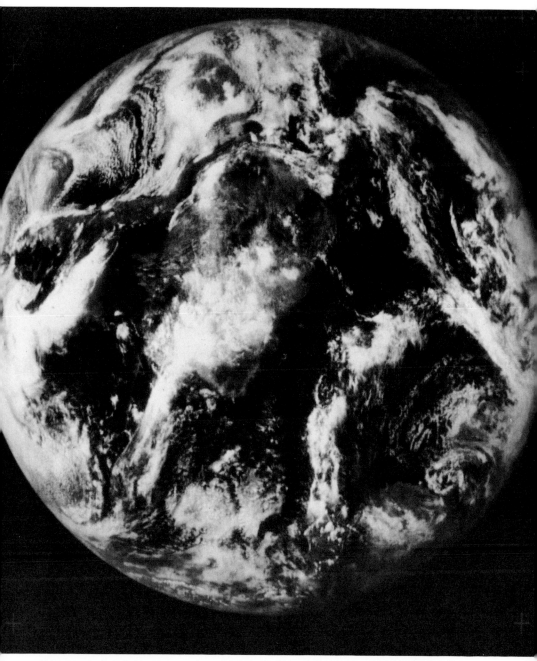

The earth, photographed by an American satellite from an altitude of 22,300 miles

is because most of the sunlit side of the earth would be visible from Venus while at such times we see only a little of the sunlit side of Venus. The moon, at such times, would appear from Venus as bright as Jupiter does to us. The earth would have a blue-white color while the moon would look yellow. Earth and moon, of course, would be less bright from the more distant planets.

Astronomers agree that it would not be possible from any of the other planets to detect evidence of human activity on the earth. The existence of life would have to be inferred from changes in appearance which indicated clouds, from changes in color indicating winter snow and summer foliage, from temperature measurements, and from spectroscopic evidence of carbon dioxide, water vapor, and oxygen.

From our own moon, the earth is an imposing sight in the sky, appearing much larger than the moon does to us. Gorgeous pictures of the earth in the lunar sky have been taken by American spacecraft. These photographs show large areas of the continents and oceans obscured by clouds.

As already mentioned, the average distance of the earth from the sun is 93 million miles. It takes 365.25 days to make one revolution around the sun. It rotates on its axis once in 23 hours, 56 minutes, and 4 seconds. The axis is tipped so that the earth's equator makes an angle of 23.5 degrees with the ecliptic, or plane, of the earth's orbit.

The average diameter of the earth is 7,913 miles. Because of its rotation, it is not a perfect sphere, but a flattened sphere, known as an oblate spheroid. As a result of its equatorial bulge it wobbles on its axis, like a toy top that is running down. Consequently, the direction of the axis in space is slowly changing. The change is known as precession. The north end of the earth's axis, when projected into the sky, defines the north pole of the heavens. Therefore, precession causes the north celestial pole to describe a circle in the sky, completing one revolution in 26,000 years. Today the North Star is Polaris in the constellation Ursa Minor (the Little Bear or Little Dipper), but in the days of ancient Egypt it was the star Thuban in the constellation Draco. The shift of the celestial pole is accompanied by a similar shift of the celestial equator, the projection of the earth's equator into the heavens. This, in turn,

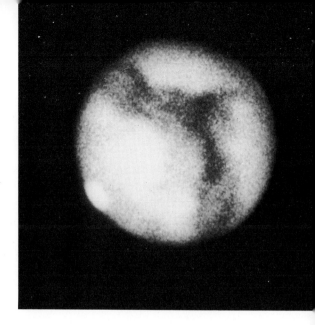

Mars, showing white polar cap

causes a slow change in the positions of the equinoxes, the two points where the celestial equator intersects the ecliptic, or apparent path, of the sun. Consequently, the phenomenon is known as the precession of the equinoxes.

The earth is discussed in detail in Part II of this book.

MARS

Mars shines with a ruddy light, appropriate for a planet named for the god of war. When, as often happens, Mars and Venus appear near each other in the evening sky, the reddish hue of Mars makes a beautiful and charming contrast to the more brilliant whiteness of Venus. The brightness of Mars varies with its distance from the earth. When it is closest, it shines like a flaming red lamp, attracting the attention of many people who rarely give a thought to the appearance of the heavens. In ancient times, and even in the Middle Ages, this periodic increase in the brilliance of the planet was regarded as an omen of war or trouble.

Mars has a diameter of 4,200 miles, a little more than half that of the earth. It is approximately 141 million miles from the sun and takes 687 days to make one revolution around the sun. It is closest to the earth every 2 years and 50 days, when the earth is between it and the sun. Because of the eccentricity of its orbit, the distance separating it from the earth at such times varies from a maximum of about 63 million miles to a minimum of slightly less than 35 million miles. The planet can be studied best at the times of these oppositions, as they are called. The time it takes Mars to rotate on its axis has been determined with exactness from observation of its surface markings. It is 24 hours, 37 minutes, and 22.5 seconds. The planet is tipped on its axis so that its equator makes an angle of 24 degrees with the plane of its orbit, only a little more than the tilt of the earth. Consequently, the planet enjoys seasons as does the earth. But since the Martian year is nearly two of ours, each season lasts twice as long.

Even a small telescope reveals that Mars is an object of great beauty, but even the largest telescope gives no better view of the planet than the view of the moon afforded by a good pair of binoculars. The planet has a reddish-orange color. However, an irregular belt running across the middle of the planet is darker in color, having bluish-gray, bluish-green, and greenish shades. The white polar caps are the most conspicuous features of the planet, and they are easily seen with a very small telescope.

The reddish-orange areas, which make up about two-thirds of the surface area, do not change with the Martian seasons. Most astronomers are inclined to regard them as rocky or sandy deserts. However, the dark areas show pronounced changes, going through a regular cycle in the course of the Martian year. The cycle appears to be related to changes in the polar caps. It is these changes that have led many astronomers to conclude that there is probably some form of life on Mars.

If we turn our telescopes on the southern hemisphere of Mars when winter is coming to an end in that hemisphere, we find that the southern polar cap is extremely large, having an area of about 4 million square miles. As spring advances in the southern hemisphere, the polar cap grows smaller and smaller. At first the shrinkage is slow. Then it becomes more rapid. About the middle of spring rifts appear in the cap and variations in brightness can be seen in it. Small areas become separated from the shrinking cap. However, they soon disappear also. The polar cap sometimes disappears entirely as spring turns into summer, but usually it remains as a tiny spot of white.

While these changes are taking place in the polar cap, the

greenish areas in the southern hemisphere are growing darker and more conspicuous.

As the summer draws to an end a strange phenomenon takes place. Dull white spots begin to appear in the polar region. They are not nearly so bright as the polar cap, and are undoubtedly bluish-white clouds or mists. They persist throughout the autumn and winter. They break up and disappear at the end of winter, once more revealing the polar cap to view.

The cycle repeats itself every Martian year with almost no variation. The polar cap may be slightly larger or smaller in some years. The rifts in the cap always appear in the same places and the last remnant of the cap is always centered at the same spot, a little distance from the planet's south pole. It is thought that this is due to the topography of the involved areas. The same sort of pattern occurs in the northern hemisphere. However, as on earth, it is summer in the northern hemisphere when it is winter in the southern hemisphere.

All observers are agreed that the polar caps cannot be mighty ice fields like our own Arctic or Antarctic. It is believed that they are layers of ice less than 10 inches thick, perhaps only an inch. Because the atmosphere of Mars is so thin and its pressure so little, this layer of ice does not melt but sublimes, going directly from the solid state to the gaseous one and becoming vapor in the Martian atmosphere.

The atmosphere of Mars apparently has about 1/30 the density of the earth's atmosphere. It is probably mostly nitrogen. Observations indicate that it contains 14 times as much carbon dioxide as the atmosphere of the earth. However, there is less than 5 per cent as much water vapor as in the earth's atmosphere and almost no oxygen at all. Two types of clouds occur in the Martian atmosphere. Bluish-white clouds, such as already mentioned, occur at high levels. They are believed to consist of ice crystals. Yellowish clouds appear at lower levels. These are believed to be dust clouds, arising from dust storms over the Martian deserts.

The temperature on Mars fluctuates violently. It can go as high as 70° F. at noon on the equator, but the average annual temperature seems to be closer to —40° F. and the nighttime temperature drops to —100° F.

Suppose that you moved the Arizona desert to the North Pole

and then lifted it 15 miles into the stratosphere. The result, some astronomers believe, would be a duplicate of conditions on Mars.

The long battle over the so-called canals of Mars is virtually over. All observers agree that there is some sort of fine detail on the planet's surface. It is difficult, however, to make out its exact nature. The markings are just at the limit of visibility with even our largest telescopes. The difficulties are increased by the turbulence of our own atmosphere. As a result photographs of Mars, which are time exposures, never show the wealth of detail that can be seen with the eye in the fleeting seconds of perfect visibility when the atmosphere above the telescope is momentarily still.

The controversy began in 1877, when the Italian astronomer Giovanni Schiaparelli announced that he had discovered a network of fine straight lines, some of them double, on the planet's surface, chiefly in the desert areas. He called them *"canali,"* the Italian word for "channels." However, it was translated as "canals" and this precipitated the battle.

In 1908 the American astronomer Percival Lowell advanced the theory that they were artificial waterways, dug by the inhabitants of Mars to convey the water of the melting polar caps across the desert regions for irrigation purposes. Lowell mapped a network of more than 400 canals and called attention to the fact that they intersected in dark spots which he called oases. Other observers, such as Edward E. Barnard, insisted that they could see only a hodgepodge of confused markings on the planet. It is not yet settled whether the canals are real features or optical illusions. However, no one today thinks that they are waterways, either artificial or natural.

The American spacecraft Mariner 4 took 21 photographs of Mars as it passed the planet on July 14, 1965, at a distance of 10,500 miles. These disclose a moonlike landscape marked with craters of all sizes. The smallest is about 3 miles in diameter, the largest about 75 miles. The clearest pictures taken by the spacecraft were the fifth to the 15th. These showed a total of 90 craters. While the existence of craters on Mars had been predicted by several astronomers, including Fred L. Whipple and Clyde W. Tombaugh, no one anticipated so close a resemblance to the lunar landscape. The photos showed no evidence of canals. However, it must be remembered that they showed only 1 per cent of the Martian surface. Additional craters were revealed by Mariner 6 and Mariner 7.

The question of life on Mars is not yet settled. Perhaps it cannot be settled until astronauts land on the Martian surface. Conditions seem much too rugged for life as we know it on earth. The temperature is too low, the atmosphere too thin, water and oxygen too scarce. There is the possibility, however, that life forms able to overcome these difficulties have evolved.

The two moons of Mars are no more than mountains on the loose. The outer one, Deimos, has a diameter of about 5 miles, the inner one, Phobos, a diameter of 10 miles.

JUPITER

Jupiter is the big brother of the solar system, the largest and most massive of the sun's family of planets. It is a brilliant object in the heavens, shining with a yellow-white light. It is surpassed in brightness only by Venus and by Mars when the ruddy planet is closest to earth.

It has a diameter of 86,800 miles, about 11 times the diameter of the earth. Its distance from the sun is 483 million miles and it takes 11.86 years to make one revolution around the sun. It rotates more rapidly on its axis than any other planet, turning once in 9 hours and 55 minutes. The flattening at its poles is very marked.

The telescope reveals Jupiter to be an object of great beauty. The planet is richly colored, various shades of red and brown predominating, with here and there an olive green. The markings, for the most part, are arranged in belts of alternate lighter and darker shades stretching across the planet parallel to the equator. The equatorial belt is bright, ranging in color from pale yellow to dull red. Above and below are darker belts ranging from reddish-brown to bluish-gray. They are known as the tropical belts. Other belts cover the planet to the north and south. The belts are not permanent but vary in number, color, and width from year to year. Small spots and other markings in them change frequently, lasting only a few weeks. It is obvious that we are not seeing a solid surface, but

Jupiter, showing the large red spot

only dense clouds in the planet's atmosphere that have been drawn into belts by the rapid rotation of the planet. It is significant that the equatorial belt turns a little faster than the others, making one turn in 9 hours and 50 minutes.

Occasionally a semipermanent feature appears in the atmosphere, lasting for a number of years. The most spectacular one, known as the Great Red Spot, appeared in 1878. It was a brick-red oval, 30,000 miles long and 7,000 miles wide. It has faded a good deal and grown a little smaller, but it can still be seen in the telescope. It does not remain stationary, but drifts in the atmosphere. Its nature is a mystery.

The Jovian atmosphere contains hydrogen, methane, and ammonia. It is thought that the atmosphere also contains helium. Measurement indicates a temperature of —226° F. for the upper portion of the atmosphere. It is thought that the clouds consist of crystals of ammonia. The colors are thought to be due to traces of various metallic compounds.

One of the mysteries of Jupiter is the more or less regular outbursts of radio noise from the planet. These apparently arise from the same spot on the planet. Each outburst equals the intensity of a billion lightning flashes in our own atmosphere. At first it was thought that the outbursts were static caused by gigantic lightning flashes in the Jovian atmosphere. However, it is impossible to explain why lightning flashes would occur only at one spot.

The low density of Jupiter makes it apparent that the planet differs greatly from the terrestrial planets. For many years it was supposed that the planet had a small rocky core surrounded by a thick layer of ice and enveloped in an immense atmosphere composed chiefly of hydrogen and helium and containing ammonia and methane. Today many astronomers are inclined to think of Jupiter, and the other major planets as well, as frozen miniature suns. It is thought that the core of each planet consists of solid hydrogen. This is surrounded by an ocean of the same elements in liquid form. Above this ocean is the atmosphere.

Temperature measurements in 1971 indicate that Jupiter radiates more energy than it receives from the sun. It is possible that the planet's core may have a considerable temperature as a result of the great pressure on it.

Jupiter has 12 moons, four of them discovered by Galileo in 1610. Three of the four are larger than our own moon while one of these three, Ganymede, is slightly larger than Mercury. Dark markings, similar to the lunar maria, have been observed on all four.

The fifth moon of Jupiter was not discovered until 1892. The next three were found early in the present century. The American astronomer Seth B. Nicholson discovered the last four between 1914 and 1951.

Saturn and its rings

SATURN

Saturn shines in the heavens with a rather dull yellow light. The ancients, who took astrology seriously and believed that the planets exercised an influence upon the affairs of men, viewed the planet with misgivings and believed that its influence was sinister.

Saturn is slightly smaller than Jupiter, having a diameter of 71,500 miles. Its average distance from the sun is almost 886 million miles and it takes 29.5 years to make one revolution around the sun.

In the telescope Saturn is the most surprising and spectacular member of the sun's family of planets. It differs from all the others, being surrounded by a great system of rings. These are easily observed with a small telescope, although it takes a large instrument to make out the finer details.

As in the case of Jupiter, the disk of Saturn is marked with belts, although they are not so well defined. The equatorial belt has a bright yellowish color, while the polar regions are greenish. These belts are obviously clouds in the planet's atmosphere, drawn into belts by the rapid rotation of the planet. The planet turns once in about 10 hours and 40 minutes but the equatorial belt turns in 10 hours and 13 minutes. The atmosphere is apparently mostly hydrogen and helium with more methane than Jupiter and less ammonia. The temperature of the upper portion of the atmosphere is $-270°$ F. The density of the planet is so low that it would float if there were an ocean big enough to hold it.

Galileo noted that Saturn was unusual in appearance, but his telescope was not sufficiently powerful to reveal the true nature of the rings. This was disclosed in 1659 by the Dutch physicist Christiaan Huygens. The planet is surrounded by a system of three thin, flat rings, lying one within the other in the plane of the planet's equator. The outer ring has a diameter of 170,000 miles. A conspicuous dark division between the outer and middle rings is known as Cassini's division after the French astronomer Jean D. Cassini, who noted it in 1675. The inner ring, because of its dark color, is called the crepe ring. The rings are not solid, but composed of millions of little moons, or moonlets, perhaps no larger than small rocks. Spectroscopic studies by Gerard P. Kuiper show that each moonlet is covered by a layer of ice. They may be composed entirely of ice and frozen gases. If so, one can think of the rings of Saturn as composed of snowballs.

Saturn has 10 moons, the largest of which, Titan, is larger than our own moon and about the size of Mercury. The tenth moon was not discovered until 1967.

URANUS

A stargazing musician created worldwide excitement by discovering Uranus. The ancients had known Mercury, Venus, Mars, Jupiter, and Saturn. But no one knew that Uranus existed until this hero of science found it on March 13, 1781. He was Sir William Herschel and his story is fascinating.

Herschel was born in Hanover, Germany. His father was the oboe player in a miltary regiment and at 14 William became oboist in the Hanoverian Guards. A few years later he left the service and settled in England. He became a bandmaster first and in 1766 the organist in the famous Octagon Chapel at Bath. He spent his spare time studying mathematics, optics, Italian, and Greek. Finally, he became interested in astronomy and decided to make his own telescope because he could not afford to buy one. He became so engrossed in polishing mirrors and lenses for his telescopes that his sister, who was his faithful assistant to the end of his days, had to feed him as he worked away on the mirrors. Before long he had built himself bigger and better telescopes than the world had ever seen. Then he launched himself upon a truly Herculean task: he proposed to make a complete survey of the heavens.

On March 13, 1781, while observing a field of faint stars in the constellation Gemini, Herschel noted an object larger and brighter than any of the stars in the field. At first he thought he had discovered a comet. But further study established that it was indeed a planet. The discovery made Herschel famous. He was appointed telescope-maker to the King, given a pension that enabled him to abandon music and devote all his time to astronomy, and in time was knighted.

Uranus is 29,400 miles in diameter. It is almost 2 billion miles from the sun, its average distance being 1.783 billion miles. It takes 84 years to make one revolution around the sun. However, it turns rapidly on its axis, rotating once in 10 hours and 45 minutes. Its axis of rotation is tipped almost at right angles to the plane of its orbit and the direction of its rotation is retrograde, or clockwise.

To the unaided eye, Uranus appears as an extremely faint star of about sixth magnitude. A telescope shows it as a sea-green disk marked with very faint belts. Spectrographic studies indicate the presence of methane and hydrogen in the planet's atmosphere. It is inferred that ammonia and helium are also present. The temperature of the atmosphere is —310° F.

Uranus has five moons, two of them discovered by Herschel. The third and fourth were discovered in 1851, but the fifth was not found until 1948. All five lie in the equatorial plane of the planet and revolve around it in the direction of the planet's rotation.

NEPTUNE

In the decades after Herschel's discovery of Uranus it became more and more apparent that the planet was not following the orbit astronomers had calculated for it. Now when someone strays from the straight and narrow path here on earth, we assume some influence has been pulling him off. Consequently, by 1840 many astronomers were convinced that there must be an undiscovered planet in the solar system whose gravitational pull was influencing Uranus.

Two young astronomers, John Couch Adams, an Englishman, and Urbain Jean Joseph Leverrier, a Frenchman, each without the knowledge of the other, set out to solve the problem. Adams, who had been graduated from Cambridge with highest honors in mathematics, left his determination of the planet's position at the Greenwich Observatory on October 21, 1845. But Sir George Airy, the Astronomer Royal, delayed the search for the planet. The next year Leverrier completed his calculations and published them. He also wrote to Dr. Johann Galle, the director of the Berlin Observatory, telling him where to look for the planet. Galle, with the aid of a new star chart, found the planet on September 23, 1846.

Airy now made Adams' paper public. Two storms arose at once, a storm of indignation in England that faster action had not been taken, and a storm of rage in France. Accusations were made that the whole thing was a British plot to rob France of glory. Adams and Leverrier refused to be drawn into the quarrel, and in the end the counsel of one astronomer that "there was glory enough for both" prevailed. The new planet, because of its pale green color, was named Neptune.

Neptune is slightly smaller than Uranus, having a diameter of 28,000 miles. It is almost 3 billion miles from the sun, its average distance being 2.794 billion miles. It takes 165 years to make one revolution around the sun. It turns on its axis once in approximately 16 hours.

Unlike Uranus, Neptune cannot be seen without a telescope. It is so far from the sun that no details can be distinguished. Spectroscopic studies indicate its atmosphere resembles that of Uranus. The temperature is about —345° F. Neptune has two moons.

PLUTO

By the start of the present century it was apparent that the presence of Neptune was not sufficient to explain the perturbations in the orbit of Uranus. It seemed, therefore, that there must be yet another planet in the solar system. Percival Lowell, who had founded the Lowell Observatory at Flagstaff, Arizona, largely to study Mars, turned his attention to the problem. By 1905 he had calculated where he thought the trans-Neptunian planet could be found. These calculations were published in 1914. Similar calculations were made by William H. Pickering at the Harvard Observatory. Lowell searched for Planet X, as he called it, but without success.

The search was resumed by Clyde W. Tombaugh, a young assistant at the Lowell Observatory, in 1929. He had the advantage of a new 13-inch telescope with a wide field and a device known as a blink microscope. His method was to make two photographs of a small area of the sky several nights apart. The two plates were than placed in the blink microscope. This device enabled him to view alternately and rapidly one plate after the other. The stars appeared to stand still as he switched from plate to plate. But any object which had moved in the interval between the two exposures would appear to blink, or jump rapidly back and forth from one plate to the other. Tombaugh worked for almost a year without success, spending a total of 7,000 hours at the blink microscope. Then on February 18, 1930, comparing plates taken on the nights of January 23 and 29, he found the jumping spot of light that proved to be Planet X. The planet was subsequently named Pluto, after the god of the underworld, a fitting name for two reasons. One is that it is so far from the sun that it moves in the outer darkness of the solar system, receiving almost no heat or light from the sun. The other is that the first two letters of its name are the initials of Percival Lowell.

Astronomers were amazed to find that Pluto bore no resemblance to Jupiter and the other giant planets. Its diameter seems to be about 3,600 miles, less than half that of the earth.

The orbit of Pluto is extremely eccentric, the distance of the planet from the sun varying from 2.8 billion to 4.5 billion miles. It takes the planet 248 years to go once around the sun. It is believed that Pluto turns on its axis in a period of about 6.5 days.

It is difficult to understand how a planet no larger than Pluto could exercise a sufficient gravitational pull to cause the observed perturbation of the orbit of Uranus, as well as the perturbation that had been noted subsequently in the orbit of Neptune. Some astronomers have suggested that Pluto originally was a moon of Neptune which escaped from Neptune's gravitational field.

four

ASTEROIDS, COMETS, AND METEORS

When beggars die there are no comets seen,
The heavens themselves blaze forth the death of princes.
—SHAKESPEARE

ASTEROIDS

AS early as Kepler's time it was noted that there was a wide gap between the orbits of Mars and Jupiter. Kepler even predicted that a planet might someday be discovered there.

A curious relationship in the distances from the sun of the six planets then known was pointed out by the German astronomer Johann Titius in 1766. Write down a series of 4s. Add 0 to the first, 3 to the second, 6 to the third, 12 to the fourth, 24 to the fifth, and so on, doubling the added number each time. The first seven numbers in the series thus formed are 4, 7, 10, 16, 28, 52, and 100. If we disregard the fifth number, the other six approximate the average distance of the six planets from the sun, if the earth's distance is taken as 10. Titius published his discovery as a footnote in a book by another author which he translated from French into German. As a result it received practically no notice.

Attention was called to it in 1772 by Johann Elert Bode, afterward director of the Berlin Observatory. Bode refused to believe that the solar system ended with Saturn or that the space between Mars and Jupiter was empty.

"Can we believe that the Creator of the world has left this space empty?" he wrote in 1772. "Certainly not!"

Herschel discovered Uranus in 1781. Its distance from the sun was found to correspond to the next number in the series, namely 196 (4 plus 192). Bode was now more certain than ever that a planet existed between Mars and Jupiter, and because he continued to press the subject, the discovery made by Titius became known as Bode's law.

In 1800 Bode and Baron Franz von Zach, director of the observatory at Gotha, Germany, called a meeting of astronomers to organize a hunt for the missing planet. They divided the sky into 24 areas and assigned each one to a European astronomer. Von Zach jokingly called them the astronomical detective police.

One area was assigned to the Italian astronomer Guiseppe Piazzi at Palermo, Sicily. However, before Piazzi had been notified that he was to take part in the hunt, he discovered by chance a tiny planet in the gap between Mars and Jupiter. On the night of January 1, 1801, the first night of the nineteenth century, Piazzi noted a moving object in the star field he was studying. At first he assumed that it was a comet. Further study proved that it was indeed a tiny planet. He named it Ceres after the traditional goddess of the island. Its distance from the sun corresponded to the fifth number in the Titius-Bode series.

Astronomers were amazed by its tiny size. But other surprises were to come. A second tiny planet was discovered in the gap in 1802, a third in 1804, and a fourth in 1807. No more of these tiny planets, or asteroids as they were named, were discovered until 1845, but since that day, the number has grown steadily. Astronomers have charted the orbits of about 1,600, but it is estimated that there are at least 50,000 faint asteroids. Trails of 90 asteroids were noted on a single photograph taken with the 48-inch Schmidt telescope on Palomar Mountain.

Asteroids are frequently discovered and then lost track of. Sometimes old ones are mistaken for new ones. When the discovery of a new asteroid is verified, the discoverer is permitted to name it. The first ones were named after mythological gods and personages.

They have since been named after nations, cities, colleges, friends of the discoverers, and even steamboats and pet dogs. A Belgian astronomer named one Hooveria, in grateful memory of Herbert Hoover's services in feeding the Belgian children during World War I.

All of the asteroids go around the sun in the same direction as do the planets, that is, counterclockwise. Orbital periods range in general from 3 to 6 years. The orbits are more eccentric, or flattened, than those of the planets. A few of the asteroids have such eccentric orbits that they cross the earth's orbit and as a result sometimes pass fairly close to the earth. Eros comes within about 14 million miles of the earth. The closest known approach of an asteroid occurred in October 1937, when a small asteroid, Hermes, came within 500,000 miles of the earth. Icarus has so eccentric an orbit that it crosses the orbit of Mercury.

Ceres, the first asteroid discovered, is the largest, having a diameter of 480 miles. Pallas has a diameter of 300 miles, Vesta 240 miles, and Juno 120 miles. Smaller asteroids range from 50 miles in diameter to less than a mile. Ceres and Pallas appear to be spherical, but the others are irregular in shape.

The first theory advanced to account for the asteroids was that they are the remnants of a planet that exploded or disintegrated for some reason. It is now thought more likely that the powerful gravitational influence of Jupiter prevented the formation of a single big planet and that a number of small ones formed instead. Ceres and Pallas, which are spherical, are believed to be two of these original planets. The others were shattered by frequent collisions into the small irregular asteroids which we see today.

COMETS

From the dawn of written history to the present the superstitious have always regarded comets with fear as the portents of disaster. In 1910 many Chinese villagers shot off fireworks in the hopes of driving Halley's comet away. In the United States many people believed that the comet of 1812 foretold the war of that year and that Donati's comet in 1858 heralded the Civil War.

Head of Halley's Comet photographed on May 8, 1910, with the 60-inch telescope of the Mt. Wilson Observatory, on the comet's most recent visit to the earth's neighborhood

The ancients misunderstood the nature of comets completely. Aristotle regarded them as some sort of exhalation coming out of the earth and becoming inflamed in the upper regions of the atmosphere. Some early astronomers seem to have regarded comets as living creatures possessed of will and purpose, swimming about in the atmosphere not unlike fish in the ocean. Exaggerated descriptions of comets as being blood-red, taking the form of a hand with a flaming sword, and the like are frequent in the Middle Ages.

It was Tycho Brahe who first proved by his observations of the bright comet of 1577 that comets were out among the planets and not in the earth's atmosphere. Tycho saw, moreover, that the orbit of the comet of 1577 cut across the orbits of the planets and upset the ancient notion that the planets were lodged in a nest of concentric crystalline spheres.

Until the invention of the telescope, only the very bright and therefore very large and spectacular comets could be seen. About 400 comets were recorded prior to 1600. Since then, the number of comets seen annually has increased with the increase in the size of telescopes. About 500 comets have been recorded since 1600. As a rule about seven comets are discovered annually. Only an occasional comet becomes bright enough to be seen with the unaided eye. Comets are named after their discoverers and many amateur astronomers spend much time hunting for them.

When a comet is first picked up with a telescope, it appears only as a fuzzy spot of light moving rapidly against the background of stars. When the comet is within 200 million miles or so of the sun, it begins to develop a tail, which grows longer as the comet drawn nearer. As the comet swings around the sun, the tail always points away from the sun. When the comet is receding from the sun, the tail streams out in front of the comet, gradually growing smaller and finally disappearing. As a rule, a comet remains visible for only a few days or at most a few weeks.

The telescope reveals that the head of a comet consists of a small bright center, known as the nucleus, and, surrounding this, a fuzzy, hazy mass called the coma. The nucleus rarely exceeds 500 miles in diameter, but the diameter of the coma is enormous, ranging from 30,000 to a million miles in some cases. The tail is even more enormous and may range from 5 million to 100 million miles in length.

But while the volume of a comet is great, the amount of matter composing it is very small. This is apparent from the large amount by which a comet is deflected from its original orbit by the gravitational pull of one of the larger planets when it happens to pass near such a planet, and from the fact that the comet has no effect upon the orbit of the planet.

A theory of the nature of a comet first proposed in 1950 by the American astronomer Fred L. Whipple is now generally accepted. According to it, the nucleus of a comet is a "dirty iceberg," a mixture of frozen water, ammonia, methane, and carbon dioxide, in which a conglomeration of stony particles, ranging in size from grains of sand to small pebbles, is embedded. As the comet approaches the sun the ices begin to warm up and vaporize, thus producing the coma. As the comet gets still closer to the sun, the pressure of sunlight and the impact of the solar wind, the streams of subatomic particles from the sun, drive the gases of the coma into a tail. In some respects the comet's tail is like the smoke from the stack of a steam locomotive. As the locomotive speeds along the rails, the trail of smoke is a continuous accompaniment. But it is continuously being created anew from smoke emitted from the stack and it is continuously being dissipated into the air. So too the tail of the comet is being continuously created from gases driven out of the coma and continuously dissipated into space. It has been calculated that a comet loses about 0.5 per cent of its mass each time it swings around the sun. Observations of two comets in 1970 with hydrogen-sensitive equipment from earth-orbiting satellites indicated that these comets were surrounded by huge spheres of hydrogen.

Comets are assumed to be composed of material that was left over when the solar system took shape. According to the theory advanced by the Dutch astronomer Jan H. Oort, there is a cloud of comets, perhaps as many as 200 billion, on the outskirts of the solar system. They go around the sun in huge orbits that take them out as far as 10 trillion miles from the sun. Because the mass of a comet is so small, this entire cloud of comets represents only about 10 to 100 times the mass of the earth.

Occasionally, a comet, when farthest from the sun, may approach fairly close to a star. The gravitational pull of the star may alter the comet's orbit so much that it now swings into a highly eccentric, or elongated, orbit that carries it down into the central portion of the solar system, across the orbits of the planets and close to the sun. We see the comet as it swings around the sun, and since it may take from 100,000 to a million years to go around its new orbit, we see it only once. Such a comet is known as a long-period comet.

If, however, the comet passes close to Jupiter, the gravitational pull of the giant planet will cause a second change in the comet's orbit. It is now confined to a much smaller orbit, taking from 3 to 200 years to make one circuit. The comet then is known as a short-period, or periodic, comet. About 70 periodic comets are known.

The most famous of them is Halley's comet, named after Edmund Halley, Sir Isaac Newton's friend who became the British Astronomer Royal. Using Newton's law of gravitation, Halley, in 1704, calculated the orbits of 24 bright comets that had been observed between 1337 and 1698. He noticed that three of the orbits, those of the comets of 1531, 1607, and 1682, were strikingly similar. Suspecting that the three were one and the same comet, going around an orbit in 75 or 76 years, he boldly predicted that it would return in about the year 1758, but did not live long enough to see his prediction come true. It was sighted on Christmas Day, 1758, but Halley had died 16 years earlier. A search of old records revealed that the comet had been seen as early as 240 B.C. It appeared in 1066 and was then regarded as the omen that heralded the defeat of King Harold by William the Conqueror at the Battle of Hastings. The comet is portrayed on the famous Bayeaux tapestry. Halley's comet was last seen in 1910 and is scheduled to return in 1986.

The inner end of the orbit of Halley's comet is within the orbit of Mercury, the outer end is beyond the orbit of Neptune.

About 45 of the periodic comets have orbits which extend out as far as the orbit of Jupiter. These are known as Jupiter's family

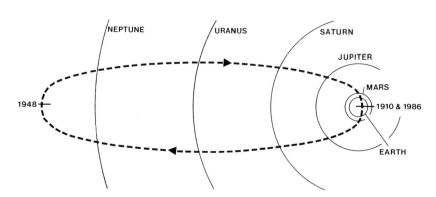

Figure 6. Orbit of Halley's Comet

of comets. Astronomers are agreed that they were originally long-period comets that were captured by Jupiter's gravitational pull. They have periods ranging from 3 to 9 years. Being continuously so near the sun damages them a great deal, causing them to dissipate their mass rapidly, growing fainter and fainter with each return to the solar neighborhood.

Biela's comet, one of these captured comets with a period of 6.6 years, was sighted a number of times as it rounded the sun between 1772 and 1832. In 1846 it was seen to split into two, the two comets traveling along parallel paths about 150,000 miles apart. The twin comets returned in 1852 but were now more than a million miles apart. They have never been seen since. Another comet, Taylor's comet, split in two in 1916. Several other comets have gradually faded from view as though they have completely disintegrated or melted away.

METEORS

If you watch the heavens on a clear night for 15 or 20 minutes, you are almost certain to see one or two starlike objects suddenly burst into view and streak across the sky, sometimes leaving a luminous trail that persists for a few seconds. They are known popularly as shooting stars. An old superstition says that if you make a wish when you see one, it will come true.

Astronomers call them meteors. They are, of course, not stars at all but incredibly tiny objects no larger than a grain of sand or the head of a pin. Such a bit of material remains invisible until it enters the earth's atmosphere. Then friction causes it to grow so hot that it begins to vaporize. The fiery trail of the shooting star is the trail of the incandescent vapors as the bit of material is consumed.

On a clear moonless night an observer will see about 10 meteors an hour. The average meteor becomes visible at an altitude of 60 to 80 miles above the earth's surface and burns itself out by the time it has descended to an altitude of 40 miles.

The number of meteors seen from any one location is only a small fraction of those entering the earth's atmosphere. It is estimated that over the whole earth 24 million meteors, visible to the unaided eye, appear each day. However, the telescope reveals a vastly greater number of meteors than can be seen by the unaided eye. It is probable that 8 billion meteors enter the earth's atmosphere every 24 hours.

It is now thought that about 90 per cent of the meteors the earth encounters are the debris of comets which have disintegrated under the destructive influence of the sun. The other 10 per cent are believed to be tiny fragments resulting from the collisions of asteroids with one another.

METEOR SHOWERS

A spectacular display of meteors, visible over the eastern half of the United States on the night of November 12, 1833, convinced many terrified witnesses that the end of the world was at hand. The display began before midnight and increased in intensity as the night wore on. In the hours before dawn the meteors were as thick as snowflakes and it appeared as though the heavens were raining fire. Some of the meteors rivaled Jupiter or Venus in brightness. One was reported to have been nearly as large as the moon. It was estimated that 10,000 flashed across the sky in an hour.

Calmer and more competent observers noted that all the meteors seemed to come from one point in the constellation Leo. Astronomers recognized that this meant that the earth had collided with a vast swarm of meteors.

It was recalled that the explorer Alexander von Humboldt had reported a similar display from South America on November 11, 1799. A search of historical records revealed similar earlier displays at 33-year intervals. It was evident, therefore, that the earth periodically collided with a swarm of meteors moving in an orbit around the sun. It was anticipated that there would be a repetition in 1866 and this proved to be the case. The 1866 event was as spectacular as that of 1833. But the 1899 and 1932 displays were disappointing. Between 1866 and 1899 the swarm had passed near both Jupiter and Saturn and its orbit had been changed enough that the earth passed only through the outer edge of the swarm. Brief displays occur annually in November when the earth crosses the orbit of the swarm, indicating that stragglers are strung out throughout the orbit.

The 1833 spectacle inspired astronomers to pay more attention to the subject of meteors and it was found that a considerable number of modest displays repeat themselves annually. They became known as meteor showers. The point from which a shower seems to come is called the radiant, while the shower is named for the constellation in which the radiant is located. Thus the spectacular shower seen in 1833 and again in 1866 became known as the Leonids. Among the more prominent showers are the Lyrids about April 21, the Aquarids about May 4, the Perseids about August 12, the Draconids about October 10, the Orionids about October 22, the Taurids about November 1, the Andromedids about November 14, and the Leonids about November 17.

The orbits of a number of these meteor swarms have been found to coincide with the orbits of known comets, supporting the view that the meteors are debris driven out of the comet when it swings around the sun. The Leonids were found to occupy the same orbit as a faint comet seen in 1866. Both the Aquarids and the Orionids share the orbit of Halley's comet. The Andromedids occupy the orbit of Biela's comet, which split in two and has not been seen since 1852. It is assumed that the meteors are all that is left of Biela's comet.

METEORITES

Occasionally an exceptionally bright fireball, or bolide, flashes across the sky, leaving a luminous trail behind it that may endure for half an hour or more. It may rival the full moon in brightness. Its passage through the atmosphere may be seen by startled witnesses along a path several hundred miles long. A rumbling sound like thunder often accompanies its passage and its journey may end abruptly with an explosion. Such brilliant fireballs are sometimes visible even in full daylight. The fireball is not completely consumed by its encounter with the atmosphere and the remnants which fall to earth are known as meteorites.

Less spectacular fireballs also give rise to meteorites. It is estimated that about 25 meteorites fall to earth within the United States each year and about 2,000 a year over the whole earth. There are about 1,600 meteorites in museums today. Some were seen to fall, others were found by accident and recognized as meteorites.

Meteorites consist of three types known as stony, iron, and stony-iron meteorites. The stony meteorites consist of minerals which are well known on earth, chiefly silicates of calcium, magnesium, and iron. The most common constituent is the magnesium-iron silicate known as olivine. The iron meteorites consist chiefly of iron and nickel with small amounts of other chemical elements. The stony-iron meteorites, as their name implies, are mixtures of the other two types.

Most of the stony meteorites contain tiny glassy globules known as chondrules. The origin of these globules is a mystery. It is possible that they formed in the cloud of gas and dust from which the solar system evolved, perhaps as the result of lightning flashes. They were subsequently incorporated into the planets and asteroids as those bodies took shape.

Meteorites vary greatly in size. Most of them are small but some extremely large meteorites have been found. Admiral Robert E. Peary, who discovered the North Pole, found an iron meteorite in Greenland that weighed 36 tons. It can be seen in the American Museum of Natural History in New York City. The largest known meteorite, a 60-ton iron meteorite, lies where it fell in Hoba in southwest Africa. Stony meteorites are smaller, because of a tendency to shatter in flight. The largest known one, found in Kansas, weighs about 1 ton.

The large fireballs which give rise to meteorites are believed to be fragments of asteroids that have crossed the earth's path and are quite different from the meteors, which are the debris of comets. The mechanism by which they become luminous is also quite different. Because of its size the fireball traps and compresses the air in front of it. Compression causes this cap to grow sufficiently hot to melt the surface of the meteorite. The incandescent droplets, swept away by the airstream, form the luminous trail. The pressure is usually great enough to shatter a stony meteorite. As the meteorite or its fragments reach the lower, denser portion of the atmosphere, they are slowed up by friction so that they fall to earth as though dropped from an airplane. Large meteorites will bury themselves in the ground. There are no records of anyone ever having been killed by a meteorite, but there are some reports of persons having been struck by small fragments. In 1954 a woman in Alabama was bruised by a small meteorite that fell through the roof of the house.

METEORITE CRATERS

In the Arizona desert, near the city of Winslow, there is a great crater 4,000 feet in diameter and 600 feet deep. Its walls rise to a height of 120 feet above the surrounding plain. It bears a striking resemblance to the craters on the moon. Thousands of small iron meteorites have been found in and around the crater. Astronomers are certain that Meteor Crater, as it has been named, was the result of the impact of a huge meteorite, perhaps 60 feet in diameter and weighing as much as 100,000 tons. It is thought to have landed about 300,000 years ago. About a dozen other craters in various parts of the world appear to be the result of meteorite impacts.

Extremely large meteorites have hit the earth twice in the present century, in both cases in Siberia. The first Siberian fall occurred on June 30, 1908, the second on February 12, 1947.

The 1908 fall occurred in the forested area of the Tunguska River in central Siberia. The meteorite was seen as a brilliant fireball in broad daylight from hundreds of miles away. The meteorite, which apparently exploded, created more than 200 craters, the largest about 150 feet in diameter. An area around the craters with a radius of 15 miles was seared and charred as if with a gigantic blowtorch. For an additional 20 miles in all directions, the trees were uprooted and lay pointing away from the center of the explosion.

More than 200 small craters were created by the 1948 fall, which occurred in the Sikhote-Alin mountain range in southeast Siberia. It was estimated that the nickel-iron fragments in the area weighed more than 100 tons.

MICROMETEORITES

A constant rain of dustlike particles, less than 1/30,000 of an inch in diameter, falls into the earth's atmosphere. Because of their tiny size, they are not vaporized by their passage through the atmosphere and settle unchanged on the earth's surface. They are known as micrometeorites.

They have been found on mountaintops, on the ice fields of the Arctic and Antarctic, and in ooze dredged up from the floors of the oceans. It is estimated that 10,000 tons of these tiny particles fall to earth daily, more than 3 million tons a year. They may be the debris of disintegrated comets or shattered asteroids, perhaps both. Unmanned satellites and planetary spacecraft have been used to evaluate the density of micrometeorites near the earth, Venus, and Mars.

THE ZODIACAL LIGHT

Just after twilight, on clear moonless nights, a faint streak of light can sometimes be seen extending into the western sky from the point where the sun has set. A similar streak can sometimes be seen in the eastern sky ahead of the sunrise. The phenomenon is known as the zodiacal light and is believed to result from the reflection of sunlight on a huge lens-shaped cloud of micrometeorites surrounding the sun in the plane of the ecliptic. Directly opposite the sun a still fainter spot of light, called the gegenschein, or counterglow, can sometimes be seen. This is believed to be the reflection of sunlight on micrometeorites in the cloud around the earth.

five

THE SUN

The glorious lamp of heaven, the radiant sun,
Is Nature's eye.
—DRYDEN

ENDLESSLY circling the sun in the grasp of solar gravitation, the earth is dependent upon the sun for its light and heat. Were the sun to go out, the earth would be plunged into darkness, relieved only by the feeble light of the stars, for the moon and planets shine only by reflected sunlight. Within a few days the tropics would be as cold as the polar regions, and not long after all plant and animal life would be frozen to death. Soon the oceans would be frozen solid. Then the atmosphere itself would freeze, forming first a layer of liquid air upon the frozen surface of the earth and then a layer of solid air.

We are dependent upon the sun for our food and fuel, for plants cannot grow without the energy of sunlight, and coal and oil are only the fossil remains of plants that grew millions of years ago.

Modern astronomy has revealed that the earth is a dwarf in comparison with the sun. The sun is a star, a huge self-luminous yellowish-white whirling sphere of heated gases, with a diameter almost 110 times that of the earth. The diameter of the earth is 7,913 miles. The diameter of the sun is 864,600 miles. The volume of the sun is so great that it would take 1.3 million globes the size of the earth to equal it.

The surface temperature of the sun is about 10,000° F. However, the interior is much hotter. Astronomers believe that the temperature at the center of the sun is about 27 million degrees. Were our earth to be suddenly thrust into the sun, it would last no longer than a snowflake in a red-hot stove.

The solar energy which falls on each square yard of the earth's surface per second in the form of light and heat is equivalent to about 1 horsepower. But the earth intercepts only about one two-billionths of the sun's radiation. Astronomers were long at a loss to explain the sun's continuous outpouring of energy. It is now believed that the center of the sun is an immense atomic furnace in which hydrogen is converted into helium. Four hydrogen atoms combine to form one helium atom. But in this thermonuclear reaction, as it is called, there is a loss of mass. The lost mass is converted into energy in accordance with the famous Einstein equation $E=mc^2$, that is, the energy produced is equal to the mass multiplied by the square of the velocity of light. It has been calculated that the sun loses 4 million tons a second. Even so, it seems certain that the sun will maintain its present rate of energy production for the next 5 billion years.

Although the sun is 93 million miles away, astronomers have assembled an immense amount of information about it. This has been possible through the use of special telescopes and delicate measuring instruments of various kinds. The chief aid has been an instrument known as the spectroscope.

THE SPECTROSCOPE

Everyone knows that if a little glass prism is held up to the sunlight, it divides the white beam of light into a rainbow. This action of the prism was first explained by Sir Isaac Newton in 1672. He showed that white light is a mixture of all the colors of the rainbow. The colors differ from one another because of the wavelength of their light. Red light consists of the longest waves, violet light the shortest. The prism bends the light of different wavelengths by different amounts and thus spreads out the sunlight into a band of its component colors.

The next important step in the study of sunlight was taken more than a century later by Joseph von Fraunhofer, whose brilliant and tragic career is one of the most interesting in the history of astronomy. On July 21, 1801, Maximilian I, Elector of Bavaria, happened to be walking along a street in Munich, when he witnessed the collapse of two old houses. All the occupants but one were killed in the accident. The one to escape was a young boy named Joseph von Fraunhofer. The Elector, moved by the obvious distress of the boy, gave him 18 ducats he happened to have in his pocket. The boy used part of the money to purchase his freedom from a manufacturer of mirrors to whom he had been apprenticed. With the rest he bought books and machinery for polishing lenses. In time the youth became the most skillful lensmaker in all Europe, grinding lenses for the largest telescopes of the day. In addition he became an authority on the subject of optics, devising and performing many experiments of the utmost importance. But the hardships of his younger days told upon him, and death cut short his brilliant career when he was only 31.

It was in 1815 that Fraunhofer performed his epochmaking experiment with sunlight. Like Newton, he admitted a tiny ray of sunlight into a darkened room through a narrow vertical slit, permitting the beam to fall upon a glass prism. However, he used the small telescope of a surveyor's theodolite to examine a magnified image of the spectrum. To his surprise, he found the spectrum crossed by hundreds of dark lines of various thicknesses and intensities. He counted more than 576 and subsequently established that they were always present in their exact relative positions. He named the more noticeable lines by letters of the alphabet, a method of designation still in use.

It was not until a half-century later, however, that the riddle of these Fraunhofer lines, as they became known, was cleared up. In 1859 a professor at the University of Heidelberg, Gustav Kirchhoff, not only explained the mystery, but laid the foundations for some of the most important advances modern science had made in astronomy, physics, and chemistry.

Kirchhoff and his colleague Robert Bunsen, the inventor of the little gas burner known as a Bunsen burner that is still used in every physics and chemistry laboratory, had been carrying on

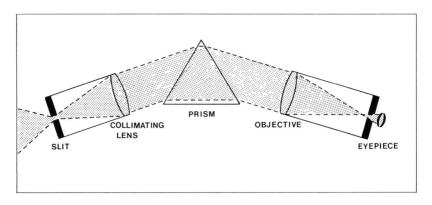

Figure 7. The spectroscope

experiments with an improvement on Fraunhofer's arrangement, which had been named the spectroscope.

Light was admitted to the prism through a slit at the end of a tube known as the collimator. A convex lens in the tube rendered the light rays parallel. A small telescope on the other side of the prism was used to view the enlarged image of the spectrum. The same scheme is still used in modern spectroscopes. However, when the eyepiece of the telescope is replaced by a photographic plate, the instrument is known as a spectrograph. (See Figure 7.)

They found that various chemical elements when vaporized in the flame of a Bunsen burner furnished spectra that consisted only of a number of isolated bright lines. There was no continuous rainbow. Each element had its own characteristic pattern of bright lines so that an element could be identified by its spectrum lines.

Kirchhoff extended these studies by permitting a strong source of white light, such as the light from a white-hot platinum wire, to shine through a flame containing some chemical element. Such a source of light by itself furnishes a so-called continuous spectrum, a continuous rainbow unmarked by any dark lines. But when the light passed through the flame, dark lines appeared in the spectrum. In every case the dark lines appeared exactly where bright lines would have appeared had the element in the flame been the only source of light. In other words, a layer of relatively cooler gas removes from the spectrum of an extremely hot source of white light the very spectrum lines which otherwise it would itself create.

Here then, at last, was the explanation of the Fraunhofer lines. Light arising in the very dense and extremely hot gaseous surface of the sun has to pass through the less dense and relatively cooler gases of the sun's atmosphere. Hence the formation of the dark lines in the solar spectrum.

But now the very secrets of the sun were unlocked. All that it was necessary to do was to form spectra from various chemical elements in the laboratory and compare the bright lines thus formed with the dark lines in the sun's spectrum. In this way it was possible to identify the chemical elements which compose the sun's atmosphere. Modern spectroscopic studies have revealed more than 30,000 lines in the solar spectrum. Each line, of course, is a different wavelength of light.

Modern laboratory experiments, moreover, have shown that the number and intensity of spectrum lines depend upon conditions of temperature and pressure, and upon electrical and magnetic conditions. Consequently, the spectroscope, in addition to disclosing the chemical composition of the sun, has yielded valuable information as to conditions of temperature and pressure and as to electrical and magnetic conditions in the sun.

The spectroscope reveals the presence of 67 of the 92 chemical elements in the sun. There is no reason to suppose that the other 25 are not there also. They are elements which are present in the depths of the sun and so do not appear in the spectrum, or elements whose spectrum lines lie in the far ultraviolet region of the spectrum and cannot be observed through the earth's atmosphere. Most recent studies indicate that the atmosphere of the sun is about 80 per cent hydrogen by volume and almost 20 per cent helium. All the other chemical elements together make up only about 0.2 per cent.

THE SOLAR SURFACE

A word of caution must be given the amateur astronomer before we discuss the sun in detail. It is sometimes possible, at sunrise or sunset, when the sun is dimmed by a hazy atmosphere, to look directly at the sun. But this should never be attempted when the sun is higher in the sky. The result might be serious damage to the eyes, even blindness. Under no circumstances should binoculars

The 150-foot-tower solar telescope of the Mt. Wilson Observatory

ever be pointed at the sun. It is equally dangerous to point a telescope at the sun unless it is equipped with a solar eyepiece.

The luminous surface of the sun is known as the photosphere. It is not a solid surface such as we have on earth, but a dense, boiling, stormy sea of white-hot gases. The density of the photosphere is such that we cannot see into it to any appreciable depth. Our best pictures of the photosphere have been obtained with telescopes carried into the stratosphere by balloons, above the dust and turbulence of the lower atmosphere. These show that it has a mottled or speckled appearance. It consists of constantly changing bright granules, ranging from 200 to 1,000 miles in diameter, separated by narrow dark lanes. The life of a granule is about 4 minutes, new ones appearing as old ones disappear. It seems certain that the granules are gigantic bubbles of gas rising from the solar interior.

SUNSPOTS

The most noticeable feature of the photosphere are the dark areas known as sunspots. They appear as dark spots only by contrast with the photosphere. Actually, a sunspot is brighter than the brightest artificial light.

A sunspot consists of two parts, a dark central portion known as the umbra, and a somewhat lighter outer portion called the penumbra. The penumbra consists of long luminous filaments radiating outward from the umbra. Spots are very irregular in shape, size, and structure. The umbra is not always at the center of the penumbra. Frequently a group of spots will have a common penumbra. The umbra varies in size from a diameter of 500 miles in the case of a small spot to a diameter of 30,000 miles or more. An extremely large spot, observed in 1947, had an umbra 52,000 miles in diameter and a penumbra 90,000 miles in diameter.

Sunspots are not permanent features. They start as small dark

Direct photograph of sun (left), *showing photosphere and sunspots. Photograph taken in one line of hydrogen spectrum* (right), *showing hydrogen clouds in the chromosphere*

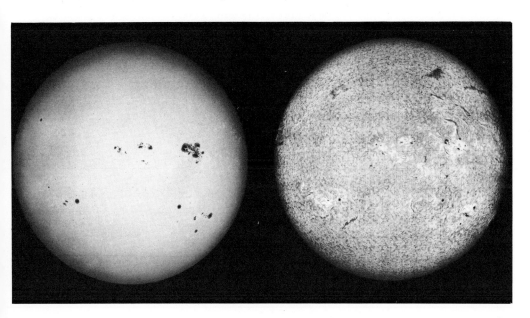

spots, known as pores, and develop to their maximum size in a few days. As a rule, they originate in groups. There are usually two large spots with a number of smaller ones between them. The life of such a group is usually a week or two, the spots finally growing smaller and disappearing.

Sunspots are gigantic whirlpools in the photosphere. Because the gases in them are expanding, they are cooler than the surrounding surface. This accounts for the fact that they appear darker than the photosphere. Spectroscopic observations show that the lines of the spectrum of a sunspot are doubled. Such a doubling, known as the Zeeman effect, after the Dutch physicist Peter Zeeman, is the result of a strong magnetic field. Streams of electrons whirling about may be the cause of the magnetic field in sunspots.

It has been known for several centuries that sunspots occur in cycles, going from minimum numbers to maximum and back to minimum in a period of about 11.2 years. The cycle can be as short as 7.5 years or as long as 16 years. A new cycle is heralded by the appearance of spots at considerable distances from the sun's equator. The end of a cycle is marked by spots near the sun's equator.

Bright patches and streaks are usually found around sunspots. They frequently appear before the spot breaks out. They are known as faculae, from the Latin for "little torches." Apparently, they are clouds of heated gases just above the photosphere. It is assumed that they are created by the same mechanism that causes sunspots.

Observation of sunspots quickly established the fact that the sun is rotating on its axis. However, the sun does not rotate like a solid body. The equatorial region has the shortest rotation period, turning once in 25 days. At 75° north or south latitude the period is about 33 days.

THE SOLAR ATMOSPHERE

For centuries it was possible to observe the solar atmosphere only during the few precious minutes of a solar eclipse. At the instant of totality the moon blots out the photosphere but leaves the solar atmosphere projecting beyond its edge. As a result the dark disk of

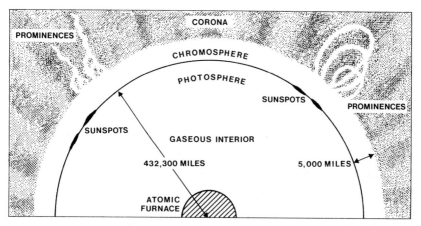

Figure 8. Cross-section of the sun

the moon appears to be rimmed by a thin border of rose-colored fire from which flamelike tongues extend. Beyond this rim a huge silvery halo of pearly streamers makes its appearance. The fiery rim is the inner portion of the solar atmosphere known as the chromosphere. The flamelike tongues are the prominences. The silvery halo is the corona. (See Figure 8.)

If a spectroscope is turned on the edge of the sun as totality approaches, the dark lines in the solar spectrum change to bright lines at the instant of totality. This bright-line spectrum is called the flash spectrum because of its sudden appearance. It is, of course, the spectrum of the chromosphere. Because the denser, lower portion of the chromosphere is responsible for the conversion of these bright lines into the dark lines of the normal solar spectrum, it is called the reversing layer.

The solar atmosphere can now be studied at any time with the aid of special instruments. One of the most important of these is the spectrohelioscope, invented independently by the American astronomer George Ellery Hale and the French astronomer Henri Deslandres. It is a combination of a telescope and a spectrograph with certain additions. The telescope forms an image of the sun on a screen which is in front of the spectrograph. There is a slit in the screen. The screen is put in motion, causing the slit to move across the sun's image, admitting light in succession from each portion of the solar image. A second screen is situated behind the spectrograph. It also has a slit in it. A photographic plate is placed behind this second screen. The slit in the second screen permits the light from only a single line in the solar spectrum to reach the photographic plate. No spectrum line is completely dark but appears so only by contrast with the adjacent bright portions of the spectrum. The photographic plate is made to move in unison with the first screen. As a result a photograph of the sun is built up on the plate in a single line of the solar spectrum. As a result this photograph shows the distribution of some particular gas in the chromosphere, depending upon the chosen spectrum line. Spectroheliograms, as they are called, are usually made in one of the prominent lines of hydrogen or calcium.

In recent years special filters of such extremely high efficiency have been developed that it is now possible to photograph the sun in a single spectrum line with a suitable telescope.

The corona can be studied at any time with a special telescope known as the coronograph, developed in 1930 by the French astronomer Bernard Lyot. This contains a number of diaphragms to eliminate scattered light and a metal disk which creates an artificial eclipse of the sun.

The chromosphere is a layer of gases, chiefly hydrogen and helium, about 3,000 miles thick. It is in a state of constant turbulence, torn by winds and storms of unbelievable fury. Bright patches, known as plages (pronounced "plahzhes"), occur in the chromosphere above the faculae. Great tornadoes appear to be blowing in the chromosphere above the sunspots. Numerous spike-shaped projections, known as spicules, constantly rise from the top of the chromosphere. They shoot up with velocities of 12 miles a second and disappear in 2 or 3 minutes. These are the tops of gaseous columns rising from the granules in the photosphere.

SOLAR PROMINENCES

Perhaps the most spectacular feature of solar activity is the formation of the prominences. The prominences are enormous clouds and streamers of gas that rise from the chromosphere to heights ranging from 10,000 to a million miles. By attaching a motion-picture camera to the spectroheliograph, astronomers have obtained

Total eclipse of sun showing solar corona

amazingly dramatic and vivid records of the behavior of prominences. There is a wide variety of prominences and they often undergo surprising changes in a matter of hours.

One type, known as an active prominence, consists of a huge, confused, tangled cloud of gas extending upward from the chromosphere to a height of 50,000 miles or so. Long curved streamers extend downward from it in great arches to the chromosphere. The material in these streamers is clearly moving downward. Often the streamers from two active prominences strike the same spot as though drawn to some center of attraction. The main mass of an active prominence appears to remain undiminished despite the constant drain of material into the streamers. Sometimes a prominence makes a sudden appearance like a cloud high above the chromosphere. These are known as quiescent prominences. Yet other prominences consist of great loops or arches. There is a constant flow of gas down both sides of the arch. At times the arch grows larger and rises higher above the chromosphere.

The downward motion in prominences is a puzzling feature. It has been pointed out, however, that here on earth rain always falls down. The rise of water vapor into the clouds is invisible. It is quite possible, therefore, that gases rising above the chromosphere may do so in a non-luminous form, becoming luminous only when the prominence takes shape.

Most startling of all the prominences are the eruptive prominences. Both quiescent and active prominences can become eruptive. The prominence begins to rise, first slowly, then more rapidly, sometimes attaining a velocity of more than 500 miles per second. The top of an eruptive prominence may attain so high a speed that it breaks away and disappears into space.

SOLAR FLARES

Sometimes a plage area grows intensely bright in a matter of minutes, then slowly fades in an hour or two. The sudden outbreak of light is known as a flare. It appears to be associated with large sunspot groups, often occurring in an area where such a group subsequently appears and continuing while the group is developing. Flares vary greatly in intensity. As many as 100 small flares have been recorded near a sunspot group in 1 day. However, very large flares occur only a few times a year.

The flare is a violent disturbance in the solar surface. Most of the energy of the flare is in the ultraviolet region of the spectrum but there is also an increase in radio microwaves. The flare is accompanied by the ejection of streams of energetic subatomic particles, chiefly hydrogen nuclei, or protons.

The appearance of a flare is reflected dramatically by conditions on the earth. The outbreak of a large flare often is accompanied by a fadeout of radio communications. This is due to the effect of the ultraviolet radiation on the ionosphere, or radio ceiling, the upper region of the atmosphere which normally reflects

THE SUN

radio waves back to the ground. The microwaves from the flare may impair television reception, causing the appearance of "snow" in the picture. Other disturbances follow a day or two later, when the proton streams reach the earth. These disturbances include magnetic storms that upset the compass and disrupt long-distance telephone and telegraph communications. There may also be increased displays of the Aurora Borealis, or Northern Lights.

It is recognized that these streams of subatomic particles can be a distinct hazard to manned space exploration. Striking the skin of a spaceship, they might generate X-rays of lethal intensity.

THE SOLAR CORONA

The silvery halo around the sun, known as the corona, is composed of delicate fan-shaped rays. Its appearance changes from

Prominence, 205,000 miles high, rising from the solar chromosphere

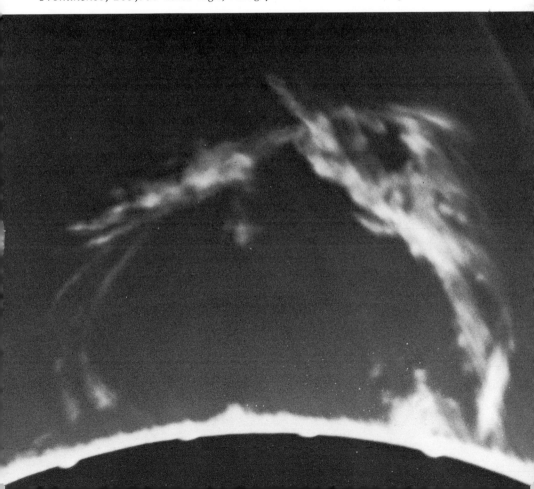

year to year in step with the sunspot cycle. At sunspot minimum there are enormous streamers, several times the solar diameter, extending from the equatorial region of the sun, while the rays at the poles are extremely short. At sunspot maximum the rays are more evenly distributed around the sun.

The corona consists of highly ionized, or electrified, atoms and free electrons. The silvery color results from the scattering of sunlight by the electrons. Spectrographic observation of the corona shows a number of bright lines. These could not be identified for many years and it was suggested that they were due to a chemical element that did not exist on earth. The name coronium was proposed for it. It is now known that these lines are the product of atoms of calcium, iron, and nickel, so highly ionized that they have lost from 9 to 14 electrons. However, such a high state of ionization requires a temperature of 1 million degrees or so. The problem is to explain how the corona can be 100 times hotter than the photosphere. It is important, however, to note that the gases of the corona are so thin that they resemble what on earth is a virtual vacuum. Temperature in the corona does not mean what temperature does on earth. Rather, it is a measure of the speed with which the ionized atoms and free electrons are moving.

THE SOLAR WIND

Satellites and planetary spacecraft have confirmed that there is a constant flow of ionized gases outward from the sun. It consists largely of hydrogen nuclei, or protons, and electrons. This has been named the solar wind. It is far less dense than our own atmosphere, and consists of only 10,000 or so protons per cubic inch. However, the particles in the wind are traveling at about 600 miles per hour.

It is possible to think of the solar wind as an extension of the corona. On this basis the earth is inside the corona. The streams of particles accompanying solar flares can be regarded as localized increases in the solar wind not unlike the winds of storms in the earth's atmosphere.

It is extremely probable that geophysicists have not yet discovered all the ways in which our earth is influenced by the activities on the sun. It is obvious that there is some deep-seated rhythmic process in the sun which manifests itself most clearly in the sunspot cycle. Its influence, however, is broader as can be seen in the association of solar flares with sunspots and the changes in the corona. Numerous attempts have been made to associate sunspots with various terrestrial phenomena, particularly the weather. To date these attempts have been unsatisfactory.

It is important to realize that the sun is a star, our local star. In many ways it is a typical star. Perhaps 10 per cent of all the stars in our galaxy are very much like our sun. The sun is the only star that we can study in detail. All the other stars are so far away that they are only points of light in the most powerful telescopes. Consequently, whatever we learn about the sun helps us understand the stars.

six

THE STARS

The sad and solemn Night
Hath yet her multitude of cheerful fires;
The glorious host of light
Walk the dark hemisphere till she retires;
All through her silent watches, gliding slow,
Her constellations come, and climb the heavens, and go.
  —BRYANT

ON a clear moonless night the heavens reveal the true grandeur of the universe. In the dark ocean of space shine the celestial beacon lights, the stars. Here and there glow the bright jewels of a familiar constellation. Scattered in lavish profusion are the hosts of lesser stars. As one gazes upon the wondrous beauty of the stars it is easy to understand the fascination they have always exercised upon mankind. It is easy to understand how they held the shepherds of ancient days spellbound, so that they wove the stories of their kings and legendary heroes into the pattern of the stars, and why the Arabs, living on the desert under the stars, gave names to the brightest of them. Today, alas, too many of us let the bright lights of the city crowd the softer lights of the heavens out of our lives. It is one of the paradoxes of modern life that while astronomers know more about the stars than ever before in the history of civilization, most people are less familiar with the stars than were the shepherds of 2,000 years ago.

Poets speak of the countless stars. But that is poetic license.

The stars visible to the naked eye are by no means countless, as astronomers proved long ago by counting them. About 2,000 can be seen from any single location at a given time. The total number visible in both the northern and southern hemispheres is about 6,000. The number increases rapidly, however, when a telescope is used. Galileo's little telescope was capable of disclosing a half-million stars. Today's giant telescopes reveal billions of stars in our galaxy.

The apparent brightness of a star is known as its apparent magnitude. We still use the system devised in ancient times to classify stars according to their apparent magnitudes. The brightest stars are called first-magnitude stars, the next brightest second-magnitude, and so on. Stars just visible to the unaided eye are sixth-magnitude. Today, when astronomers measure the apparent magnitude of stars with great exactness, the scale has been standardized so that a difference of 5 magnitudes represents a ratio of 100 times in brightness. This means that the stars of each magnitude are approximately 2.5 times brighter than those of the magnitude which follows them. Twenty-one stars are rated as first-magnitude stars. These are listed in Table II of the Appendix. There are about 60 second-magnitude stars. The faintest stars visible with the 200-inch telescope are 21st-magnitude. It is obvious that the apparent magnitude of a star is a combination of its real, or intrinsic, brightness and its distance from the earth. Astronomers call the actual brightness of a star its luminosity. Stars differ greatly in luminosity: the most luminous are a million times brighter than the sun; the least luminous a million times fainter than the sun.

STELLAR DISTANCES AND MOTIONS

Because the stars are so very far away, astronomers failed to determine stellar distances until the nineteenth century. In the 1830s three astronomers, Friedrich Bessel, Thomas Henderson, and Friedrich Struve, working independently, succeeded in measuring the parallax of three of the nearest stars. It will be recalled from Chapter 2 that in determining the lunar parallax, the distance between two observatories was used as a base line. A much longer

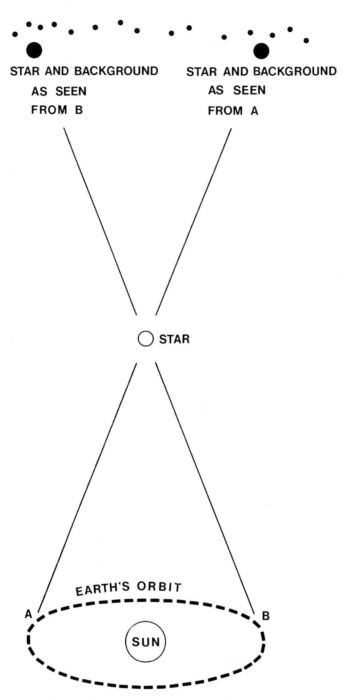

Figure 9. Determination of stellar parallax

base line is needed to determine stellar parallax. Observations of the apparent shift of a nearby star against the background of distant stars are made 6 months apart from a given observatory. The base line, therefore, is the diameter of the earth's orbit. (See Figure 9.)

The nearest star, Alpha Centauri, is 4.3 light-years away. (A light-year, you will recall, is approximately 6 trillion miles.) Sirius, the Dog Star, is 8.7 light years away. Fifteen stars are known to be within 12 light-years of the earth. Using the parallax method, astronomers have determined the distances of about 6,000 of the nearer stars. The most distant of these is about 300 light-years away. Once we know the distance of a star, we can calculate its luminosity from its apparent magnitude.

The parallax of stars more than 300 light-years away is so small that it cannot be determined with any reliability. Their distances, however, can be estimated by comparing their apparent magnitudes with those of stars with similar spectra whose distances are known. It is assumed that stars with similar spectra have the same luminosity.

The term absolute magnitude is used to designate the apparent magnitude a star would exhibit if placed at a distance of 10 parsecs from the sun. The parsec, a unit of distance equal to 3.26 light-years, gets its name from the fact that a star at this distance would have a parallax of 1 second of arc. The absolute magnitude of a star, it will be seen, is a measure of its luminosity.

The ancients spoke of the "fixed stars." It was assumed by later astronomers that the stars remained fixed in their exact positions in the constellations. However, Edmund Halley showed in 1718 that this was not the case. By comparing ancient records with observations of his own day, he proved that some of the brighter stars had moved slightly. Today we know that all the stars are in motion.

The angular motion of a star across the line of sight is known as its proper motion. Because the stars are so very far away, proper motion can be detected only for the nearer stars.

The motion of a star along the line of sight, that is, toward or away from the earth, is known as its radial motion. This is determined from the spectra of the star. You have undoubtedly noticed that the pitch of an automobile horn rises as the vehicle approaches you and falls as it recedes. This is the so-called Doppler effect, named for the Dutch physicist Christian Doppler, who first explained the phenomenon in 1842. As the auto approaches, the sound waves are crowded on each other. More of them reach your ear per second and so the pitch rises. As the auto recedes, fewer waves reach your ear per second and the pitch goes down. The same thing happens to light waves. If a star is approaching the earth, the light waves are crowded on each other. This increase in frequency causes a decrease in wavelength and the spectrum lines are shifted toward the violet end of the spectrum. If the star is receding, there is a decrease in frequency and a consequent increase in wavelength. The spectrum lines are now shifted toward the red end of the spectrum. In either case the radial velocity of the star can be calculated from the amount of shift in the spectrum lines.

If the distance of a star is known, its velocity across the line of sight can be calculated from observations of its proper motion. By combining this transverse velocity, as it is called, with the radial velocity, it is possible to determine the exact direction in space in which a star is moving and how fast it is going.

The stars in our part of the galaxy are found to be moving in all directions like a swarm of bees, with velocities ranging, in general, from 5 to 20 miles a second. But some are moving faster. Arcturus, for example, has a speed of 84 miles per second.

Because the stars are so very far away, these speeds result in only a very slow change of the appearance of the heavens. We see the constellations today just as the Greeks did 2,500 years ago. But in 50,000 years the constellations will be completely out of shape.

Analysis of the proper motions of the nearer stars show certain systematic trends. Stars in the direction of the constellation Hercules seem to be opening out while stars in the opposite direction appear to be closing in. Analysis of radial velocities shows that the stars in Hercules, on the average, seem to be approaching us at a velocity of about 12 miles a second. It is obvious that these are apparent effects due to the sun's own motion. The sun is moving toward Hercules at a velocity of 12 miles a second, carrying, of course, the whole solar system with it. The point toward which the sun is moving is called the solar apex.

THE HERTZSPRUNG-RUSSELL DIAGRAM

If you examine the night sky carefully, you will note that the stars differ in color as well as apparent magnitude. Many stars, like Sirius or Vega, the bright star in the constellation Lyra, are a brilliant white. Other stars, like Betelgeuse in Orion, or Antares in Scorpio, are red. The colors of stars range from red, through orange, yellow, and white, to blue. These colors are more noticeable in the telescope. The colors of the first-magnitude stars are given in Table II of the Appendix.

As one might suspect differences in color are the result of surface temperature. The red stars are red-hot, with surface temperatures in the neighborhood of 5,000° F. The blue stars are blue-hot, with surface temperatures of 100,000° F. or more.

Stars of different colors show characteristic differences in spectra. A system of classification devised at the Harvard College Observatory at the end of the nineteenth century, in which spectral types are designated by letters of the alphabet, is still used. The system has been somewhat modified so that the classes are now designated by letters in the following order: O, B, A, F, G, K, M, N. Types O and B are blue; A stars are white; F, yellow-white; G, yellow; K, orange; M and N, red. A small number of stars which do not fit well into this sequence have been designated as types R and S.

Stars differ not only in luminosity and surface temperature, but in diameter and mass. Diameters range from 3,000 times that of the sun to about 1/400. Some of the smallest stars are smaller than the moon. The range in mass is less since the largest stars are less dense than the smaller ones. The most massive stars have about 50 times the mass of the sun, the least massive about 1/25.

It was realized early in the present century that all conceivable combinations of luminosity, size, mass, and surface temperature do not occur. The combinations which do occur are most easily shown on a chart which astronomers call the Hertzsprung-Russell, or H-R, Diagram. The chart is named for the Danish astronomer Ejnar Hertzsprung and the American astronomer Henry Norris Russell who pioneered in the study of these relationships. (See Figure 10.)

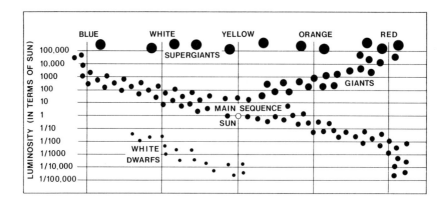

Figure 10. The Hertzsprung-Russell Diagram

You will note that in Figure 10 luminosity, in terms of the sun, is plotted along the side of the diagram, while color is plotted along the base with blue at the left. About 85 per cent of the stars fall into an elongated S-shaped track that runs diagonally across the chart from the upper left to the lower right. It is called the main sequence. Another band of stars slopes upward toward the right-hand corner of the diagram from near the middle of the main sequence. This is known as the giant sequence. There is a scattering of stars across the top of the diagram. These are the supergiants. Finally, there are some stars in the lower-left-hand corner of the diagram known as the white dwarfs.

The main sequence represents a simultaneous downward progression in luminosity, temperature, size, and mass. It begins with supergiant blue stars that are 100,000 times as luminous as the sun. They have surface temperatures of about 100,000°, their diameters are about 20 times the sun's, and they are about 30 times as massive. Proceeding down the main sequence, we find the stars grow less luminous, cooler, smaller, and less massive. As the temperature decreases, the colors of the stars change to white, then yellow, orange, and finally red. The stars in the lower half of the main sequence are known as yellow dwarfs and red dwarfs. Our sun is a typical yellow dwarf. Its position in the main sequence is marked in the diagram.

The giant stars in general are about 1,000 times as luminous as stars of similar color on the main sequence. Since surface temperatures are the same, this means that the giants are very much larger than the yellow and red dwarfs. The diameter of a giant star is about 30 times that of a main-sequence star of the same color.

The supergiants, ranging in color from red to blue, are not only the most luminous stars but are also the largest. They are about 100 times as luminous as the giants and have diameters 10 times larger. If one of the supergiants, like Antares or Betelgeuse, was placed at the center of the solar system, it would extend almost to the orbit of Saturn.

The white dwarfs are extremely hot, as their color indicates, but very small. They are the smallest stars known, some of them no larger than the moon. They possess about the same mass as the sun. This means that they are so dense that a spoonful of the material composing them would weigh several tons.

DOUBLE STARS

The second star in the handle of the Big Dipper is Mizar. If you look carefully at it on a clear moonless night, you will note a small star near it. Its name is Alcor. The Arabs many centuries ago called Mizar and Alcor the "horse and rider," and considered the ability to see Alcor a test of good eyesight.

In 1650 the Italian astronomer Giovanni Battista Riccioli turned his little telescope on Mizar and found to his surprise that the telescope separated it into two stars. If you own a small telescope you can repeat his discovery. Other similar instances were found subsequently but attracted little attention. It was assumed that these were merely cases of accidental alignment, the one star being actually far behind the other.

It was not until 1803 that Herschel discovered that this was not so in the great majority of cases. His observations established that in most cases the two stars formed a physical system, revolving in an orbit around a common center of gravity. Such pairs are known as double stars, or visual binaries. Many double stars have been under observation for more than a century and the orbits of more than 100 have been calculated. Periods range from a little less than two years to more than 700. By applying Kepler's third law, it is possible to calculate the mass of the system from the orbit. Many stars have been found to consist of more than two components. Astronomers are now certain that more than half of all the stars in our galaxy are twins, triplets, quadruplets, or even sextuplets.

The components of a double star are sometimes identical twins, but this is usually not the case. In some cases one component is a giant or supergiant star, the other a white dwarf. Sirius is such a double star. The brighter component is a white giant about 100 times as bright as the sun. The fainter component, which astronomers have nicknamed "the Pup," is a white dwarf less than 1/100 as bright as the sun.

Early in the nineteenth century it was realized that Sirius must be a double star, although existing telescopes did not show the faint companion. This was deduced from the fact that Sirius was not moving in a straight line but had a wavy motion. This could be explained only on the assumption that Sirius and a companion were revolving around a common center of gravity which was moving in a straight line. The companion was seen for the first time in 1862 by Alvan G. Clark with the 18.5-inch refracting telescope which his father had built.

In some cases the components of a double star are so close together that the telescope cannot separate them. However, they can be detected with the spectroscope. Once more the story begins with Mizar. In 1889 the American astronomer Edward C. Pickering was studying the spectrum of the brighter component of Mizar. He was amazed to find that the spectrum lines were sometimes double, at other times single, and that they went through a regular cycle. He reasoned that the component was a double star. As the two stars revolve around their common center of gravity, one at times is moving toward the earth while the other is moving away. As a result the lines of one are shifted toward the violet end of the spectrum while those of the other are shifted toward the red end. Consequently, the lines in the spectrum appear double. At other times, when both stars are moving across the line of sight, the lines appear single. Such double stars are known as spectroscopic binaries. Because in most cases one component is so much brighter than the other, its spectrum "drowns out" that of the fainter component. But the regular changing shift of the spectrum lines, first toward the violet and then toward the red, indicates that the star is a spectroscopic binary. More than 1,000 spectroscopic binaries have been catalogued. They have periods ranging from less than 10 days to about a year.

Several hundred spectroscopic binaries are known as eclipsing binaries. These not only exhibit the expected shift in their spectrum lines but periodically show a drop in apparent magnitude. This is the result of the fact that the orbit of such a double star is seen edgewise from the earth. Consequently, the two components eclipse each other in each revolution around the orbit. If one component is considerably brighter than the other, the drop in brightness is particularly marked when the faint star gets in front of the bright one.

The first eclipsing binary to be recognized was Algol in the constellation Perseus. There is a drop in brightness that lasts several hours every 2 days and 21 hours. It is believed that the Arabs had noted this peculiar behavior of the star, although they had no explanation for it. The name Algol is Arabic for "the demon."

Among the most interesting of the eclipsing binaries are five in which one component is a supergiant star. Unlike other eclipsing binaries, they have periods ranging from almost 3 to 27 years. In each case the atmosphere of the supergiant star is so extensive that the light of the smaller component shines through it for several weeks before and after it is eclipsed by the supergiant. This is shown by the increase of absorption lines in the composite spectrum of the two components.

Peculiarities in the spectra of some eclipsing binaries can be explained only on the assumption that the components have a disturbing effect upon each other so that great streamers of gas are being pulled out of one or both components. The gas in these streamers is constantly being lost into space. In some cases, however, the gas lost by one component appears to be absorbed by the other.

VARIABLE STARS

Back in 1596, before the invention of the telescope, the Dutch astronomer David Fabricius noted a third-magnitude star in the constellation Cetus, the Whale, where none had previously been known. A few months later, to his great surprise, the star faded and disappeared from view. A few years later the star was seen again by another astronomer. In succeeding years it was found that the star had a habit of appearing and disappearing. Astronomers named it Mira, from the Latin for "wonderful."

When astronomers had fairly large telescopes, the mystery was solved. Mira does not shine with a constant light but alternately grows brighter and dimmer. At its faintest, it is a ninth-magnitude star and can be seen only with a fairly large telescope.

As time went on it was found that other stars also fluctuated. About 20 were known by 1800. The number was increased to about 1,000 by 1900. Since then, more than 10,000 have been catalogued. There must be many millions of them in our galaxy.

Variable stars are classified into groups on the basis of their periods, that is, the interval of time from minimum magnitude to maximum and back to minimum. It has been found that there is a relationship between period and spectral type or color. Practically all of the variable stars are giant or supergiant stars.

Among the most important group is that of the Cepheid variables, so called because the first one discovered was the star Delta in the constellation Cepheus. They are divided into cluster-type Cepheids, so named because they are found frequently in globular clusters, and classical Cepheids.

The Cepheids go through their light variations with clocklike regularity. Study of their spectra disclosed that they are pulsating stars, alternately expanding and contracting, growing hotter and cooler by turns. The cluster Cepheids have the shortest periods, showing light changes of about 1 magnitude in intervals ranging from 2 to 24 hours. They are white giant stars. They are also known as RR Lyrae stars after the one in the constellation Lyra which is typical of the class.

Classical Cepheids include yellow giants with periods from 2 to 10 days, and orange and red giants with periods from 10 to 45 days. About 500 classical Cepheids are known in our galaxy. About a dozen are visible to the unaided eye. The best known one is Polaris, the North Star.

In 1911 Miss Henrietta Leavitt of the Harvard Observatory noted that there was a direct relationship between the periods and the apparent magnitudes of the Cepheid variables in the nearby galaxy known as the Small Magellanic Cloud. Magnitude increased with period so that Cepheids of periods around 30 days were about 2 magnitudes brighter than those with periods of 2 days. Since all the stars of the cloud are essentially the same distance from the earth, this showed that the period of a Cepheid was an indication of its absolute magnitude, or luminosity. This is now known as the period-luminosity relationship.

Once astronomers had established the absolute magnitude scale of relationship by observation of Cepheids at known distances in our galaxy, it became a useful indicator of the distances of more distant Cepheids. The period reveals the star's luminosity. By comparing this with the apparent magnitude, its distance can be calculated. This technique has proved of immense value in determining the dimensions of our own galaxy and the distances to neighboring galaxies in which Cepheids can be observed.

Another class of variables are the long-period variables. The star Mira is a typical member of this class. They are giant red stars with periods ranging from about 100 to more than 1,000 days. Still other classes of variables are known as semiregular and irregular variables because of variations in their periods.

A small number of dwarf red stars comprise a class of variables known as flare stars. They have sudden outbursts of light, lasting from a few minutes to half an hour, which increase their luminosity from 1 to 3 magnitudes.

EXPLODING STARS

In November 1572 all Europe was startled by the sudden appearance of a bright star in the constellation Cassiopeia. In a few days it grew brighter than Venus and could be seen in broad daylight. The superstitious were certain that it heralded some dreadful

The Crab Nebula in Taurus, a remnant of the supernova seen in A.D. *1054*

disaster, perhaps the end of the world. However, nothing happened and the star gradually began to fade. By March 1574 it had disappeared from view. At the time of the star's appearance the great Danish astronomer Tycho Brahe was dabbling with alchemy and neglecting astronomy. The event turned his attention back to the heavens. His study of the star convinced him that it was out among the other stars and not an apparition in the earth's atmosphere as many university professors insisted. He wrote a book about it titled *De nova stella* (Concerning the New Star). Astronomers still refer to the star as Tycho's Nova.

Another bright star appeared in 1604. It did not become as bright as Tycho's Nova but equaled the brilliance of Jupiter and did not fade from view until 1606. Kepler wrote a book about it and so it became known as Kepler's Nova.

Although the name "nova" is still used, we know that these are not new stars but only faint stars that have suddenly flared up with new brilliance. Since 1890, about a dozen have been visible to the unaided eye. More than 100 have been noted on photographic plates taken with big telescopes. It is thought that about 20 or 25 appear annually in our galaxy but most of them escape notice. Novae have also been noted in some of the nearer galaxies.

A nova is a small, very hot, very dense blue star which undergoes a sudden explosion. If placed on the H-R Diagram, it occupies a place between the main sequence and the white dwarfs. For this reason it is sometimes spoken of as a sub-dwarf.

Apparently some explosion in the star causes it to blow off a great shell of gas. This shell expands rapidly, attaining the diameter of a supergiant star. As a result the luminosity of the nova may increase 500,000 times. It may take anywhere from 2 or 3 days to a month for the nova to reach its maximum brightness. Then the shell grows cooler as it continues to grow larger, and the nova begins to fade. While the temporary increase in brightness is spectacular, the nova loses only about 1/100,000 of its mass in the explosion. A number of novae, designated as recurrent novae, are known to have undergone more than one explosion.

Extremely brilliant novae, such as Tycho's Nova and Kepler's Nova, are extremely rare in our galaxy and are now called supernovae. Only one other is known to have occurred in historic times. This was the supernova which appeared in 1054 and was recorded in the Chinese annals. However, supernovae have been detected in other galaxies. A supernova may increase 100 million times or more in brightness. In some cases it attains the luminosity of an entire galaxy. It loses about 1/10 of its mass in the explosion.

Astronomers have succeeded in locating the supernova of 1054. It is a white dwarf star in the constellation Taurus, surrounded by a huge cloud of tangled gaseous filaments. Because of its general appearance, the cloud has been named the Crab Nebula. It has been found to be a strong source of radio waves.

Strangely enough, the spot where Tycho's Nova appeared in 1572 is a strong source of radio waves. But the star itself cannot be found. It seems as though the ghost of Tycho's Nova haunts the sky. Almost as strange is the fact that astronomers have found a tangled nebula which is the remnant of Kepler's Nova, but no radio waves have been detected from it.

THE BIRTH AND DEATH OF STARS

Discoveries in the field of atomic energy during the last few decades have enabled astronomers to construct a satisfactory theory of the life history of a star. Earlier attempts had failed because there were only vague notions of a star's source of energy. It is obvious that an acceptable theory of stellar evolution must account for the stellar types revealed by the H-R Diagram.

From a study of the rate at which stars are radiating energy it has become apparent that stars differ in age as well as physical characteristics. Some of the stars we see in the heavens are young stars. Others are very old stars. Stars are being born in our galaxy today. Other stars are dying.

A star is born in the contraction of a huge cloud of gas and dust, under the influence of the force of gravity. In time the cloud is transformed into a gigantic dark sphere known as a globule or protostar.

As contraction continues the protostar grows hotter as a result of the release of gravitational energy. Finally it becomes hot enough to be luminous. It is now a star. Convection currents in the contracting star cause it to become more luminous than it will be in later life. Gradually, as these currents subside, the star becomes less luminous.

In time the center of the star reaches a temperature of 10 million degrees or more. When this happens, the center of the star becomes an atomic furnace in which energy is created by the transformation of hydrogen into helium. This can take place in two ways. One process, known as the proton-proton reaction, takes place in the sun and the cooler stars of the main sequence. Two protons, or hydrogen nuclei, unite to form an atom of deuterium, or double-weight hydrogen. This, in turn, unites with a third proton to form an atom of lightweight helium. Finally this unites with a fourth proton to form a normal atom of helium. A more involved process, known as the carbon cycle, takes place in hotter stars. This involves a number of steps which start with the union of a carbon nucleus and a proton. Astronomers speak of these reactions as hydrogen burning. But it must be realized that these are nuclear transformations, quite different from ordinary burning.

Soon after the production of atomic energy has started, the star reaches a state of equilibrium in which the outward push of the heated gas and the radiation balance the inward pull of gravity. The star has now reached maturity and settles down on the main sequence in the steady state which it will retain for most of its life.

How long it takes a star to arrive at the steady state depends upon its mass. A massive star, evolving from a very large cloud of gas and dust, contracts more rapidly because of the greater gravitational pull. It reaches the steady state in about 100,000 years and settles down on the main sequence as a highly luminous blue supergiant or white giant. Less massive stars take 50 million or more years to arrive on the main sequence as yellow or red dwarfs.

How long a star remains on the main sequence depends upon how rapidly it uses up the hydrogen at its center. When this happens, changes take place in the star which cause it to move off the main sequence. The more massive stars consume their hydrogen most rapidly and so are the first to depart from the main sequence.

A highly luminous blue or white star is a spendthrift star, using up its hydrogen at a furious rate. Consequently, it may remain on the main sequence only 10 million years. A yellow or red dwarf, consuming its hydrogen at a far more modest rate, may remain on the main sequence for 10 billion to 20 billion years. The faint red dwarfs at the bottom of the main sequence may stay there for a trillion years.

The life patterns of giant and dwarf stars differ. Let us turn first to the pattern of dwarf stars whose mass is in the neighborhood of the mass of the sun.

As such a star consumes its hydrogen, a core of helium forms at its center. In time, depending on the mass of the star, all the hydrogen in the center is converted to helium. Energy production now ceases in this core, but hydrogen-burning begins in a layer, or shell, around the core. This leads to a series of changes which cause the star to move off the main sequence.

The helium core begins to contract, but the hydrogen-burning around the core causes the outer portion of the star to expand. The star grows more luminous, but expansion of the surface causes the surface to grow cooler. The star is now moving to the right and upward in the H-R Diagram. For a time it is unstable and becomes a variable star, but eventually it becomes a red giant with a diameter 30 to 50 times its original diameter.

As hydrogen-burning continues around the helium core, the core grows larger. When it represents about 40 per cent of the star's mass, another change takes place. Contraction of the core causes it to reach a temperature of more than 250 million degrees F. At this temperature, a thermonuclear reaction begins in the core. Three helium atoms unite to form one atom of carbon. Further reactions also convert some of the carbon into neon and oxygen. The star now has two sources of nuclear energy, hydrogen-burning and helium-burning. This heralds the end of the red giant stage.

The subsequent evolution of the star is not yet clearly understood. It is possible that the star exists for a time as a long-period red variable or as a Cepheid variable. But with helium-burning the star grows hotter and more luminous, probably moving to the left in the H-R Diagram, moving across the main sequence to become a white giant. As it moves to the left, it goes through an unstable stage in which it becomes an RR Lyrae variable.

Eventually, the star exhausts both its supply of helium in the core and of hydrogen around it. The star now has no source of nuclear energy. It begins to contract, but as it grows smaller, it

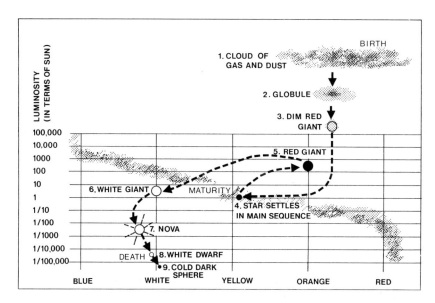

Figure 11. Life and death of a star

grows hotter as a result of the release of gravitational energy. It is now on its way to become a white dwarf. There is some uncertainty about the exact path the star now follows, but as it grows smaller, it is wracked by one or more explosions which blow off the outer portion. It becomes a nova, or perhaps a repeating nova.

Finally, it is a white dwarf, in some cases completely devoid of hydrogen. It is now about the size of the moon. Its atoms are so squashed together that a spoonful of its material weighs several tons. In time contraction will cease, the star will grow cooler and cooler, until it is a cold dead star. (See Figure 11.)

Giant stars leave the main sequence early in life and, like dwarf stars, develop a core of helium and move into the region of the red giants on the H-R Diagram. A star very high on the main sequence will move into the realm of the supergiants.

The giant star, when it leaves the main sequence, will move upward and to the right on the H-R Diagram. Like the dwarf star, it will develop a helium core and eventually will have two sources of energy, helium-burning in the core and hydrogen-burning in the layer around the core. In time an inner core of oxygen and neon forms in the center of the helium core. The star continues to grow hotter and eventually a thermonuclear reaction begins in the

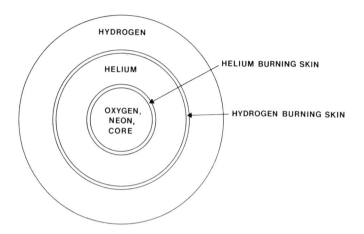

Figure 12. Star with three energy sources

oxygen-neon core. The star now has three sources of nuclear energy, as shown in Figure 12.

If the star is sufficiently large, this process can repeat itself until it has seven zones in which nuclear reactions are taking place. It will be seen, therefore, that the stars are not only nuclear furnaces, but nuclear stoves in which hydrogen is gradually "cooked," or converted, step by step, into the heavier atoms of the chemical elements.

The seven-zone star has iron in its innermost core. The creation of atoms heavier than those of iron is a complex process by which some atoms begin to disintegrate and the neutrons thus released combine with the atoms of the iron group.

Finally at a central temperature of 5 million degrees, catastrophe overtakes the star. The iron atoms disintegrate into helium atoms, the core shrinks, and the star collapses. But the gravitational energy thus released causes the star to undergo a gigantic explosion. The giant star becomes a supernova, blowing off as much as 10 per cent of its mass. Like the dwarf star, the giant is now on its way to becoming a white dwarf.

PULSARS

In the summer of 1967 a point source of radio waves emitting sharp pulses at regular intervals of 1.33 seconds was discovered at the Mullard Radio Observatory of Cambridge University, England. Several more were found soon after. Some scientists thought they might represent signals from distant inhabited planets or some sort of celestial beacons to guide interstellar navigation. Martin Ryle, director of the observatory, facetiously called them LGM's, an abbreviation of "Little Green Men." About 60 such pulsars, as they are now called, are known today. They have sharp and stable emission periods ranging from a few hundreths of a second to a few seconds.

In 1968 astronomers succeeded in identifying one pulsar with the white dwarf star at the center of the Crab Nebula, the remnant of the supernova seen in 1054. It was found that its light pulsates in exact synchronism with the radio pulse. A second pulsar was identified in 1970 with a white dwarf star in the Gum Nebula.

It is now believed that the pulsars are rapidly rotating neutron stars. These represent a state of collapse beyond the white dwarf stage. A neutron star may be only a few miles in diameter.

Many astronomers believe that the final stage of a star is the result of its size. A star whose mass is 1.4 that of the sun or less eventually becomes a white dwarf. The electrons are stripped from its atoms and the nuclei are squashed together so that a spoonful of the star weighs many tons, perhaps as much as 1,000 tons.

If the mass of the star is between 1.4 and 2 times that of the sun, the greater gravitational attraction causes a greater collapse in the star whose nuclear fuel is exhausted. The atomic nuclei are shattered and the electrons jammed into the protons to form neutrons. Some astronomers believe that a spoonful of such a star would weigh a billion tons.

Finally, if the mass of the star is more than 2 times that of the sun, its collapse will be so great and its concentrated gravitational field so intense that no radiation can escape from it. The star is now completely invisible, a "black hole" in the heavens.

seven

THE GALAXY

> *A broad and ample road whose dust is gold*
> *And pavement stars.*
> —MILTON

ONE of the striking features of the night sky is the magnificent arch of the Milky Way, ascending from one horizon, sweeping grandly across the sky, and descending to the other horizon, a luminous pathway across the heavens. It is best seen on a clear moonless night in late summer, out in the country and away from city lights. You will note at once that its width and brightness vary greatly from section to section. Starting, for an observer in the northern hemisphere, on the northeastern horizon in the constellation Cassiopeia, it is a single silvery band until it reaches the Northern Cross, or Cygnus, high overhead. From there to the southern horizon it is split lengthwise by a dark streak known as the Great Rift. It attains its maximum width and brightness in the constellation Sagittarius, low in the south. As summer gives way to autumn and other constellations come into view, it becomes apparent that the Milky Way encircles the entire heavens.

The ancients had no way of knowing the nature of the Milky Way, although some of the Greek philosophers guessed that it consisted of faint stars. Ovid called it a celestial road that led to the home of the gods. When Galileo turned his little telescope on the heavens in 1609, he confirmed that it did indeed consist of thousands upon thousands of faint stars. But more than a century passed before some notion of the structure of the Milky Way and its relationship to the solar system was attained. The true state of affairs was first suggested by the British astronomer Thomas Wright about 1750 and confirmed by Sir William Herschel in 1787.

In the course of his survey of the heavens Herschel counted the number of stars his telescope revealed in small areas scattered over the heavens. He found that the number increased steadily as he approached nearer and nearer to the Milky Way. From this he rightly inferred that the stars form a vast flattened system shaped like a grindstone or a watch. The Milky Way is the perspective view we get from our place inside the system. When we look at the Milky Way, we see so many stars because we are looking into the depths of the system. We are looking along the hands of the watch. When we look away from the Milky Way we see so few stars because we are looking out through the face or the back of the watch.

Herschel thought that the sun was situated near the center of the Milky Way system. This opinion still prevailed at the start of the twentieth century. It is obvious that the sun lies in the plane of the system since the Milky Way appears as a great circle dividing the heavens in two. However, we now know that the sun is far from the center of the system.

Astronomers refer to the Milky Way system simply as the galaxy or as our galaxy since we are a part of it. Our present notions of the structure of the galaxy and our position in it had their start in the studies made by Harlow Shapley in 1916. Twenty-five years ago it was suspected that our galaxy was a spiral galaxy very much like the Andromeda galaxy. This has been confirmed in recent years. If we could station ourselves in the Andromeda galaxy, our galaxy would appear to us from it as the Andromeda galaxy does to us from here.

Our galaxy is a highly flattened lens-shaped system containing 100 billion stars and enough gas and dust to equal some billions of more stars. It is about 100,000 light-years in diameter. The sun is situated far to the outer edge of the galaxy, about 33,000 light-years from the center.

Astronomers today recognize five chief structural features of the galaxy. These are the nucleus, the central bulge, the disk, the spiral arms, and the halo.

The nucleus, the center of the galaxy, is hidden from our view by the immense star clouds in the constellation of Sagittarius. We know very little about the nucleus because of the obscuring clouds of gas and dust between us and it. The nucleus appears to be about 5,000 light-years in diameter and is extraordinarily dense. It may consist of stars that are unusually close together. Recently it has been suggested that it may consist of some sort of gigantic superstar.

Surrounding the nucleus is the central bulge. This is a flattened sphere, or ellipsoid, of stars about 20,000 light-years in diameter and about 6,000 light-years from top to bottom. Its stars are two or three times as close to one another as the stars in the solar neighborhood.

The central bulge, in turn, is surrounded by a large flat disk of stars with a diameter of 100,000 light-years. It is about 3,000 light-years in thickness.

Embedded in this disk are the spiral arms, consisting of more stars and enormous clouds of gas and dust. The spiral arms are irregularly shaped and astronomers have not yet succeeded in determining their outline with exactness. The exact number is not known. Each arm appears to be about 2,500 light-years in width,

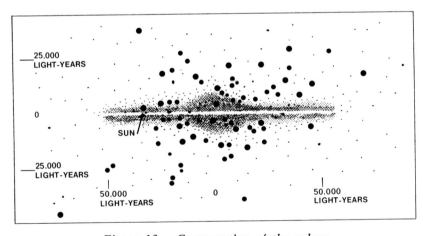

Figure 13. Cross-section of the galaxy

and adjacent arms are separated by about 6,500 light-years. The sun is located in one of the spiral arms often referred to as the Orion arm because it passes through the constellation Orion. The dust clouds in the Orion arm constitute a kind of local fog, reddening the light of distant stars. Where the dust is thickest, the clouds obscure our view of large portions of the galaxy.

Finally, we come to the halo which surrounds the galaxy. It is a gigantic tenuous sphere of gas, containing more than 100 globular clusters and a sprinkling of individual stars. A cross-section of the galaxy and its halo is shown in Figure 13.

STELLAR POPULATIONS

One of the problems to which astronomers have given much attention in the twentieth century is the distribution of different types of stars.

A list of the 21 brightest, or first-magnitude, stars is given in Table II of the Appendix. Eleven of the 21 are blue supergiants or white giants. All but one of the 21 is more luminous than the sun. All but four are at immense distances from us, ranging from 23 to 650 light-years. The blue star Rigel in the constellation Orion has an intrinsic brightness, or luminosity, 23,000 times that of the sun. But it is important to note that Rigel is 650 light-years away. This means that there is only one supergiant the size of Rigel in a volume of space with a radius of 650 light-years. It is obvious that supergiant and giant stars are rare in our galaxy.

If we turn now to an exploration of the solar neighborhood, as astronomers like to call relatively nearby space, we find 41 stars within a distance of 16.5 light-years from the sun. Only four of them—Sirius, Altair, Procyan, and Alpha Centauri—are among the 21 brightest stars. Only these four are more luminous than the sun. Fewer than a dozen of these nearby stars can be seen with the unaided eye. The majority of them are red dwarfs. There are also some white dwarfs. The faintest star in the group, known by its catalogue number of Wolf 359, has only 1/60,000 the luminosity of the sun. Astronomers are now convinced that the great majority of stars in our galaxy are dwarf red stars.

A tendency of certain types of stars to be more prevalent in certain parts of the galaxy was first noted in the 1920s by the American astronomers Harlow Shapley and Robert Trumpler. In 1944 Walter Baade established that the stars could be divided into two groups which he named Population I and Population II. The stars of Population I occur in the outer spiral regions of the galaxy, while those of Population II occur in the central regions and in the globular clusters. The brightest stars of Population I are blue supergiants. It also includes white giants and red supergiants. The brightest stars of Population II are red giants, much less luminous than the supergiants of Population I. Large amounts of gas and dust are associated with Population I, particularly in the spiral arms. There is much less gas and dust, perhaps almost none, in the Population II regions.

Recently some astronomers have suggested the recognition of three populations. They would confine Population I stars to the spiral arms, calling the stars of the disk the Disk Population.

From the types of stars they contain it seems evident that Population II consists of very old stars, Population I of young stars, and the Disk Population of stars of intermediate age. Many astronomers today think that this division into three populations is an oversimplification and that there are at least five different populatons.

GALACTIC CLUSTERS

It is impossible to view the sky on a clear winter night without noting the little cluster of twinking stars in the constellation Taurus. Called the Pleiades, they have stirred the imagination of man since ancient times. The little cluster is known also as the Seven Sisters, although only six stars are easily seen. However, a person with sharp eyes may count 11 or 12 on a clear night. Binoculars reveal about 60 stars in the cluster, while a big telescope shows more than 200 stars embedded in luminous clouds of gas and dust. The Pleiades constitute a physical system and not a chance grouping of stars. All of them are moving through space in the same direction and with the same speed.

Astronomers are aware today of more than 400 such clusters in the galaxy. They were formerly called open clusters, but the preferred name today is galactic clusters. It is probable that many more exist but are lost to view against the star clouds of the Milky Way. Each cluster contains from 20 to 30 to more than 1,000 stars. Five of the seven stars of the Big Dipper belong to a cluster of more than 100 stars known as the Ursa Major cluster. Because the other two stars of the Big Dipper are moving in a different direction, the familiar dipper will be completely out of shape in another 100,000 years.

It seems obvious that all the stars of a galactic cluster must have evolved together and would be the same age. Consequently, an analysis of the stars in a cluster is an index of its age. It will be recalled from the preceding chapter that the brightest and most massive stars evolve the fastest. Since the Pleiades contains many giant white stars, it cannot be more than 50 million years old. Another cluster, containing no stars of the upper end of the main sequence, must be more than a billion years old.

GLOBULAR CLUSTERS

About 100 globular clusters are known, although it is quite possible that another 100 may be hidden from our view. They differ from the galactic clusters in size, structure, composition, distribution, and motion. Each globular cluster is a huge globular formation containing from 10,000 to a million stars. The density increases toward the central part of the cluster, while the center is lost in a great blaze of light in which it is impossible to distinguish individual stars.

It was Shapley who first showed that the globular clusters were distributed around a center which coincided with the center of the galaxy. This enabled him to show that our sun was not at the center of the galaxy but far from it. The globular clusters are most numerous within 10,000 light-years of the galactic center but a considerable number are scattered out to 30,000 or more light-years at all angles to the plane of the galaxy.

An analysis of the stars of a globular cluster reveals that

there are no blue supergiant or white giant stars. The upper portion of the main sequence is missing. The brightest stars are red giants and hence well removed from the main sequence. There are also numerous yellow sub-giants which lie on the H-R Diagram between the main sequence and the red giants. The globular clusters appear to be nearly free of gas and dust. Astronomers conclude that the globular clusters are extremely old systems, perhaps more than 10 billion years old, which have been free from outside influences since their formation. The stars are thought to be all extremely old stars. The blue supergiants and white giants which originally existed in the globular clusters have long since lived out their lives. The yellow sub-giants and red giants represent stars that were once on the main sequence. Only the dwarf stars are still on the main sequence.

GAS AND DUST

The galaxy contains a tenuous mixture of gas and dust equivalent to many billions of stars. The mixture is about 98 per cent gas and 2 per cent dust. The gas is chiefly hydrogen, although it contains some helium and atoms of other chemical elements, notably calcium and potassium. The dust consists of microscopic crystals of ice and simple compounds of the lighter elements, including carbon, nitrogen, potassium, sodium, and calcium. There is a thin all-pervading haze of this interstellar medium in the plane of the galaxy, but most of it is concentrated in vast clouds in the spiral arms. The amount of dust in these clouds varies greatly. Some contain very little dust, others are heavily charged with it.

These clouds are for the most part invisible, but their presence is revealed indirectly. When the spectrum of a distant star is recorded, it is found to contain a few lines that do not belong to the star. It will be recalled from the previous chapter that stellar spectra exhibit a so-called Doppler shift due to the motion of the star in the line of sight. The extraneous lines, which do not show this shift, are called stationary lines. They result from the absorption of some wavelengths of the star's light by the intervening gas cloud. The dust causes the light of distant stars to be redder than

The Great Nebula in Orion

it would otherwise be. Finally, because the neutral hydrogen emits radio waves in the 21-centimeter band, the clouds can be charted with the aid of radio telescopes. This emission is due to changes in the direction of spin of the electron relative to the direction of spin of the proton in the hydrogen atom.

In numerous spots in the sky the interstellar medium reveals itself to visual observation as bright or dark patches. These are known as diffuse nebulae from the Latin for "cloud." It has been noted that the bright nebulae are always associated with stars either within or near them. In other words, these bright nebulae are patches of larger clouds which have been rendered luminous by the stars associated with them. Two types of bright nebulae are known. If the star is a very hot blue supergiant or white giant, its intense radiation of ultraviolet light causes the gas of the nebula to become fluorescent, so that it glows like the gas in a neon lamp. Such a nebula is known as an emission nebula. If the star is a cooler type, the nebula shines only by the reflection of starlight from its dust particles. Such a nebula is known as a reflection nebula.

One of the most spectacular sights revealed by a big telescope

Globular star cluster in Hercules

is the emission nebula in the constellation Orion, known as the Great Nebula in Orion. It is visible to the unaided eye as a starlike fuzzy spot of light. The telescope reveals its true glory. It is a vast luminous cloud of spectacular beauty, soft green in color, brightest in the center. Majestic streamers, duller in color, swirl upward from the center like the petals of a gigantic flower. A cluster of four blue supergiant stars, known as the Trapezium, is embedded in its center. The nebula is about 30 light-years in diameter. It contains a considerable number of stars in addition to the Trapezium. Some of these are blue supergiants, but others are red variable stars, known from their prototype as T Tauri stars. These are believed to be extremely young stars in which the nuclear reaction of hydrogen-burning has not yet begun.

Time-exposure photographs made with large telescopes reveal that many nebulae are also magnificent objects. Astronomers have catalogued more than 1,000 bright nebulae of various sizes and luminosity. All of them fade away at various distances from their central stars. Recent studies have disclosed many extended, faint filmy nebulae. A great faint wispy nebula extends over almost all of the constellation Orion.

The dark patches, known as dark nebulae, are areas in the interstellar region where the concentration of dust is high. They reveal their presence when they occur between us and a dense starfield, or bright nebula. The most spectacular one is the Great Rift in the Milky Way. A spectacular dark nebula in the southern hemisphere is known as the Coal Sack. At one time it was thought these were regions of space devoid of stars. It is now known that there are dark nebulae of various degrees of concentration. The brighter stars show through the obscured areas but the fainter stars do not. No dark nebula is totally opaque.

Some bright nebulae contain small round very dark nebulae. These have been named globules and are believed to be stars in the process of formation. They are called protostars.

About 500 objects in the galaxy are known as planetary nebulae. The name was given to them many years ago because their appearance in the telescope is somewhat like the disk of a planet. In each case the planetary nebula consists of a very hot bluish-white star surrounded by a shell of luminous gas. It is probable that there actually are thousands of them in the galaxy. The best known is the Ring Nebula in the constellation Lyra.

ROTATION OF THE GALAXY

The flattened structure and spiral arms of the galaxy strongly suggest that the system must be in rotation. This was confirmed in the 1920s by the researches of the Swedish astronomer Bertil Lindblad and the Dutch astronomer Jan Hendrik Oort. It was stated in the previous chapter that the sun was moving toward the constellation Hercules at a velocity of 12 miles a second. This is the motion of the sun with reference to the nearby stars. But all the stars, including the sun, are revolving about the center of the galaxy. We are passengers on a cosmic carrousel.

The galaxy does not rotate like a solid wheel. The rate of rotation is fastest near the nucleus of the galaxy and becomes progressively less with increasing distance from the nucleus. This means that stars closer to the galactic center than the sun are catching up to the sun and passing it, while stars farther out from the center are lagging behind. The sun is going around the galactic center at the incredible speed of 140 miles a second. But because the galaxy is so huge, it takes the sun about 200 million years to make one trip around the galaxy. If our sun is 10 billion years old, as astronomers think, it has made only 50 trips around the galaxy since it came into existence.

The globular clusters and the individual stars of the halo do not share in the rapid rotation of the galaxy. They are moving in long elliptical orbits that extend at various angles from one side of the galactic plane to the other, passing through the galactic center.

THE EVOLUTION OF THE GALAXY

Astronomers have not had the same success in arriving at a theory of the evolution of the galaxy that they have had in depicting the life history of individual stars. From the differences in ages of the stellar populations, it seems obvious that the galactic nucleus, the

Horsehead Nebula in Orion

central bulge, and the globular clusters and individual stars of the halo formed first, then the stars of the galactic disk, and finally the spiral arms. But just why things happened this way is not yet understood.

It is generally assumed that the galaxy began as an immense cloud of hydrogen gas which contracted under its own gravitation. Irregularities in the cloud caused it to begin to rotate as it contracted. Perhaps a gravitational collapse caused a considerable amount of gas to fall to the center. The collapse caused this central mass to grow hot and turbulent and it separated into smaller fragments which became protostars and then stars. The gas left behind became the galactic halo and condensation in it led to the formation of the globular clusters. It is now believed that stars do not evolve in isolation but in "showers" containing from 100 to a million stars.

The first generation of stars included many huge blue supergiants which led brief lives, ending as supernovae. But during their

Trifid Nebula in Sagittarius

lives they transformed hydrogen into helium and created the heavier chemical elements including the metals, as explained in the previous chapter. When they exploded, they injected these newly formed elements into the interstellar gas. It is thought that star formation went on faster as dust grains formed in the gas. The second and subsequent generations of stars contained more short-lived supergiants but also long-lived yellow and red dwarfs.

Star formation eventually came to an end in the central region. It is significant that the spectra of the Population II stars show them to be "metal poor." Their metal content is only about one tenth that of the Population I stars. It seems certain that the stars of Population II are old stars which formed in the early days of the galaxy.

We assume that the stars of the galactic disk, sometimes called the Disk Population, formed next from gas and dust that streamed out of the central bulge. Their metal content lies between that of Population I and Population II.

Most of the gas and dust in the galaxy is concentrated in the spiral arms. Just how these arms took shape is not clearly understood. One explanation bears the startling name of the "coffee-cup theory." You can perform an amusing and fascinating experiment that shows how this works. Stir a cup of black coffee near the middle of the cup. The coffee will rotate fastest at the center. At the rim of the cup it will not be rotating at all. Now pour a little cream into the center of the surface. It will be drawn into beautiful spiral arms by the differential rotation of the coffee. You can make a great number of different spiral structures in this fashion, some of which bear a startling resemblance to photographs of familiar spiral galaxies.

Ring Nebula in Lyra

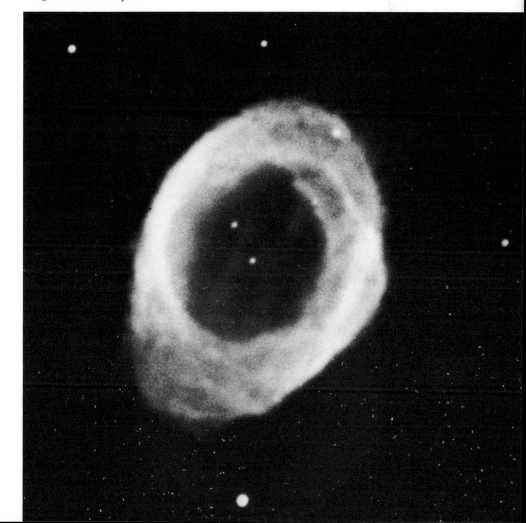

While this experiment is amusing, it is probably much too simple. Astronomers are inclined to believe that the spiral arms are not permanent features but are constantly forming and dissipating. There is evidence that gas is still streaming out of the central bulge. There may be a circulation of gas from the halo into the central bulge, then into the disk, and finally from the circumference of the disk back into the halo.

The young supergiants of the galaxy are found in the spiral arms of the galaxy. It seems certain that star formation is still going on in these arms.

Perhaps in another decade astronomers will have more positive information about the evolution of the galaxy.

eight

THE UNIVERSE

> *Lo, these are but the parts of his ways: but how little a portion is heard of him? but the thunder of his power who can understand?*
>
> —Job xxvi: 14

A LITTLE over 200 years ago, in 1755, the great German philosopher Immanuel Kant hazarded the guess that small elliptical patches of light visible in the telescopes of the day, called nebulous stars, were actually great collections of stars at immense distances from our own Milky Way system in the vast ocean of space. He called them "island universes." A similar view was advanced in England by Thomas Wright.

In 1771 the French astronomer Charles Messier made a catalogue of 103 such fuzzy objects, but his telescope was not powerful enough to reveal the nature of any of them. Today we know that his catalogue was a jumble of diffuse nebulae, galactic clusters, globular clusters, and galaxies.

When Sir William Herschel made his survey of the heavens near the end of the eighteenth century, he noted more than 2,500 such fuzzy spots of light. For a time he inclined toward the island universe theory, but later changed his mind. Although his telescope was the best of its day, it was not good enough to distinguish between a diffuse nebula and a galaxy.

The nature of the galaxies was still in doubt as late as 1920.

The Great Andromeda Galaxy with two small satellite galaxies

By then big telescopes had given clearer views of the nearer galaxies. But many astronomers still thought that they were diffuse nebulae within our own galaxy and that what appeared to be stars in them were only knots of nebulous matter.

The question was finally settled in 1923 when Edwin P. Hubble, using the 100-inch telescope on Mt. Wilson, located a Cepheid variable star in what was then called the Great Nebula in Andromeda. This made it possible to determine the distance of the Andromeda nebula, establishing that it was far out in space beyond our own galaxy. Subsequently Hubble located more Cepheids in the Andromeda galaxy, as we now call it.

Today's giant telescopes reveal a great deal of detail in the nearer galaxies, but the more distant ones are still only fuzzy spots of light. The most amazing revelation has been the incredible number of galaxies. It is now known that there are about 100 billion galaxies within the range of the 200-inch telescope. It is impossible to estimate the total number in the universe. It may be many trillions. They do not thin out with distance as do the stars in the Milky Way. If the universe is infinite, their number may be infinite.

On the large scale the galaxies are distributed almost uniformly in space. However, there are local groupings within this large-scale distribution. Frequently galaxies appear in pairs, in groups of a half-dozen or so, and in clusters that in some cases contain as many as 10,000 galaxies. There are also clusters of clusters. The French-American astronomer Gerard de Vaucouleurs has suggested that there are still larger groupings of galaxies and clusters of galaxies in flattened supersystems which he calls supergalaxies.

Galaxies show a considerable variation in size, ranging from small ones containing 10 million or so stars to giants that are estimated to contain a trillion stars.

TYPES OF GALAXIES

Three general types of galaxies are known. These are the elliptical galaxies, the spiral galaxies, and the irregular galaxies. Each class is further divided into a number of subclasses.

About 20 per cent of the galaxies are ellipticals. They range from forms which are spherical through various forms which are more and more flattened. Astronomers designate the spherical ones as E0 galaxies, the next group as E1, and so on to the most flattened ones, which are called E7. In some ways the elliptical galaxies resemble supergiant globular clusters. They are brightest at the center, gradually fading off with no sharp outer boundary. They show no evidence of internal structure.

About 75 per cent of the galaxies are spirals. In a general way their appearance, when we see them full-face, is reminiscent of the familiar Fourth of July fireworks known as pinwheels. There is a bright center, resembling a small elliptical galaxy, surrounded by spiral arms. There are usually two coiled spiral arms emerging from the center, although there may be some minor intermediate arms. We see many spirals edgewise. These exhibit a prominent central bulge and are split lengthwise by a dark streak.

About three-fourths of the spirals are known as normal spirals. These are classified in turn as Sa, Sb, and Sc spirals. The Sa type has the largest central region and the shortest, most closely coiled spiral arms. The Sb has a smaller center and more fully developed spiral arms. The Sc has a very small center and widely separated spiral arms.

The other fourth of the spirals are known as barred spirals, because the spiral arms, instead of emerging from the center, originate at either end of a bright bar that extends across the center. The barred types are known as SBa, SBb, and SBc, depending upon how tightly the arms are coiled.

Our own galaxy is classed as an Sb type. It is thought that it is very much like the Andromeda galaxy and that from that galaxy would look very much as the Andromeda galaxy looks to us.

The Sa galaxies merge into a type known as the S0 which resemble the most flattened of the ellipticals, the E7s, in some ways. They contain no gas or bright stars, but slight traces of the dust of spiral arms can be detected in them.

The irregular galaxies, as their name implies, are ragged and irregularly shaped aggregations of stars. They are designated by the letter I.

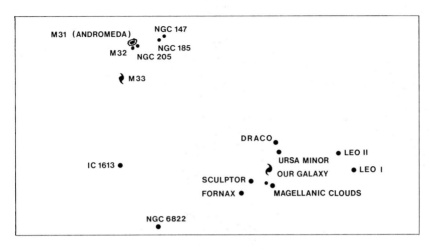

Figure 14. The Local Group

THE LOCAL GROUP

Our galaxy and the Andromeda galaxy are senior members of a cosmic club, a group of 17 galaxies to which astronomers have given the prosaic name of the Local Group. It occupies an egg-shaped volume of space with our galaxy and the irregular galaxies known as the Magellanic Clouds at one end of the long axis, and the Andromeda galaxy and three smaller ones at the other end. The group consists of three spirals, four irregular galaxies, four ellipticals, and six dwarf ellipticals, so called because of their low luminosity. The distance from our galaxy to the Andromeda galaxy is about 2 million light-years. (See Figure 14.)

The three spirals are our own galaxy, the Andromeda galaxy, and one still known as M33 from the number Messier gave it in his catalogue. The Andromeda galaxy is larger than our own, having a diameter of about 120,000 light-years. It is surrounded by a halo containing some 250 globular clusters.

During World War II, when Los Angeles was blacked out, Walter Baade took advantage of the unusually dark sky to launch an intensive study of the Andromeda galaxy with the 100-inch telescope on nearby Mt. Wilson. It is interesting to note that he made his discovery of the two stellar populations not in our own galaxy but in the Andromeda galaxy in the course of this study. As

Spiral galaxy in Ursa Major

in our galaxy Population II stars compose the central region of the Andromeda galaxy and the globular clusters around it. The Population I stars are in the spiral arms.

In determining distances in our own galaxy, astronomers had made use of the cluster-type Cepheids, or RR Lyrae, stars which occur in the globular clusters and also as single stars in the galactic halo and in the central region of the galaxy. These are the faintest of the pulsating variable stars as well as the ones with the shortest periods. Baade had been unable to locate any RR Lyrae stars in the Andromeda galaxy. This was not surprising since it was agreed that the 100-inch telescope was not sufficiently powerful to reveal them. But when in 1950 he could not find them with the 200-inch telescope, he realized that the situation needed explaining.

The RR Lyrae stars were obviously Population II stars, while the longer-period Cepheid variables noted in the spiral arms of the Andromeda galaxy by Hubble and subsequent observers were Population I. It had been assumed that all the pulsating variable stars fell on the same absolute magnitude scale which enabled the astronomer to infer the absolute magnitude of the star from its period. However, this was challenged in 1944 by the French astronomer Henri Mineur, who felt that the longer-period Cepheids fell on a scale that was 1.5 magnitudes brighter. Baade concluded rightly that there was this difference of 1.5 magnitudes between Population I and Population II pulsating variables.

But this conclusion had a profound effect upon the whole picture of the universe. The distance to the Andromeda galaxy had been calculated from the comparison of the observed apparent magnitude of the Cepheids in the spiral arms with their inferred absolute magnitudes. A change in the absolute magnitude scale of 1.5 magnitudes meant that these Cepheids were four times brighter than previously supposed. Since the apparent magnitude of a star is inversely proportional to its distance, this meant that these Cepheids were twice as far away as previously supposed. It meant also that the diameter of the Andromeda galaxy was twice as big as previously supposed. It had been thought that the Andromeda galaxy was smaller than our own and about a million light-years away. It became apparent that the Andromeda galaxy is larger than our galaxy and about 2 million light-years away. Baade's work did not affect the accepted views of distances in our galaxy since these had been based on Population II stars, the RR Lyrae variables. But every distance and diameter beyond our galaxy had to be doubled. Our notion of the size of the visible universe was doubled.

THE MAGELLANIC CLOUDS

Our nearest neighbors in intergalactic space are the Magellanic Clouds. These are still classed as irregular galaxies although recent studies have led some astronomers to think that they show some evidence of a partial spiral structure. They can be seen only from the southern hemisphere and to the unaided eye look like luminous patches that have broken loose from the Milky Way. They were first noted by the explorers of the fifteenth century as their ships rounded the Cape of Good Hope. The Portuguese navigators called them Cape Clouds, but in later centuries they were named after the great circumnavigator Ferdinand Magellan who sailed around the world in 1618–20. They are about 150,000 light-years away.

The Large Magellanic Cloud is about 30,000 light-years in diameter. While it is smaller than our own galaxy, it is larger than the average galaxy in the universe. The Small Magellanic Cloud is 20,000 light-years in diameter. Both clouds appear to consist chiefly of Population I stars. They are rich in blue supergiants, white giants, Cepheid variables, and diffuse nebulae. They contain scores of clusters like our own galactic clusters. Many of them are like the Pleiades in which the stars are enmeshed in a tangle of gaseous material. One cluster in the Large Magellanic Cloud contains more than 100 giant and supergiant stars. It includes one star, S Doradus, which is the brightest star known to astronomers. It is a million times more luminous than our sun. The most conspicuous diffuse nebula in the Large Magellanic Cloud is known as the Loop Nebula because of its shape. It is far larger than the Great Nebula in Orion in our galaxy. In fact, if the Loop Nebula were put in place of the Orion nebula, it would fill the entire constellation of Orion and be bright enough to cast shadows. The Large Magellanic Cloud contains so much dust that more distant galaxies cannot be seen through it. The Small Magellanic Cloud contains less obscuring material. A number of globular clusters are included in the Large Magellanic Cloud. RR Lyrae variables, Population II stars, have been detected in them. Recent observations indicate that a faint bridge of stars may connect the two Magellanic Clouds. It is also thought that they are surrounded by a common gaseous halo and that this may merge into the halo around our own galaxy.

THE MAFFEI GALAXIES

In 1968 the Italian astronomer Paolo Maffei detected the existence of two massive galaxies which had escaped attention because they are almost completely obscured from our view by the gas and dust of one of the spiral arms of our own galaxy. These galaxies have been named Maffei 1 and Maffei 2. It appears that Maffei 1 is a large elliptical galaxy about 3 million light-years away, while Maffai

Spiral galaxy in Canes Venatici

2 is a spiral galaxy farther away. Some astronomers think that both these galaxies should be considered part of the Local Group, thus raising the number of members in our cosmic club from 17 to 19.

EVOLUTION OF THE GALAXIES

Hubble showed that the galaxies could be arranged in an orderly sequence on the basis of their shape, starting with the spherical E0 and proceeding through the more flattened ellipticals to the S0. At this point the line separated into two branches, one consisting of the normal spirals, the other of the barred spirals. He did not include the irregular galaxies in this sequence, which is shown in Figure 15.

While Hubble had been careful not to claim that this represented an evolutionary sequence, some other astronomers concluded that it did. The thought was that an elliptical E0 galaxy contracted under its own gravitation and that this contraction caused it to

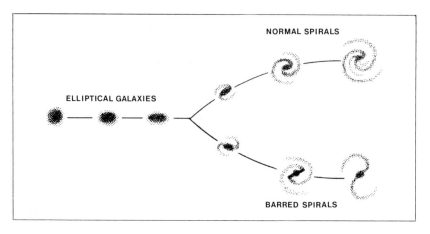

Figure 15. Hubble's sequence of galaxies

rotate more rapidly. The galaxy grew more flattened, finally developing spiral arms.

However, as more information accumulated about the nature of the various types of galaxies, it became apparent that this could not be the case. The elliptical galaxies are redder than the spirals. The brightest stars in them are red giants. They are virtually free of gas and dust. It is evident that they are very old aggregations of Population II stars. Apparently they used up their gas and dust quickly and star formation stopped in them many billions of years ago.

The spiral galaxies, like our own galaxy, have central regions of old Population II stars, but their spiral arms are full of gas and dust and contain many young Population I stars, blue supergiants and white giants. The irregular galaxies contain great numbers of supergiant stars entangled in vast clouds of gas and dust. They appear to be the youngest of all the galaxies.

These considerations led Harlow Shapley some years ago to suggest that the Hubble sequence proceeded in just the opposite direction. According to this view, a galaxy, taking form from a gigantic cloud of gas, would start its life as an irregular galaxy. In time rotation would produce a symmetrical and orderly structure, resulting in a spiral galaxy with a small center and widespread arms. As time went on, and one generation of stars followed another, the central region would grow larger while the arms would begin to wind up, closing the gaps between them and the central bulge. Eventually all the gas and dust would be exhausted, and the galaxy would become an elliptical.

Astronomers today are inclined to accept the picture of the evolution of spirals from irregular galaxies. However, no way is now seen by which a spiral galaxy could become an elliptical one. The reason is the disparity in mass. The American astronomer Thornton Page has shown that elliptical galaxies average 30 times more massive than spiral galaxies. It is interesting to note that the spherical E0 galaxies are the most massive of all and that there is a decrease in mass from the E0 to the E7 and S0 types.

It has been suggested that just as the life history of a star depends upon its mass, the same is true of galaxies. This view pictures an original state of the universe in which space was filled with a homogeneous fog of hydrogen. Local irregularities developed in this fog, causing it to break up into fragments of various sizes with different rates of rotation. These clouds have been called protogalaxies. It is assumed that a very large slowly rotating protogalaxy became an E0 galaxy. Smaller clouds, rotating more rapidly, formed other types. The density of the protogalaxy and the existence of magnetic fields in it would also play an important part in its evolution. It is thought that the moment at which star formation took place in a protogalaxy has much to do with its type. It is obvious that we do not yet know as much about these things as we should like to know.

RADIO GALAXIES AND SUPEREXPLOSIONS

A new and exciting phase of astronomy began in 1931, when the American engineer Karl Jansky of the Bell Telephone Laboratories discovered that faint radio waves were entering the earth's atmosphere from the center of our galaxy. This led, a decade later, to the construction of radio telescopes. A radio telescope is a directional antenna connected to an extremely sensitive radio receiver. The incoming waves are recorded on a moving tape. They can be fed into a loudspeaker, but produce only hissing noises of various intensities.

A large variety of radio telescopes are in use today. Some are huge complexes of steel masts and wire nets. One in Australia, known as the Mills Cross, consists of two wires, each 1,500 feet long, in the form of a cross. The most spectacular type consists of huge steel or aluminum parabolic reflectors mounted on tall masts, and so arranged that they can be pointed at any spot in the sky. The radio receiver is placed at the focus of the reflector, or dish, as radio astronomers call it. The largest of this type, located at Jodrell Bank near Manchester, England, has a dish 250 feet in diameter.

It was discovered early that radio waves came from the sun and some of the planets. Subsequently, it was found that strong radio waves were coming from some 2,000 discrete spots in the heavens. These became known as radio stars. With improvements in radio telescopes it became possible to pinpoint some of these sources with sufficient accuracy to turn the 200-inch telescope on them. It became evident that they were not stars. One of the first sources identified turned out to be the Crab Nebula, the remnant of the supernova whose explosion was recorded in the Chinese annals in 1054. Other radio stars were found to be turbulent diffuse nebulae, perhaps the remnants of unrecorded supernovae. However, the most amazing development has been the fact that most of the radio sources identified so far with visible objects are galaxies. These are now known as radio galaxies. About 100 of them have been identified.

The first radio galaxy identified by the 200-inch telescope was in the constellation Cygnus. It looks very much like two galaxies that had crashed into each other head on. Some other radio galaxies have similar appearances and for a time astronomers were inclined to accept them as examples of galactic collisions. Today, however, astronomers lean to just the opposite view, namely, that these are single galaxies which have been torn apart by superexplosions. It has been found in many cases that the radio waves originate not in the visible portion of a radio galaxy but in two areas beyond it, one at either end of a diameter of the galaxy. It is assumed that the radio sources are huge invisible clouds of electrons trapped in magnetic fields. Such clouds of electrons, it is argued, could result only from tremendous explosions within the galaxy, explosions more violent than the simultaneous detonation of trillions of hydrogen bombs. Such an explosion would be the

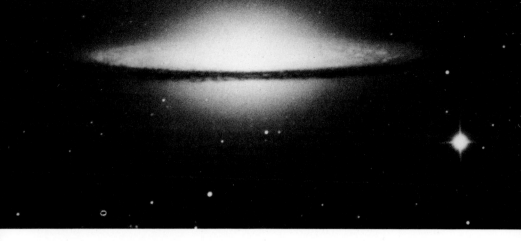

Spiral galaxy in Virgo seen edge on

equivalent of the annihilation of 100 million stars and the conversion of their total mass into energy.

Strong evidence for the explosion theory has been supplied from the study of the radio galaxy M82, a huge saucer-shaped elliptical galaxy, by Alan Sandage at the Palomar Observatory. Sandage equipped the 200-inch telescope with a filter which admitted only red light. His photographs showed a vast structure of hydrogen gas above and below the central portion of the galaxy. Further studies by C. R. Lynds at the Lick Observatory revealed that the hydrogen was streaming out of the galaxy's center at a speed of about 600 miles a second.

Photographs of another radio galaxy, the elliptical galaxy M87, revealed a great jet of hydrogen gas emerging from its center.

It has been suggested that some sort of chain reaction in the center of a galaxy, in which one star after another in rapid succession became a supernova, might account for such an explosion. However, Sandage believes that it is not possible to explain the size of the explosion by any conventional mechanism. The British astronomer Fred Hoyle and the American physicist William A. Fowler have advanced a possible explanation. According to their theory, the energy is released by a sudden gravitational collapse at the center of the galaxy. Hoyle thinks such a collapse may have taken place in our own galaxy and that the material hurled out in the subsequent explosion accounts for the globular clusters and stars of our galaxy's halo. Such explosions are thought to account for the cosmic rays, the rain of high-energy atomic particles which enters the earth's atmosphere from outer space.

Further evidence for the explosive nature of galaxies is contained in the unique *Atlas of Peculiar Galaxies*, compiled by Halton C. Arp of the Mt. Wilson and Palomar Observatories. This contains a total of 338 galaxies which are distinguished by many strange features. They are twisted into weird corkscrew shapes and some are linked together by luminous bridges of stars, gas, and dust. The luminous bridges, in some cases, are more than 100 light-years long. A number of the strange galaxies resemble spiral galaxies with a small or dwarf irregular galaxy attached to the end of one arm. In some cases, the little galaxy is attached to a greatly elongated arm which extends far beyond the spiral.

What might be called an explosion theory of galactic evolution has been advanced in the last few years. According to this theory, the original fog of hydrogen in the universe broke up into protogalaxies which became giant spherical E0 galaxies. Instability in these galaxies caused explosions which ejected fragments of various sizes. Larger fragments formed other elliptical galaxies. It will be recalled that there is a steady decrease in mass from the E0 to the E7 types. The smaller fragments of lower density evolved into spirals. For a time after each explosion, the exploding galaxy emitted radio waves. According to this theory, the radio galaxies now observed in the universe are the most recent ones to have exploded.

In support of this theory it is pointed out that there are many small groups of galaxies in which a massive elliptical galaxy is associated with one or more spiral or irregular galaxies. Many such groups appear to be flying apart as though they were fragments of a gigantic explosion.

THE EXPANDING UNIVERSE

It is possible to obtain the spectrum of a galaxy which represents the integrated light of all the stars in it. During the 1920s Vesto

Melvin Slipher at the Lowell Observatory and Milton Humason at the Mt. Wilson Observatory determined the motions of a number of galaxies in the line of sight, the so-called radial velocity, from the shift of the lines in such spectra. (It will be recalled from Chapter 6 that this shift is a Doppler effect, revealing motion in the line of sight.)

In 1929 Hubble announced a relationship between the radial velocities and the distances of 20 galaxies whose distances he had determined. Every spectrum showed a shift to the red, and the farther away a galaxy was, the greater was the shift. In other words, every galaxy was moving away from the earth, and the farther off it was, the faster it was running away. Hubble boldly predicted that this relationship would be found true for more distant galaxies. The relationship became known as Hubble's Law of Red Shifts and is regarded as the most spectacular discovery of the last 50 years.

Humason continued the laborious task of determining the red shift of more distant galaxies and by 1936 had succeeded in measuring the speeds of galaxies up to a billion light-years away. He found the most distant ones receding at the incredible speed of more than 13,000 miles a second. By 1956 Humason and his colleagues, working with the 200-inch telescope, had found that galaxies 2.5 billion light-years away are receding at a speed of 38,000 miles per second, a speed equal to one-fifth the speed of light. Since then, more distant galaxies have been found to show even greater speeds.

A superficial view of this situation might lead to the conclusion that our earth is some center of repulsion from which all the galaxies are running away and that the farther away they are, the faster they are running. This is an obviously impossible conclusion.

It was quickly realized that the correct explanation was that the universe is expanding. You would, therefore, observe the same state of affairs from a planet in any galaxy of the universe. A simple analogy will help you picture this. Imagine that you have pasted a large number of small paper dots on the surface of a toy rubber balloon. If you blow more air into the balloon, it will grow larger. As its surface expands, the distance between dots will increase. In addition, the farther apart any two dots are, the more rapidly will they move apart. Any dot will constitute a center from which all the other dots are receding.

It will be seen, however, that with our experiment we have introduced a new idea into our concept of the universe. Until now we have been thinking of stars and galaxies as moving through static space. Now we must think of space itself as expanding and carrying the galaxies along with it. Apparently gravitation is strong enough within a galaxy or even a group or cluster of galaxies to hold it together. There is no evidence of expansion within the Local Group. But the distances between isolated galaxies and clusters of galaxies is constantly increasing as a result of the expansion of space.

While the great majority of astronomers accept the theory of the expanding universe, some attempts have been made to explain the red shift as something other than a Doppler effect. One suggestion, not well received by most astronomers, is the theory of "tired light." This holds that the light waves from distant galaxies slow down with the passage of time. At the moment the only known physical principle that will explain the Law of Red Shifts is the Doppler effect.

CURVED SPACE

Until astronomers began the exploration of the distant reaches of the universe, there was no reason to be concerned with the nature of space. For measurements of distances within our own galaxy or to the nearer clusters of galaxies it could be assumed that the laws of Euclidean geometry held true. This is the ordinary view of three-dimensional space, which has been called "flat" or "Euclidean." In more extensive investigations the possibility that space possesses a curvature must be taken into consideration.

The idea of curved space was first advanced by the German mathematician Bernard Riemann in 1854. But it attracted no particular attention until Einstein introduced it into his general theory of relativity in 1915. Three possibilities are recognized today concerning the nature of space. It may have positive curvature, zero curvature, or negative curvature. Zero curvature, of course, means no curvature at all. If space is curved, it must be curved into a fourth dimension. This can be expressed by mathematical equa-

Spiral galaxy in Pisces

tions, but it is hopeless to try to visualize it. Some simple analogies, however, will help you understand it.

The surface of the earth can be thought of as a two-dimensional space curved into a third dimension. It is finite but unbounded. If you travel long enough in any direction, you eventually find yourself back at the starting point. If space has a positive curvature, it is also finite but unbounded.

If space has a negative curvature, its properties resemble those of a saddle. Mathematicians call such a surface a hyperbolic paraboloid. It does not close on itself like the surface of a sphere. It can be extended indefinitely. If space has a negative curvature, it is open and infinite.

Einstein concluded that space had a positive curvature. His model of the universe, known in the 1920s as the "Einstein universe," gave space a radius of 84 billion light-years. Another model of the universe was proposed by the Dutch astronomer Wilhelm de Sitter. Both the Einstein and de Sitter models were contradicted by astronomical observations and soon abandoned.

Early in the 1920s three scientists independently pointed out some cosmological consequences of the theory of relativity. They were the Russian mathematician Alexander Friedman, the Belgian mathematician Abbé Georges Lemaître, and the American mathematical physicist Howard P. Robertson. They showed that positive, zero, and negative curvature of space were all compatible with relativity. Their most startling conclusion was that the universe cannot be static and that it must be either expanding or contracting. They could not tell, however, which was the case. In 1929, as already related, Hubble found that the universe is expanding. It is an amazing fact that mathematical reasoning disclosed the possibility of an expanding universe before astronomical observations indicated it.

THE MYSTERIOUS QUASARS

By 1962 the radio astronomers at Jodrell Bank in England had located some extremely small, or compact, radio sources. Subsequently, the exact positions of some of them were pinpointed by radio astronomers in Australia who recorded the exact instant when one was occulted, or eclipsed, by the moon. Since the moon's position is known with great exactness, this revealed the precise location of the radio source. Sandage was then able to turn the 200-inch telescope on them. To his surprise, they appeared to be faint blue stars. However, a study of their spectra revealed enormous red shifts. It became apparent that they were not stars in our own galaxy at all, but are possibly the most distant objects yet recorded in the universe. They became known as quasi-stellar radio sources, a name subsequently contracted to "quasars." About 120 quasars have been located. The red shift of about half of them has been determined. Dr. Maarten Schmidt, who has made many of these measurements, has found that a half-dozen of them are apparently more than 8 billion light-years away. The most distant one appears to be receding at a velocity of 149,000 miles a second, a rate equal to 80 per cent of the speed of light.

If a quasar is as far away as the red shift appears to indicate, it means that it is actually about 100 times brighter than the brightest galaxy. The situation is complicated by the fact that quasars have been found to exhibit variations in brightness, sometimes changing as much as 40 per cent in a few weeks. Some quasars vary in brightness and radio output in periods of a few days. These variations indicate that a quasar cannot be more than a few light-years in diameter. The problem is to explain how an object far smaller than a galaxy can radiate more energy than an entire galaxy.

Recognizing that a quasar is too large to be a star and too small to be a galaxy, Hoyle and Fowler have suggested that it is a superstar equal in mass to 100 million suns. They think that such a superstar would result from the sudden gravitational collapse of a large cloud of gas.

An attractive theory put forward in 1967 by W. H. McCrea of the University of Sussex, England, pictures a quasar as the first observable state of the formation of the nucleus of a galaxy. According to this theory, evolution has proceeded to the point where the protogalaxy has fragmented into protostars, some of which have already become supergiant stars. McCrea thinks that some of the

Barred spiral in Eridanus

Part of cluster of galaxies in Coma Berenices

supergiants have already lived their short lives and exploded into supernovae. At a given time four or five such supernovae would be sufficient to account for the luminosity of a quasar. The rapid fluctuations in brightness which are so baffling a feature of the quasars are explained by McCrea as a "searchlight and shutter" effect. When the light of the supernova emerges from the quasar, we have the searchlight effect. When one is occulted, or eclipsed, by a protostar, we have the shutter effect. McCrea has also shown that the effect would be the same if all the light of the quasar came from one superstar. This would be partially eclipsed by the random motion of the dark protostars.

Not all astronomers feel that the red shift of the quasars is an index of their distance. Among those who think that the quasars are much closer to us is Halton C. Arp. He believes that the quasars are small compact galaxies fairly close to the earth as distances go in the universe. He thinks they may be only 30 million to 300 million light-years away. He bases this opinion on his *Atlas of Peculiar*

Galaxies. He found that in a considerable number of cases the peculiar galaxy is situated between two powerful sources of radio waves. In some cases they are radio galaxies, but in others they are quasars. He believes that these radio galaxies and quasars were formed from immense masses of gas expelled by an explosion in the center of the peculiar galaxy. Two explanations for the very great red shift in the spectra of quasars are offered by Arp. One is that the red shift is a measure of the rate at which the hot gases in the quasar are collapsing. The other is that the quasar is a compact galaxy with an exceptionally high gravitational field. Such a field would cause a red shift in the light leaving the galaxy.

The existence of a gravitational red shift was first predicted by Einstein in his general theory of relativity in 1915. It was subsequently verified in the case of the sun by the American astronomers Walter S. Adams and Charles E. St. John. The German astronomer Karl Schwarzschild calculated that in a gravitational collapse the density would finally become so great that the light could not fight its way out against the gravitational field. As a result a star or even an entire galaxy would disappear from view. Such a supposed disappearance is known as a Schwarzschild singularity, or "black hole."

Quasars emit not only strong radio waves, but also very large amounts of ultraviolet light. Sandage used this fact to locate additional quasars. To his surprise, he found a number of faint blue starlike objects which radiated large amounts of ultraviolet light but no radio waves. They show red shifts comparable to quasars. Sandage called them blue stellar objects. They present the same puzzle as the quasars. It is possible that they might be either an earlier or a later stage in the development of a quasar.

THE ORIGIN AND NATURE OF THE UNIVERSE

We come finally in this survey of astronomy to the biggest and most fundamental of all problems, the origin and nature of the universe. The branch of astronomy concerned with these problems is known as cosmology.

In 1931 Lemaître proposed a theory of the universe which was subsequently extended by the Russian-born American scientist George Gamow. It was known at first as the theory of the exploding universe, but astronomers today, with a delightful display of humor, call it the "big-bang" theory. The fact that the universe is expanding means that the galaxies are drawing farther and farther apart. Lemaître pointed out that this meant that if we could go back in time, we would find the galaxies closer and closer together. From this Lemaître concluded that originally the whole universe was crowded together into one single mass. He called this the primeval egg or the primeval atom. It is also known today as the primeval fireball. One must not think of this superatom as existing somewhere in the midst of infinite space. According to Lemaître, this was the whole universe. He estimated that it was about the size of Mars, about 4,000 miles in diameter. All the matter of the universe, all the energy of the universe, and space itself was packed into this superatom. A cubic inch of it weighed about 2 billion tons.

The universe which we know today began with a big bang, the explosion of the superatom. It is estimated that this took place about 12 billion years ago. Gamow calculated that the temperature of the exploding superatom was about 25 billion degrees F. But as space expanded, the temperature fell rapidly. It dropped to about 1 billion degrees in the first 5 minutes. By the end of the first 250 million years it had dropped to $-200°$ F. At the moment of explosion radiation had the upper hand, and the universe was a huge blaze of light. But as expansion continued the universe grew dimmer. Finally space was filled with a great dark fog of hydrogen gas. But in time turbulence and gravitation were able to cause local condensations in the fog. These became the protogalaxies which eventually fragmented into showers of stars. With the formation of stars light returned to the universe and it gained its present majesty. With the passage of time the galaxies will draw farther and farther apart. If the expansion goes on forever, the time will come when it will be impossible to see another galaxy outside the Local Group. In time all the stars of every galaxy will leave the main sequence, degenerate into white dwarfs that will finally cool off and die. The universe will again be cold and dark. It will be a dead universe.

Astronomers speak of this type of universe as an evolving universe. It requires that the curvature of space be negative.

However, another possibility was suggested by Howard P. Robertson and the American physicist Richard C. Tolman. They reasoned that if space had a positive curvature, the rate of expansion would eventually begin to slow up and finally stop. Gravitation would then gain the upper hand and the universe would begin to contract. After some billions of years it would again be consolidated into a superatom. Then there would be another big bang and expansion would begin again. The kind of universe postulated by Robertson and Tolman is called an oscillating universe.

An alternative to the big-bang theory was proposed by the British astronomers Fred Hoyle, Herman Bondi, and Thomas Gold. It is known as the "steady-state" or the "continuous-creation" theory. It holds that the universe had no beginning, that it will have no end, and that it always looked essentially as it does today. The theory recognizes that space is expanding and the galaxies drawing apart. It contends, however, that as the galaxies draw apart, new galaxies come into existence in the space between them. Consequently, while the universe changes in detail, its large-scale aspect is unchanging. The new galaxies take shape out of the accumulation of hydrogen atoms which are constantly coming into existence at the rate of one atom a year in a volume of space equal to a 20-story building. This sounds like very little, but actually it would mean the creation of trillions of hydrogen atoms in a cubic light-year of space. This continuous creation of hydrogen atoms is an arbitrary assumption not subject to direct test.

In 1965 Hoyle startled the astronomical world by stating in a communication to *Nature*, the British scientific weekly, that he felt that the idea of a steady-state universe would have to be discarded. His new view was that oscillations take place in finite regions within an otherwise infinite universe. Hoyle calls these oscillating regions "bubbles" and considers the observable portion of the universe as one such bubble, a pretty big bubble to say the least.

Astronomers must turn to the most distant galaxies and quasars in attempting to distinguish among the various theories of the origin and nature of the universe. This is because the study of distant objects is an exploration in time as well as space. When we

Radio galaxy in Centaurus

turn the 200-inch telescope on a galaxy that is 5 billion light-years away, we are not seeing it as it is today, but as it was 5 billion years ago.

If the steady-state theory is correct, we should see the same proportion of old and young galaxies at all distances from us. But if the big bang theory is correct, we should find more old galaxies in our neighborhood and more young galaxies at great distances. Unfortunately, we do not yet know enough about the evolution of galaxies to apply this criterion with any confidence.

Astronomers have had more success in drawing conclusions from the distribution of galaxies with distance and from changes in the rate of recession with distance.

The big-bang theory requires that the galaxies were closer together 5 billion years ago than they are today. The study of distant radio galaxies indicates that those at immense distances are closer together than nearby ones. Since we observe these galaxies, not as they are today, but as they were billions of years ago, the observations support the big-bang theory.

The most impressive evidence at present for the nature of the universe comes from the determination of the rate of recession of galaxies and quasars at immense distances. Calculations show that this should differ for a steady-state universe, for an evolving universe, and for an oscillating universe.

From a study of the red shifts of the most distant known galaxies and quasars, Sandage has concluded that the rate of expansion was greater 8 billion years ago than it is today. This means that the expansion is slowing up. He concludes, therefore, that the curvature of space is positive and that we live in an oscillating universe. He calculates that the present cycle of the universe began with a big bang 12 billion years ago and that it will continue for another 29 billion years. Then expansion will stop, gravitation will get the upper hand, and in another 41 billion years the universe will again be concentrated in a superatom ready for the next big bang. In other words, each cycle of the oscillating universe occupies 82 billion years.

The most recent evidence in favor of the big-bang theory was the discovery of a low-energy cosmic radio radiation that fills all space and enters the earth's atmosphere from all directions. It was discovered by Arno A. Penzias and Robert W. Wilson of the Bell Telephone Laboratories in 1965. Its existence had been previously predicted by Robert H. Dicke of Princeton University. It is sufficiently strong to be received by conventional radio telescopes and probably accounts for some of the "snow" seen on television screens. Study of the radiation has been carried on at Princeton by a team headed by P. J. E. Peebles and David T. Wilkinson. They have concluded that the radio waves are the result of the original big bang, the explosion of the superatom which they prefer to call the primeval fireball. The radiation corresponds to a temperature of 3 degrees above absolute zero (absolute zero is $-459°$ F.).

As Hubble once said, the history of astronomy is a history of receding horizons. Every improvement in telescopes and auxiliary instruments has enabled astronomers to reach further into the depths of space, first to the edge of our own galaxy, then to the nearby galaxies, finally to the most distant ones. It has been an exploration in time as well as space. With the study of the all-pervading low-energy cosmic radio radiation, we may be back almost to the instant when the big bang started the present cycle of the universe 12 billion years ago.

PART 2

****

The Earth

nine

THE RISE OF GEOLOGY

In the beginning God created the heaven and the earth.
—Genesis 1: 1

IT is one of the paradoxes of man's intellectual development that his acquaintance with the distant stars progressed more rapidly than his knowledge of the earth under his feet. He gave names to the constellations, he charted the courses of the planets, he learned to predict eclipses, all before he came to realize that there was an equally fascinating record locked within the earth itself. Even within comparatively recent times the study of the earth has lagged behind the study of the stars. Copernicus ushered in the era of modern astronomy in 1543. But it was not until 1795 that James Hutton laid the foundations of modern geology.

Man's tardiness in deciphering the earth's record must be blamed upon the geography of the ancient world. Babylon was a flat country in a dry region, the great flood-plain of the Tigris-Euphrates. The sky was clear and cloudless and the horizon free. It was possible to watch the majestic rise of planets and constellations on the eastern horizon and to follow their solemn march across the sky and down to the western horizon. And so the Babylonians laid the foundations of astronomy.

But they developed no geology because the rocks of Mesopotamia are out of sight, buried deep under the huge flood-plain. The earth they saw seemed static and unchanging. And so they took it for granted that the earth as they knew it was the earth which had existed since the time of creation. The Hebrews inherited their science from the Babylonians and therefore there is little geology in the Old Testament.

THE GREEKS

The Greeks, living in a hilly, rainy district, were brought face to face with the record in the rocks. In addition, they were sailors and they saw more of the record on their travels. As a result they made a beginning at the deciphering of that record. The Greeks were our first geologists.

They saw that many of the rock layers, the so-called stratified rocks, were precisely like beaches and deltas and sea floors, and therefore must have been ancient beaches and deltas and sea floors which had hardened and dried up. They also identified the fossils in these rock layers as creatures that had once lived in the sea.

From this they reasoned that many regions which were dry land must have been at the bottom of the sea once upon a time. Aristotle, in his *Meteorics* written about 330 B.C., sets forth the idea that the land must sink at times, letting in the sea, and that later it must rise again out of the sea. Strabo, writing in Rome in about 7 B.C., states the same idea in his monumental *Geography*. When the Barbarians overthrew Rome, Greek knowledge passed into the hands of the Arabs. A tenth-century Arabian scientist named Omar wrote a book titled *The Retreat of the Sea*.

MEDIEVAL EUROPE

Medieval Europe turned chiefly to the Bible for knowledge, and because the Babylonians and the Hebrews had developed no geology, the study of geology for all practical purposes came to a standstill. Medieval Europe went back to the account of the earth's beginning which is to be found in Genesis. Some of the Greeks had placed the age of the earth at a million years, but medieval theologians placed its age between 4,000 and 6,000 years.

It took many centuries for the world to get back to the geological notions of the Ancient Greeks. Leonardi da Vinci, Italian painter, engineer, inventor, and genius, who was born in 1452, was commissioned in his youth to build some canals in the northern part of Italy. The excavations cut through stratified rocks, that is, rocks which lay in layers. Leonardo noticed the rock layers. He noticed the fossils of crabs and clams and snails and other marine creatures in them, and he revived the idea of the Greeks that the stratified rocks were old sea floors. Some of his countrymen agreed, but most learned men, not only of the fifteenth century, but of the sixteenth and seventeenth centuries as well, disagreed.

Two explanations were advanced to account for fossils and these explanations held sway even into the eighteenth century. The Bible says "the earth brought forth abundantly the moving creatures that hath life." The fossils, therefore, were assumed to be creatures which formed in the earth at the time of creation, but which never came forth in possession of life. They were tentative models, as it were.

The other theory of fossils was bound up with an equally erroneous biological theory, namely, that of spontaneous generation. According to this theory, living creatures came suddenly to life out of the air or out of water or mud. Such a creature in the process of formation might fall into a crevice in the rocks and consequently fail to develop completely. The partially developed form would constitute a fossil.

NOAH'S FLOOD

In the eighteenth century the favorite method of explaining the stratified rocks and their fossils was by recourse to Noah's flood. This method had started earlier. For example, Dr. John Woodward published in 1695 *An Essay toward a Natural History of the Earth and Terrestrial Bodies, especially Minerals; as also of the Sea, Rivers, and Springs, with an account of the Universal Deluge, and of the Effects that it had upon the Earth.* Vast numbers of theories based upon the idea of Noah's flood developed during the eighteenth century. It was soon seen that a single flood could not explain all the stratified rocks and so a whole series of floods was invented.

The flood theories reached their climax with Abraham Gottlob Werner, who became professor of mineralogy at the University of Freiberg in 1775. Werner was small in stature. He had a pug nose and a shy disposition. But he was a brilliant man and a gifted speaker and he made the school at Freiberg famous. Distinguished scientists came from all over Europe to hear him lecture.

The followers of Werner's theory became known as the Neptunists, after Neptune, the mythological god of the ocean, because Werner insisted that all the rock layers had formed from chemical materials precipitated out of a universal ocean. At first the ocean covered the entire earth, so he taught, but later parts of the earth protruded so that later rock layers were deposited only in certain regions. This theory gives us what has been called an "onion coat" earth, since rock layers would be deposited all around the earth at once, the layers eventually resembling the coats or skins of an onion. Werner's theory was also known as that of catastrophism, since it held that each of the floods which deposited a rock layer came suddenly and was, in fact, a great catastrophe at the time.

MODERN GEOLOGY

As the eighteenth century drew to its close, the flood theory met its Waterloo. Modern geology was founded in 1795 with the publication of James Hutton's *Theory of the Earth, with Proofs and Illustrations*. Hutton introduced the idea of "uniformitarianism" as it is sometimes called. It was the rational idea that the earth's past history was to be explained in terms of what could be seen happening in the earth at the present time. The ideas of sudden and great catastrophes were abandoned, and because Hutton insisted that many rocks were solidified lavas and not deposits out of water, the followers of his theory became known as Vulcanists, after Vulcan, the mythological god of fire. For a time a heated battle raged between Neptunists and Vulcanists, but the Neptunists, the followers of Werner, were fighting a losing fight.

Hutton was a Scot, educated at the University of Edinburgh. He became a skilled chemist and made a fortune by inventing a process for manufacturing sal ammoniac from coal soot. Then he studied medicine in Edinburgh, Paris, and Leyden. A large inheritance of land came to him and he settled down to farming. It was during this time that his interest in geology grew. In 1768 he became financially independent and took up his residence in Edinburgh. One of his close friends was Adam Smith, the great economist.

In his *Theory of the Earth*, Hutton set forth in broad outline the general principles of present-day geology. The forces of nature, he said, wear away the land. The sediment is swept into the sea. There it forms a layer which in time hardens and solidifies into rock. Eventually, change in level of the earth's surface raises the sea floor and the layer of rock is once more exposed to view. Hutton also called attention to the volcanoes that pour forth molten lava. Many rocks, he said, were formed by the solidification of such streams of lava.

His theories were further clarified in 1802 by his friend John Playfair, who published *Illustrations of the Huttonian Theory*. The clinching arguments were supplied by William Smith, British engineer and surveyor, who lived from 1769 to 1839. He became known as "Strata Smith," for he made an elaborate study of the rock strata of Great Britain, determining their geological order and showing how the age of a layer could be determined by the type of fossils in it. His monumental work was published in four volumes titled *Strata Identified by Organized Fossils*. Hutton's theory was strengthened further later in the nineteenth century by the researches of the British geologist Sir Charles Lyell.

While Werner was quite wrong in his theory of floods, he deserves to be remembered as the pioneer student of the composition and structure of rocks. He devised a system of classification based on mineral composition. A rock usually does not consist of a single mineral. As a rule it is an intricate mixture of a number of minerals.

THE TWENTIETH CENTURY

With the start of the twentieth century the science of geology was enhanced by the rise of geochemistry and geophysics, the application of chemistry and physics to the study of the earth. Pioneer studies were carried out by the U. S. Geological Survey and by the Geophysical Laboratory of the Carnegie Institution of Washington. Laboratory experiments were performed to augment field observation. Researches were carried on to determine the order in which minerals crystallized from molten material and how rocks behaved under conditions of high pressure and temperature. Such experiments are still being conducted. Geology gained greatly from the development of seismology, which is concerned with the behavior of earthquake waves.

A new era of earth exploration was set off by the International Geophysical Year, actually a period of 18 months from July 1, 1957, through December 31, 1958. This was a gigantic cooperative effort in which some 60,000 scientists in 66 countries joined hands to attack the secrets of the earth. Cooperative research programs were organized to study the mysteries of the earth's interior, its crust, its oceans and atmosphere, and the effect of solar and cosmic forces upon the earth. Plans for the use of rockets and artificial satellites led to the dawn of the Space Age. During the IGY, as it became known, simultaneous worldwide observations were made of the weather, the aurora, cosmic rays, and other phenomena. A large number of expeditions concentrated on researches in the Arctic and Antarctic.

Today's earth scientists are no longer content to study the chemical composition of a rock or the structure of a geological formation. They are seeking to reconstruct the whole story of the earth's history and the forces which directed it.

How did the earth begin? What is its present internal structure? Why is its surface divided between continents and oceans? How are mountains created? These are some of the basic questions which today's geologists are striving to answer.

ten

THE EARTH'S BEGINNING

*This world was once a fluid haze of light,
Till toward the center set the starry tides
And eddied into suns, that wheeling cast
The planets, then the monster, then the man.*
—Tennyson

THE AGE OF THE EARTH

FROM his study of the genealogies in the Old Testament the Irish Archbishop James Ussher came to the conclusion in 1650 that the earth had been created in 4004 B.C. However, a more venerable age was assigned to the earth as the nineteenth century drew to a close. The British scientist Lord Kelvin, on the basis of what he thought was the rate at which the earth was losing heat, estimated that the earth was between 20 million and 40 million years old. But the Irish geologist John Joly, from calculations of the rate at which salt was washed out of the rocks and concentrated in the sea, placed the age of the earth at 90 million years. Similar estimates were arrived at from calculations of how long it had taken the sedimentary rocks to accumulate.

At the start of the twentieth century the great British physicist Lord Rutherford realized that radioactivity offered a geologic clock for the determination of the age of the earth. As uranium slowly disintegrates it goes through a series of transformations, ending as a particular isotope of lead. Since the rate of disintegration is known, the age of a rock can be determined from the ratio of uranium to lead in it. From this and other radioactive transformations geophysicists are now agreed that the earth is about 4.6 billion years old. This makes the earth almost as old as the sun and about one-third as old as the universe, according to the big-bang theory.

THE NEBULAR HYPOTHESIS

It seems evident that the earth must have come into existence as part of the evolution of the solar system. All the planets and thousands of asteroids go around the sun in the same direction. Their orbits lie approximately in the same plane, which is very close to the plane of the sun's equator. The sun and all of the planets, with the exception of Uranus and possibly Venus, rotate on their axes in this same direction. So do the large majority of the 32 moons in the solar system.

It was already realized by some thinkers as early as the middle of the eighteenth century that the members of the solar system must have had a common origin. In 1750 the French naturalist Georges Buffon suggested that the planets and their satellites took shape from material knocked out of the sun by a comet which collided with it. At the time, of course, there was no knowledge of the nature of comets.

A few years later a different theory was advanced independently by the Swedish theologian Emanuel Swedenborg and the German philosopher Immanuel Kant. According to their theory, the solar system originated in a nebula, or great cloud of gas, which broke up as it contracted, the central portion becoming the sun, the outer portions the planets and their satellites.

A similar theory was presented in more scientific detail by the French mathematical astronomer Pierre Simon de Laplace in 1796. He was familiar with Buffon's suggestion and pointed out why such an event could not account for the nearly circular orbits of the planets. But apparently he was unaware of the theories of Swedenborg and Kant.

Laplace pictured the solar system as beginning in a rotating globular cloud of gas. As it contracted it rotated more rapidly due to the conservation of angular momentum. On a number of occasions, when the centrifugal force at the equator balanced the gravitational attraction, the shrinking globe left rings of material behind. They were not thrown off, or "cast" as Tennyson's poem, "The Princess" has it. Laplace spoke of them as "abandoned." These rings broke up and then consolidated to form the planets. But in their turn they shed equatorial rings which became the satellites.

Laplace's theory attracted much attention and under the name of the nebular hypothesis dominated scientific thinking for more than a century. In time, however, it became evident that it possessed fatal defects. The sun, which accounts for about 99.9 per cent of the mass of the solar system, possesses only about 2 per cent of the angular momentum of the system. If things had happened as Laplace imagined, the sun would have the major portion of the angular momentum. The British mathematician James Clerk Maxwell showed that a ring of gaseous material might condense into a formation like the rings of Saturn, but that it could never coalesce into a planet.

THE PLANETESIMAL HYPOTHESIS

To overcome the objections to Laplace's theory, Thomas C. Chamberlain and Forest R. Moulton of the University of Chicago in 1916 advanced a theory which in some respects went back to Buffon. According to it, the sun originally had no planets. Random motions in the galaxy caused a star to pass near the sun. Its gravitational pull raised gigantic tides in the sun, causing immense waves of gaseous material to be ejected from the solar surface. The tendency of this material to follow the passing star provided the angular momentum which the planets now possess. The material quickly condensed into small fragments which the two scientists named planetesimals. For this reason, their theory became known as the planetesimal hypothesis. Mutual gravitation caused the larger planetesimals to sweep up the smaller ones. In this way the planets and their satellites grew by slow accretion.

This theory was modified by the British scientists Sir James Jeans and Harold Jeffreys into the so-called tidal theory. According to it, the passing star drew one long filament out of the sun. This quickly broke up into segments that became the planets and their satellites.

However, physicists raised serious objections to these theories. The principal one was that gaseous material pulled out of the sun would have been extremely hot and under very high pressure. It would explode violently once the pressure was released. As a result, the planetesimals would never have been formed, since the material would have been lost in space.

THE PRESENT VIEW

To be acceptable today, a theory must do more than account for the motions of the solar system and the strange distribution of angular momentum. It must also explain the variations in the chemical compositions of the various members of the system. These are very considerable, as you will recall from Chapters 3 and 4. It is generally agreed that we do not yet possess enough knowledge to construct a completely satisfactory picture of the evolution of the solar system. A number of scientists have offered theories. Needless to say, they do not agree in detail. It is interesting to note that they contain elements of both the nebular and the planetesimal hypotheses.

It is thought that the sun and its family of planets were born about 5 billion years ago in the collapse of a vast cloud of gas and dust like the globules which were discussed in Chapter 7. This globule may have had a diameter of 6 trillion to 9 trillion miles and its collapse may have been hastened by the pressure of radiation from nearby exploding supernovae. The globule may have contained 20 per cent more material than does the sun today. It consisted chiefly of hydrogen and helium, but contained considerable amounts of dust. It was cool, marked by turbulence, and rotating slowly with a period measured in millions of years. This slow rotation was a consequence of the rotation of the galaxy.

It collapsed under its own gravitation in a matter of a few million years. Most of it was drawn to the center to form a rapidly rotating protosun with a diameter of about 70 million miles. However, 10 or 20 per cent was left behind. Rotation caused this outer material to form a huge disk around the equator of the protosun with a diameter of about 10 billion miles.

Up to this point the modern view differs very little from that of Laplace. The big problem is to explain how the angular momentum of the system was transferred from the protosun to the disk. The German astronomer Carl F. von Weizsäcker thought it could be explained by the interaction of turbulent eddies which formed in the disk. This is now felt to be inadequate.

It is thought that the hurdle can be cleared with the aid of the new knowledge of magnetohydrodynamics. This branch of physics deals with the behavior of plasmas, that is, extremely hot ionized, or electrified, gases in a magnetic field. Very hot gases are ionized by frequent collisions of their atoms in which electrons are knocked out of the atoms. If the plasma is in motion, it creates its own magnetic field.

It is thought that as the infant sun, or protosun, increased its rate of rotation, the rapid motion of ionized gases in it set up strong magnetic fields near its surface. As the disk formed, the magnetic lines of force were carried into it by gases shed from the sun's equator. Since the outer portion of the disk was rotating more slowly than the inner portion, these magnetic lines of force were wound up into a spiral in the disk, as shown in Figure 16.

These lines acted as a magnetic brake to slow the rotation of the sun. Thus the angular momentum was transferred from the sun to the disk. As the sun slowed down, the disk speeded up. The magnetic lines also imparted a wiry rigidity to the disk, making it possible for the gases in the disk to move along the lines of force but not across them. In this way the lines created a "magnetic bottle" which prevented the gases of the disk from dissipating into space.

It is thought that temperature gradients played a major role in the evolution of the sun's family. It is obvious that as the proto-

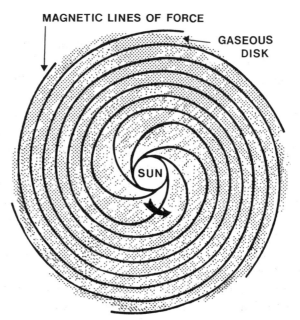

Figure 16. Magnetic lines in the solar disk

sun warmed up, the temperature in the disk would be highest in the vicinity of the solar surface, decreasing rapidly at greater distances from it.

Harrison Brown pointed out that the chemical elements can be divided into three groups and that the terrestrial planets and asteroids, the major planets, and the comets differ in their percentage of these groups. Whipple calls them the gaseous group, the icy group, and the earthy group.

Hydrogen and helium, the most abundant elements in the sun, constitute the first group. These two remain gaseous at extremely low temperatures. The icy group consists of compounds of carbon, nitrogen, and oxygen, such as ammonia, methane, and water. They freeze at moderately low temperatures. The earthy group includes the remaining heavier elements with the exception of the inert gases. The three most abundant are silicon, magnesium, and iron. They remain solid until very high temperatures are reached.

Whipple points out that the earthy elements constitute essentially all the mass of the terrestrial planets and asteroids, but rather small fractions of the major planets and comets. Comets consist almost entirely of the icy group. Jupiter and Saturn consist almost entirely of the gaseous group with some amounts of the icy group in their atmospheres. Uranus and Neptune appear to be a mixture of the gaseous and icy groups.

THE PROTOPLANET HYPOTHESIS

One present-day theory, the protoplanet hypothesis, formulated by the Dutch-born American astronomer Gerard P. Kuiper, assumes that the solar system began as a cloud of gas and dust which contracted into a protosun surrounded by a great disk in the fashion already outlined. He points out that the inner portion of the disk must have rotated more rapidly than the outer portion in accord with Kepler's third law of planetary motion.

According to Kuiper's theory, this differential rotation created eddies or whirlpools in the disk. Large eddies swallowed up small ones and in time became sufficiently dense to contract under their own gravitation. In this fashion the disk broke up into a series of rotating globes, one for each of the planets in the solar system. These were very much larger than the present planets. Jupiter's protoplanet was 10 or 20 times as massive as Jupiter is now. Protoearth may have been 1,000 times as massive as the earth today and may have had a diameter of several million miles.

The formation of these protoplanets took place in the darkness of space before the protosun had grown hot enough to begin to shine.

As the protoplanets which formed the major planets, Jupiter, Saturn, Uranus, and Neptune, continued to contract, their rotation increased and they developed equatorial disks in the same fashion that the protosun had done. These disks broke up into smaller globes which became satellites. It will be recalled that Jupiter has 12 moons, Saturn 10, Uranus five, and Neptune two.

With the passage of time, continued contraction increased the temperature of the protosun to the point where it became luminous, becoming a red variable star of the T Tauri type. This increase in temperature had no great effect upon Jupiter and the other major planets because of their great distance from the sun. However, it had a very marked effect upon the protoplanets of the terrestrial planets, Mercury, Venus, Earth, and Mars. The heat and the streams of atomic particles, the solar wind, drove the hydrogen and helium and most of the icy elements out of them, leaving behind the earthy elements which formed the solid bodies we see today.

The earth-moon system represents a special case. While Jupiter and Saturn both have satellites larger than our moon, they are small in comparison to those giant planets. The diameter of our moon is one-fourth that of the earth. Kuiper believes that the protoearth developed a double nucleus. As contraction continued, the earth formed around one nucleus, the moon around the other.

The gravitational pull of Jupiter prevented the formation of a large protoplanet in the asteroid belt. It may be that originally there were only a few fairly large asteroids. Collisions fragmented them into the present aggregation.

Whipple believes that Jupiter and Saturn may have taken shape in the fashion postulated by Kuiper. However, he suggests other mechanisms for the formation of the other planets. He thinks that the icy materials condensed on earthy nuclei to form what he calls "cometesimals." These accumulated to form the comets and the planets Uranus and Neptune. He believes that the earthy materials within the orbit of Jupiter condensed into planetesimals and that these accumulated to form the terrestrial planets.

eleven

THE STRUCTURE OF THE EARTH

And the earth was without form, and void; and darkness was upon the face of the deep.

—GENESIS I: 2

TODAY'S geophysicists know the shape and size of the earth with considerable exactness. The earth, as stated in Chapter 3, is not a perfect sphere, but a flattened sphere known as an oblate spheroid. It has an equatorial bulge which is the result of its rotation. Its diameter from North to South Pole is 7,900 miles, but its equatorial diameter is 7,927 miles. The average diameter is 7,913 miles. Because the poles are some 27 miles closer to the center of the earth than the equator, the force of gravity is greater at the poles than at the equator. Other variations in the earth's gravitational field were revealed by a study of the orbit of Vanguard 1, the satellite launched on March 17, 1958. This showed that in addition to bulging at the equator, the earth is slightly pear-shaped. It is somewhat more pointed at the North Pole and somewhat more rounded at the South Pole.

The surface area of the earth is 197 million square miles. Its volume is slightly more than 250 billion cubic miles. Its mass is estimated to be 6.6 billion trillion tons.

Gravity measurements indicate that the earth as a whole has

a density 5.5 times that of water, more exactly, 5.52 grams per cubic centimeter. The rocky surface of the earth has a density of only 2.7, making it obvious that the interior of the earth is much denser.

Pressure increases rapidly in the earth's interior because of the weight of the overlying rocks. It is calculated that the pressure at the center of the earth is 3 million times greater than the pressure of the atmosphere on the surface.

The temperature in mines increases about 1° F. for every 60 feet of depth. However, geophysicists are certain that this rate cannot continue at great depths. If it did, the temperature at the center of the earth would exceed 180,000° F. It is thought that the temperature at the center of the earth is nearer 18,000° F.

Five chemical elements are believed to account for 94.4 per cent of the earth by weight. They are iron, 34.6 per cent; oxygen, 29.5 per cent; silicon, 15.2 per cent; magnesium, 12.7 per cent; and nickel, 2.4 per cent. The only other elements which contribute more than 1 per cent to the composition of the earth are sulfur, 1.9 per cent; calcium, 1.1 per cent; and aluminum, 1.1 per cent.

IN THE BEGINNING

As we gaze upon our earth as we know it today, with its green-covered fields and hills, its pleasant lakes and streams, its majestic forests and towering mountains, it is difficult to imagine that it ever appeared different. However, it must have been very different in the beginning.

Scientists are not yet agreed as to whether the earth formed from the contraction of a protoplanet or the slow accretion of planetesimals. In either case, if the earth formed quickly in the course of 5 million years or so, the heat generated by its formation would have caused it to become molten. However, if it took shape more slowly, it would have begun as a solid body. In time, however, the release of heat from the radioactive elements in the earth, chiefly uranium, thorium, radium, and potassium, would have caused most of the earth's interior, if not the entire earth, to become molten. Thus, while geologists disagree about the earth's beginning, they agree that at least most of the earth's interior must have been molten at a very early stage. There is no other reasonable explanation of how the earth attained what is now believed to be its present structure.

THE EARTH'S INTERIOR

One of the greatest mysteries in the whole universe lies right under your feet. It is the composition and structure of the earth's interior. The difficulty is its inaccessibility. The world's deepest mine, a gold mine in Boksburg, South Africa, is approximately 2 miles deep. The deepest hole ever drilled is in the Pecos oilfield in western Texas: it is a little over 4.5 miles deep. It is nearly 4,000 miles to the center of the earth. Consequently, the deepest mines tell us no more about the interior of the earth than a pin scratch on the shell of an orange reveals about the interior of the fruit. Scientists have had to turn to indirect methods to form a picture of the earth's interior. In view of the difficulties involved, it is greatly to the credit of geophysicists and geochemists that they have been able to put together by indirect methods a reasonable picture of the interior of the earth.

The present view of the structure of the earth is based in large part upon the behavior of earthquake waves. These consist of three types. One type travels along the surface of the earth. It is these surface waves which shake the earth and do the damage. However, they die out rapidly at some distance from the center of the quake. They are known to seismologists as long waves, or L-waves.

The other two types of waves travel through the deep interior of the earth and are registered on seismographs at vast distances from the quake. They are known as primary and secondary waves. The primary waves, or P-waves, are longitudinal, or compressional, waves, resembling sound waves in this regard. They produce alternate compression and rarefaction in the direction in which they are traveling. The secondary waves, or S-waves, are shear, or transverse, waves, like the waves on a lake. The vibrations are at right angles to the direction in which the wave is traveling.

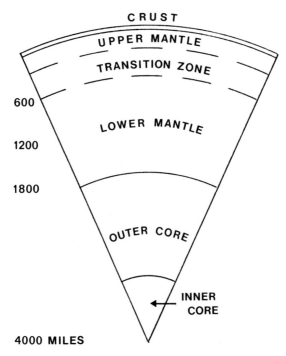

Figure 17. The earth's interior

S-waves travel through the earth only half as fast as P-waves and therefore arrive later at a distant seismological station. Experiments have shown that the speed of both waves depends upon the rigidity of the material through which they are traveling. Moreover, it has been found that P-waves can travel through both solids and liquids, but that S-waves can travel only through solids.

The behavior of seismic waves, as well as other considerations discussed below, lead to the conclusion that the earth consists of three main divisions. These are known as the crust, the mantle, and the core. The mantle is subdivided into the upper mantle, the transition zone, and the lower mantle. The core consists of an outer and inner core. This is shown in Figure 17.

An abrupt change in the speed of earthquake waves indicates the division between the crust and the mantle. This dividing line is known as the Mohorovicic discontinuity after Andrija Mohorovicic, the Yugoslav seismologist who discovered it at the start of the present century. Geologists usually refer to it simply as the Moho. The discontinuity between the mantle and the core was discovered in 1914 by Beno Gutenberg at the California Institute of Technology.

The crust is relatively thin, having an average thickness of about 10 miles. It constitutes only about 0.4 per cent of the mass of the earth. The crust is discussed in detail in Chapter 12.

THE MANTLE

The mantle accounts for about 85 per cent of the volume of the earth and more than two-thirds of its mass. It is about 1,800 miles thick. The increase in the speed of seismic waves as they enter the mantle makes it evident that it is composed of rocks more dense and rigid than those of the crust. Geochemists think that more than 90 per cent of the mantle consists of compounds of four chemical elements, magnesium, iron, silicon, and oxygen.

It is now thought that the upper mantle consists chiefly of a rock known as peridotite. This is a dense dark rock made up largely of the mineral olivine, a complex silicate rich in magnesium and iron. This rock, which is not abundant on the earth's surface, has the right density and rigidity to match the seismic observations. In addition, chunks of peridotite are frequently thrown out in volcanic eruptions, indicating that they have originated below the earth's crust. Finally, laboratory studies of the behavior of molten mixtures of silicates show that olivine is the first mineral to crystallize out as the mixture cools.

The upper mantle extends from the base of the crust to a depth of about 240 miles. At some depth in the upper mantle seismic waves are slowed up. This low-velocity zone, as it is called, is believed to represent a region where the rocks of the mantle are close to the melting point and therefore somewhat plastic.

The speed of seismic waves rises in the transition zone, which is about 360 miles thick. It becomes still greater in the lower mantle. It is thought that the increase in pressure in these regions results in changes in the crystal structure of the minerals so that the atoms are packed closer together.

Minerals which have been created in the laboratory under extreme high temperature and pressure may be some indication of the magnesium and silicon compounds which exist in these regions.

The rocks of the mantle are hot enough to be molten. They are kept solid by the tremendous pressure of the rock layers over them. As a result the mantle exhibits some of the characteristics of a liquid as well as of a solid. Many geophysicists believe that the rocks of the mantle are slowly turning over, like oatmeal cooking in a pot, in gigantic convection cells. The theory was first proposed by the Dutch geologist F. A. Veneg-Meinez. The convection currents may start where the mantle comes in contact with the hotter core of the earth. More likely they are started by the heat of the radioactive decay of potassium in the lower mantle. The heated rocks expand and rise toward the surface of the earth by solid flow. This can be compared to the flow of a glacier, which consists of solid ice. As the heated rock nears the bottom of the crust, it spreads laterally, loses some of its heat, and descends again into the depths of the mantle. The convection currents move slowly, perhaps only 4 inches a year. It may take 30 million years for a mass of heated rock to rise from the depths of the mantle to the bottom of the crust. It is now thought that many of the features of the crust are the results of these currents in the mantle. This is discussed in Chapter 12.

THE CORE

The core of the earth has a diameter of 4,400 miles, a little more than half the diameter of the earth. It represents one-sixth of the volume of the earth, but accounts for nearly one-third of its mass. It has a density of 11 grams per cubic centimeter, almost five times the density of the earth's crust. About 85 per cent of the core is iron and about 7 per cent nickel. The remaining 8 per cent consists of silicon, cobalt, and sulfur. The separation of the earth's interior into a rocky mantle and an iron core is believed to have occurred very early in the earth's history when all or most of the earth's interior was molten. The heavy iron sank to the center of the earth, leaving behind the other chemical elements which crystallized, as the earth cooled, into the minerals which form the rocks of the mantle.

The core consists of two parts, an outer core, which is still molten, and an inner core, which is solid. The outer core is about 1,400 miles thick. The fact that transverse seismic waves, or S-waves, cannot get through the outer core is ample evidence that it is molten. However, the behavior of the longitudinal, or P-waves, which go through the core indicate that the inner core is solid. It is thought that the immense pressure on the inner core explains why it is solid despite the fact that it is probably hotter than the outer core.

It is believed that the earth's magnetic field results from electric currents in the molten outer core which circulate around the solid inner core, thus converting the earth's core into a gigantic electromagnet. These currents are thought to arise from chemical irregularities in the outer core. They are put into circulation by the rotation of the earth.

The theory that the interior of the earth consists of a rocky mantle surrounding an iron core is based to a great extent upon the study of the meteorites which fall to earth. These, as explained in Chapter 4, are believed to be fragments of asteroids that were shattered by collision. You will recall that they consist of three types: stony meteorites, which are a mixture of silicates; iron meteorites, which are chiefly iron and nickel; and stony-iron meteorites, which are a mixture of the other two types.

Beach of black volcanic sand on the island of Hawaii

twelve

CONTINENTS AND OCEANS

And God said, Let the waters under the heaven be gathered together unto one place, and let the dry land appear: and it was so.

—Genesis 1: 9

THE crust of the earth, the only portion of our globe that we know with any certainty, is only a thin veneer surrounding the vast bulk of the earth. It got its name in the days when it was thought that the entire interior of the earth was molten. It constitutes only 0.4 per cent of the mass of the earth. It contains more silicon and oxygen than the mantle and less magnesium and iron. Silicon and oxygen account for almost 75 per cent of it. These two, with six other elements, namely, aluminum, iron, calcium, sodium, potassium, and magnesium, compose 98 per cent. These and the other chemical elements form more than 2,000 minerals which in their turn are organized into an almost infinite variety of rocks.

We have no evidence of what the original crust of the earth was like. The earth is 4.6 billion years old. The oldest known rocks are only 3 billion years old. We can only theorize about the previous 1.6 billion years. It is generally assumed today that the rocks of the continents, the waters of the oceans, and the gases of the atmosphere were squeezed out of the mantle as the once molten mantle solidified. The oceans cover 71 per cent of the surface of

Surf breaking on lava rock, Maui, Hawaiian Islands

the earth, but constitute only 0.025 per cent of the mass of the earth. The atmosphere is only about a millionth of the earth's mass.

One of the major unsolved geological puzzles is the division of the crust into continents and oceans. There are marked differences between the rocks which form the continents and those which form the floors of the ocean. There is every reason to believe that the ocean beds were always ocean beds and the continents always continents, although both have undergone changes in the course of geologic time.

The crust of the earth is about 20 miles thick under the continents, but less than 5 miles thick under the oceans. The continents are composed chiefly of granitic rocks, rocks made of silicates containing the lighter elements such as aluminum, sodium, and potassium. They are generally light in color and have a relatively low density, about 2.6 grams per cubic centimeter. The ocean floors are composed of basaltic rocks. These are heavier rocks, black in color, consisting of silicates which are rich in magnesium and iron. Their density ranges from 2.9 to 3.2.

Some geologists think that the oceans are older than the continents and that at one time the entire crust of the earth consisted of a thin veneer of basaltic rock above which there was a universal ocean. The continents are believed to have formed from magmas, or molten solutions, which welled up from the mantle through fractures in the basaltic layer. There is no agreement among geologists as to the composition of the bottom of the continents. Some think that the basaltic layer which forms the floors of the oceans extends under the continents. Others believe that the foundations of the continents are vast granitic plates resting directly on the mantle.

THE CONTINENTS

While geologists are aware of more than 2,000 minerals, the continental rocks are composed of surprisingly few minerals. One of the most important is quartz, a chemical compound of silicon and oxygen, known to chemists as silica. Most of the others are combinations of silicon, oxygen, and one or more other chemical elements. These are known as silicates. The most abundant of the rock-forming minerals are the silicates called the feldspars, combinations of potassium, sodium, or calcium with silicon, oxygen, and aluminum. Another important group is the micas. The micas in general are different crystal arrangements of the same elements that compose the feldspars, but some contain magnesium or iron. Still another important group are the pyroxenes. These are dark-green or black silicates containing iron and magnesium.

The continental rocks are classified into three general types on the basis of their origin. They are known as the igneous, sedimentary, and metamorphic rocks. The igneous rocks were formed from the cooling and crystallization of magmas. If the process occurs deep in the earth, plutonic rocks, so-called after Pluto, the mythological god of the underworld, are formed. If it occurs on the surface of the earth, as in the eruption of a volcano, the rising magmas are known as lavas and volcanic rocks are formed. The most abundant plutonic rock is a mixture of interlocking crystals of three minerals, quartz, feldspar, and mica. If one examines a sample of such a granitic rock, it is easy to identify the clear glassy grains of quartz, the smooth lighter-colored crystals of feldspar, and the scales or plates of mica, either white or dark brown.

The second type of rocks are the sedimentary rocks which occur in layers or strata of various thicknesses. The greater part of the continents is covered with sedimentary rocks. Common types are shales, sandstones, and conglomerates. They were formed by the cementing together of the fragments of disintegrated rocks. The forces of erosion cause exposed rocks to decay and crumble. The debris is carried away by wind and running water and spread out into layers of sand and mud which are eventually compacted into sedimentary rocks. Other types of sedimentary rocks, the limestones, were formed by the accumulation of the shells and skeletons of countless organisms on the floors of lakes and shallow seas. Yet other types resulted from the precipitation of minerals in solution from evaporating lakes and bays. In many areas the layers of sedimentary rocks are only a few hundred feet thick, but in some they are tens of thousands of feet thick.

The third type of rocks are the metamorphic rocks. They were formed from either igneous or sedimentary rocks that have been subjected to pressure or heat or both, often at considerable depths in the earth's crust. As a result they have undergone recrystallization or other mineralogical or structural change. Movements of the earth's crust may subject rocks to great pressures, altering them in various ways. The intrusion of hot magmas may cause rocks to heat up and recrystallize in new forms. In these ways limestone is changed into marble, shale into slate, and granite into gneiss. Chemical activity also plays a role in some type of metamorphism.

In most places we have to go a little distance beneath the earth's surface to find the various types of rock. The surface of the earth, as we come in contact with it, consists as a rule of a thin layer of loose materials such as soil, clay, sand, and gravel. This loose material is the result of the most recent decay and disintegration of rocks.

The oldest rocks, about 3 billion years old, are found near the centers of the continents, forming the so-called continental shields. These are low plateaus, gently arched up at the center. They get their names from their resemblance to ancient battle shields. They consist largely of huge masses of granitic rocks. At one time geologists thought that they were remnants of the earth's original crust. However, it is known today that these granitic blocks, or batholiths, are younger than the metamorphosed rocks surrounding them. The shields were once great mountain ranges, long since worn flat by the forces of erosion. The granites are intrusions which welled up into the roots of the mountains. As one proceeds outward from the shields to the rims of the continents, younger rocks are encountered, indicating that the continents grew outward as the result of countless upwellings of magmas through geologic time.

THE OCEANS

The rims of all the continents extend into the oceans. These submerged margins of the continents, varying greatly in width, are known as the continental shelves. Their depth below sea level varies, but averages about 400 feet. From the edge of the continental shelves vast slopes, descending gently through a distance of 100 miles or so, lead to the floors of the deep oceans. Deep canyonlike trenches, extending outward in some cases from the mouths of rivers, mark these slopes. These submarine canyons were cut by currents of mud, veritable avalanches, that swept across the continental shelves and down the slopes to the deep sea.

The average depth of the oceans is about 12,500 feet. It is estimated that they contain more than 300 million cubic miles of water.

The ocean floors are the last frontier of the earth to be explored. We have better maps of the moon than of the ocean floors. The old idea that the ocean floors were vast monotonous flat plains has been disproved by studies of the last few decades. There are great plains, the so-called abyssal plains, but they are only part of the story. There are also submarine mountain ranges and volcanoes.

A large mountain range, known as the Mid-Atlantic Ridge, bisects the Atlantic Ocean. It is 10,000 miles long and practically equidistant from the continents on either side. The ridge is about 900 miles wide and varies in height from 3,000 to 6,000 feet. But in places it rises above the surface of the Atlantic to form islands, among them Iceland, the Azores, Ascension Island, St. Helena, and Tristan da Cunha.

The ridge curves around Africa into the Indian Ocean, where it becomes the Carlsberg Ridge. In the Indian Ocean it joins another range of submarine mountains which extends eastward south of Australia into the Pacific Ocean, where it becomes the East Pacific Rise. Extending northward, it disappears under Lower California and reappears in the ocean just north of the state of California. These ridges, which are young compared to the earth, perhaps only a few tens of millions of years old, form a worldwide system 40,000 miles long. (See Figure 18.)

Our present knowledge of the Mid-Atlantic Ridge is in large part the work of Henry Menard of the Scripps Institution of Oceanography of the University of California and Maurice Ewing and Bruce Heezen of the Lamont-Doherty Geophysical Observatory of Columbia University. The ridge is dotted with volcanoes and the islands which mark its emergence above sea level contain many extinct or active volcanoes. Earthquakes are frequent along the ridge. Its most remarkable feature is a series of clefts, or rifts, along its crest. In the South Atlantic the rift broadens into a trench 5 miles deep.

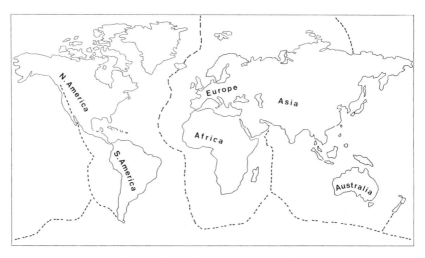

Figure 18. The mid-ocean ridges

A startling theory, first advanced by the American geologists Harry Hess and Robert S. Dietz, holds that the ocean floor is in motion, spreading east and west from the Mid-Atlantic Ridge. It is thought that the floor is being continuously formed by lavas which are rising from the rifts in the ridge, solidifying, and moving laterally. Convection currents rising in the mantle under the ridge are believed to be the driving force which creates the motion of the ocean floor. These currents, moving to the east and west, drag the ocean floor along.

A similar situation is believed to exist in the Pacific Ocean. On the western edge of the Pacific there are chains of islands, known as island arcs because of their shape. These include the Japanese, the Kurile, the Marianas, and the Aleutian Islands. Paralleling the outer, or convex, sides of these arcs are deep rifts, or valleys, in the floor of the Pacific, known as trenches. The deepest spot on earth is at the bottom of the Marianas trench, 35,810 feet below sea level, more than a mile deeper than Mt. Everest is high. It is thought that the floor of the Pacific consists of enormous sheets of lava which are emerging from the East Pacific Rise and are slowly disappearing down the trenches or under the continents. Richard Von Herzen of the Scripps Institution found that the flow of heat from the earth's interior was 40 times greater at the Mid-Pacific Rise than at the trenches. (See Figure 19.)

It is significant that no rocks more than 135 million years old have been found in the ocean basins. This is less than 2 per cent of the age of the earth.

One of the strongest pieces of evidence for belief in the spreading of the ocean floors comes from magnetic measurements. In both the Atlantic and the Pacific magnetic reversals are found in wide stripes of the floors paralleling the ridges. The stripes on either side of the ridges match each other with amazing symmetry. It is now generally agreed that the earth's magnetic field has reversed itself at least 16 times in the past 4 million years, the north magnetic pole becoming the south magnetic pole and vice versa. It is thought that as the lava flowing from the ridges solidified, the magnetic pattern of the period was frozen into it. Consequently, the pattern reversed itself each time the earth's field was reversed. Just

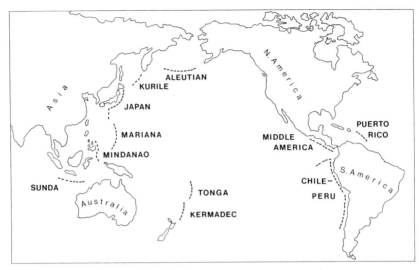

Figure 19. Island arcs and ocean trenches

why the earth's magnetic field undergoes these reversals is not yet understood.

A spectacular feature of the oceans is the many islands which are volcanic in origin. Chief among these are the Hawaiian Islands, which rise from a depth of more than 17,000 feet below sea level. The largest of the Hawaiian Islands, which bears the name of Hawaii, is composed of the union of five volcanoes. Dominating the island are the giant summits of Mauna Kea, which rises to a height of 13,784 feet above sea level, and Mauna Loa, which is 13,680 feet high. Mauna Loa is the largest single mountain mass on earth. Its total height, if measured from the floor of the Pacific, 18,000 feet down, is a little over 31,000 feet. The volcano is built entirely of lava which started to well up from the ocean floor about 500,000 years ago. The bulk of Mauna Loa is estimated to be 10,000 cubic miles.

In addition there are many truncated volcanic cones in the oceans whose flat tops are now thousands of feet below sea level. These are known as guyots. It seems apparent that these were once above sea level and that they have since submerged. Some geologists believe that an area of several millions of square miles in the central Pacific is slowly sinking.

CONTINENTAL DRIFT

The recent study of ocean-floor spreading has given new support to the theory of continental drift. According to this theory, all of the present continents originally were united in one supercontinent that was formed in the early days of the earth from granitic magmas brought to the surface by convection currents in the mantle.

The idea of a supercontinent arose more than a century ago when there was no way of marshaling scientific evidence for it. It occurred to a number of geologists because the continents could be fitted together like the pieces of a jigsaw puzzle. The east coast of South America fits into the west coast of Africa. Antarctica can be fitted against the east coast of Africa and Australia against it. The east coast of North America can be placed against Europe and the northwest corner of Africa.

It is thought that the original supercontinent, to which the name Pangaea has been given, existed until about 225 million years ago. It was surrounded by a universal ocean, the ancestral Pacific Ocean, which has been named Panthalassa. About 200 million years ago Pangaea divided into Laurentia and Gondwanda. In time Laurentia split into North America, Europe, and Asia. Gondwanda became South America, Africa, India, Australia, and Antarctica. (See Figure 20.)

The first detailed theory of continental drift was advanced in 1912 by the German geophysicist and meteorologist Alfred L. Wegener. He based it not only on the fit of the continental outlines, but on the striking similarity of rocks and fossils on opposite sides of the Atlantic. His theory gained relatively few adherents until the theories of convection currents in the mantle and ocean-floor spreading were advanced. A majority of today's geologists are convinced of the reality of continental drift.

Wegener's notion that the continents had drifted like icebergs floating on the sea could not be maintained. The present view is that the convection currents in the mantle provide the driving force. It is thought that the continents were united in one supercontinent until some 225 million years ago and that they first split apart at what are now the rifts in the mid-ocean ridges.

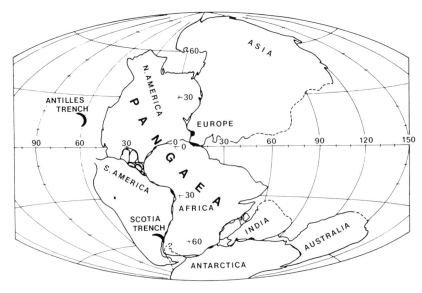

Figure 20. Pangaea

Rise of the Mid-Atlantic Ridge under Laurentia split America and Europe apart. Until then there was no Atlantic Ocean. Lava flowing from the rift in the ridge solidified and moving outward pushed the parts of the supercontinent apart. At first the Atlantic was no wider than a river. With the passage of time the ocean floor spread, pushing America and Europe apart. Many geologists believe that the process is still going on and that the spread of the Atlantic floor is pushing America and Europe still farther apart. If the Atlantic is growing wider, the Pacific must be growing narrower. This could explain the motion of the Pacific floor into the deep trenches that border the outer sides of the island arcs.

You will recall that the East Pacific Rise passes under part of Mexico and California. It is thought that motion arising in the ridge created the Gulf of California by splitting Lower California from Mexico. The San Andreas fault in California lies over the ridge, and many geologists believe that part of California will eventually be pushed into the Pacific and that the Gulf of California will extend northward, separating California from the rest of the United States.

Just as magnetic studies strongly support the theory of ocean-floor spreading, they also support the theory of continental drift.

The younger continental rocks show magnetic orientation which agrees with the earth's present magnetic field. However, this is not the case with older rocks. The magnetic orientation of these older rocks can be explained by assuming that the continents have drifted apart and rotated into new positions in the last 225 million years.

THE DEEP SEA DRILLING PROJECT

In 1968 the National Science Foundation launched the Deep Sea Drilling Project, a joint venture of five American oceanographic institutions under the management of the Scripps Institution of Oceanography. Its object was to learn more about the origin and history of the crust of the earth through the study of long cores of sediment and rock brought up from the deep ocean bottom. The *Glomar Challenger*, the project's drill ship, drilled 246 holes at 154 sites in the Atlantic and Pacific Oceans during the first 30 months of activity. In some cases cores more than 3,000 feet long were obtained where the sea floor was 20,000 feet below the surface of

The Glomar Challenger, *drill ship of the Deep Sea Drilling Project*

Floating Instrument Platform, or "Flip," of the Scripps Institution of Oceanography. As its ballast tanks are filled the ship "flips" from a horizontal to a vertical position, proving an extremely stable platform for oceanographic studies.

the water. This required a drill string more than 4 miles long.

It is now felt that the most important achievement of the Deep Sea Drilling Project has been the verification of the theory of seafloor spreading and continental drift. If the theory is correct, sediments overlying the sea floor should be thinner and younger at the mid-ocean ridges and progressively thicker and older as the distance from a ridge increases. The cores obtained by the *Glomar Challenger* establish that this is indeed the case. The cores have also shown that the present sea floors are young, no more than 200 million years old.

The project has also given strong support to the newest geologic theory, known as plate tectonics. According to this theory, the crust of the earth consists of 10 or more gigantic plates which are in motion, sliding over the mantle like sleds on a field of ice. The plates constituting the ocean floors are moving outward from the mid-ocean ridges. Lava, welling up from the rifts in these ridges, renews the edges of the plates as they move away from the ridges. The continents can be visualized as granitic structures riding high on crustal plates like wooden rafts embedded in the ice floes on a pond in winter.

It is thought that many geological phenomena can be explained in terms of the interaction of these crustal plates, for example, the collision of two plates or the movement of the edge of an ocean plate under the edge of a continental plate. It is suggested that coastal mountain ranges were buckled up by such processes. Plate tectonics is proving a unifying concept that integrates seemingly unrelated geological events into a coherent picture.

There is as yet no agreement as to the number of size of plates. Some geologists believe that North America and the western Atlantic rest on a single plate. It is also thought that major seismic regions are the areas where crustal plates meet.

THE ATMOSPHERE

Life on earth would be impossible without the atmosphere. Plants create the sugars and starches of their tissues from the carbon dioxide of the air and the water of the soil. We need oxygen to breathe. In addition, the blanket of air holds in the sun's heat and at the same time protects us from the sun's shorter ultraviolet rays and X-rays which would otherwise destroy all life on earth.

The atmosphere is 78 per cent nitrogen and 21 per cent oxygen. The remaining 1 per cent is made up of traces of a few other gases. Chief among them is the inert gas argon, which composes 0.93 per cent. Carbon dioxide accounts for only 0.003 per cent. The amount of water vapor in the atmosphere varies from place to place and day to day.

Scientists recognize five layers in the atmosphere. At the bottom of the ocean of air is the troposphere. It extends to a height of 10 miles in the equatorial region but only 5 miles in the polar areas. It is the turbulent region of the atmosphere in which the clouds occur and our weather originates. It contains about 90 per cent of the mass of the atmosphere and virtually all of the water vapor. The temperature drops about 1 degree for every 300 feet of altitude, falling to about $-67°$ F. at the top of the troposphere.

The stratosphere starts at the top of the troposphere and goes

Electron-microscope photo of skeletons of various forms of Radiolaria in Pacific sediment core obtained at a depth of 9,383 feet by the Deep Sea Drilling Project

up to a height of 20 miles. It is the home of continuous powerful winds. At the bottom of the stratosphere two vast swiftly flowing rivers of air, the jet streams, race around the world. The temperature falls to —80° F. at a height of 12 miles.

The ozonosphere overlaps the stratosphere, starting at 12 miles and going up to 30. It gets its name from the presence of ozone in it which absorbs the shorter, lethal radiations from the sun. The temperature rises to 30° F. as a result of this.

The ionosphere begins at 30 miles and rises to a height of 500 miles or more. It is a strange region of vacuumlike thinness. Subatomic particles arriving from the sun, the so-called solar wind, ionize, or electrify, the ionosphere, creating the "radio ceiling," the layers of ionized air which reflect radio waves back to earth, and the displays of the Northern and Southern Lights, the Aurora Borealis and Aurora Australis.

The temperature in the ionosphere rises to more than 1,000° F. It must be remembered, however, that this temperature is a measure of the motions of the sparse atoms and molecules in the ionosphere. It does not have the same meaning that temperature does on the surface of the earth.

Beyond the ionosphere is the exosphere, where the atmosphere fades away into empty space.

THE VAN ALLEN BELT

The first U.S. satellite, Explorer 1, launched on January 31, 1958, revealed the existence of the Van Allen belt, named after the American physicist James A. Van Allen who established its existence from an analysis of the satellite data. It was first thought that there were two belts of intense deadly radiation lying in the plane of the earth's equator like two doughnuts, one within the other. It is now known that there is a single belt starting at 400 miles above the surface of the earth and extending to an altitude of 40,000 miles. It consists of protons and electrons from the solar wind which have been trapped in the earth's magnetic field. The lower portion also contains protons and electrons knocked out of atoms at the top of the earth's atmosphere by the impact of cosmic rays.

thirteen

THE EARTH'S CHANGING FACE

> *The waters wear the stones: thou washest away the things which grow out of the dust of the earth....*
> —JOB XIV: 19

THE poet speaks of the everlasting hills. But that is poetic license of the widest latitude. For the hills are not everlasting. The history of our earth has been one of continuous and ceaseless change. The earth as we know it today is far different from the earth as it appeared when the continents took form. At the start the continents or the supercontinent, if they were originally all one, consisted of nothing but bare, hard igneous rocks. The scene must have been one of terrifying monotony. There were no green hills and pleasant valleys, only masses of rugged rocks.

But immediately forces came into play whose function it was to change the face of the earth. These same forces have been at work ever since. They can be divided into two sets of opposing forces. The first is the forces of erosion. They wear away the land and sweep the debris into the oceans. If they were the only forces at work, they would plane down the land until the continents were flat plains stretching from ocean to ocean. There have been times in

Mt. Gould and a lake in Glacier National Park, Montana

the history of the earth when the continents were worn very low by these forces and much of them was covered by vast shallow seas as the waters of the oceans invaded them.

But fortunately another set of forces is at work that at times elevates the land, buckling the continents up into mountain ranges and returning the shallow seas to the oceans. This second set, the forces of diastrophism, will be discussed in Chapter 14.

THE FORCES OF EROSION

The forces of erosion are sometimes called the forces of weathering. The first one is that of temperature change, the changes from day to night and from season to season. Heat causes the rocks to expand; cold causes them to contract. But the action is not uniform throughout a rock. The expansion and contraction take place only at the surface. Consequently, stresses and strains are set up in the rock and it begins to crack and crumble.

The second force to attack the rocks is the atmosphere. The action of the atmosphere is both chemical and mechanical. Two active constituents of the air are oxygen and carbon dioxide, in addition to much water vapor. These act chemically upon the rocks. Rocks, it will be remembered, are made up of interlocking masses of crystals of various minerals. Some minerals are more easily attacked than others.

A typical granitic rock consists of crystals of quartz, feldspar, and mica. The atmosphere has no effect on the hard and insoluble crystals of quartz, but the other constituents give way quickly. The gases and water vapor gradually cause chemical changes in the feldspar and it turns into a kind of clay called kaolin. The mica also begins to decompose and soften. The atmosphere also has a mechanical effect on rocks, for the winds blow particles of crumbled rock against the larger masses of rocks, causing an action like that of a file or a sandblast.

The next force to attack the rocks is the rain. The rain washes out the softer parts of decaying rocks, thus exposing new portions to the action of the atmosphere. The rain is also a powerful factor in leveling the land. In a heavy rain the water collects in

Yentna Glacier, Alaska, a typical valley glacier. The dark streaks are rocks and debris being transported by the ice.

Tikke Glacier, British Columbia

rivulets that at times cut deep channels into the soil. As the rivulets find their way into streams and rivers, they carry much of the soil with them. The amount of rain falling annually on all the continents is computed by Sir John Murray to be about 29,350 cubic miles. Of this amount one-fourth finds its way into the rivers and is in turn poured into the oceans by them. The remainder either evaporates or else soaks into the ground from which it later reappears in springs.

RIVERS

From a geological point of view, the rivers of the earth serve a triple purpose. First of all, they act as the main trunk drainage lines, carrying the runoff of the rain to the ocean. Second, they act as carriers for the crumbled bits of rock and sand and dust that the rain washes into them. Part of this material is also carried in solution, that is, part of the mineral matter is dissolved in the water. The rest of it is carried in suspension. If a glass of river water is allowed to stand, this suspended material sinks to the bottom as a layer of sediment. Heavier bits of rock are rolled along the bed of the stream by the current of the river.

THE EARTH'S CHANGING FACE   191

The third function of the river is to wear away the rock layers through which it flows. This function is known technically as corrasion, or downcutting. A river can be compared to a sinuous, flexible, and endless file. As it flows along it scrapes the sand and gravel it is carrying against its banks and bed, wearing them away just as sandpaper wears away a piece of wood.

It is difficult to realize how much land is worn away by the action of a river and carried into the sea. It is estimated that the Mississippi River annually delivers into the Gulf of Mexico about 340 million tons of mud carried in suspension, 136 million tons of mineral matter carried in solution, and 40 million tons of sand rolled along its bed. This makes a total of 516 million tons of land swept each year out into the Gulf of Mexico.

The Ganges River is estimated to carry 356 million tons of material into the Bay of Bengal annually, while the Yangtze is estimated to sweep away three times as much.

It is estimated that the level of the United States as a whole is

Hot springs, Yellowstone National Park

Kaibob Trail, Grand Canyon, Arizona

lowered at approximately the rate of 1 foot in about 7,500 years. At this rate it would take 15 million years to wear down the entire country to sea level. As a matter of fact, other forces would prevent this. And, the present rate could not be maintained anyway, because the rate of erosion depends upon the velocity of the current in the rivers, and as the land would become lower, the river currents would become slower.

Most of the wearing-down action of the rivers occurs in highlands, where the grades are steepest and as a result the current in the river is the swiftest. In such regions the river trough is often a narrow V-shaped gulch or canyon with steep walls.

While the chief action of a river is to wear away the land, it reverses its function in the lowlands. The amount of material that the river can transport depends upon the speed of the current. A change in current, therefore, will lessen the river's ability to carry sediment. For example, a river may come tumbling down a mountainside at a high rate of speed. The current diminishes as the river enters the broad lowland. As a result, the river will deposit much of its sediment at the point where it leaves the mountain slope and enters the plain. For smaller, ephemeral streams the deposit may take a fan shape, like a section of a cone with its apex at the foot of the mountain. Such a deposit is known as an alluvial fan or alluvial cone.

A river will also deposit much sediment when it meets with some obstruction in its course. Bars are formed in rivers in this way. Most of these formations, however, are only temporary.

Another type of deposit occurs where a river flows through a wide flat valley. During a flood season, the river will overflow its banks and spread out over the valley. But the outer reaches of the flood will not have a velocity equal to the central portion of the stream. Consequently, the sediment will be deposited there, and after the flood recedes the valley will be covered with a thick layer of sediment. This is known as a flood-plain. The flood-plains are composed of fine sediment and are usually very fertile tracts, extremely valuable for agricultural purposes.

When a river enters a lake or bay, its velocity is checked. Consequently, large deposits of sediment form at the mouths of many rivers. In time this sediment grows into a large flat plain through which the river will cut itself a number of channels. This plain is roughly triangular, resembling the Greek capital letter *delta* ($\triangle$). These formations are known, therefore, as deltas. The delta where the Mississippi River enters the Gulf of Mexico is a conspicuous example.

Grand Canyon, Arizona. View across O'Neill Butte to Bryant Angel Creek

WATERFALLS

Falls develop in a river valley where there is a change in the resisting power of the underlying rocks to the downcutting action of the current. A river will flow along over a layer of hard rock, which does not wear away very quickly. Then it will come to a layer of soft rock, which wears away very rapidly. In time a waterfall will form at the place where the hard and soft layers meet.

The famous Niagara Falls is a good example. At the top of Niagara Falls the rock is a hard limestone. The river has made little progress in wearing this away. But below the falls the rock is a soft shale. This has worn away rapidly, during the passing centuries, so that we now have the magnificent falls leaping from the edge of the hard limestone and falling upon the soft shale more than 150 feet below. The actual drop is 158 feet at the Horseshoe Falls and 167 feet at the American Falls.

Geologists have revealed an interesting fact about Niagara Falls. The falls are slowly moving up the river. Originally, the falls were at Youngstown, where the Niagara River flows into Lake Ontario. Gradually the falls have moved up the river to their present location, a distance of 7 miles. As the falls moved back they

Terrace Mound, Minerva Hot Spring, Yellowstone National Park

Giant saguaro cacti on the Arizona Desert

A variety of cacti on the Arizona Desert

created the magnificent gorge which now exists between their present location and Youngstown. The falls are still receding at a rate of about 5 feet a year. What happens is that as the water tumbles over the edge of the falls, it eats into the softer rock that lies under the layer of hard limestone that forms the edge of the falls. As a result, a ledge of limestone is left protruding with no support underneath it. From time to time the weight and action of the water break off some of the protruding ledge and as a result the falls move a little upstream.

Geologists have undertaken to estimate the age of Niagara Falls from the time it must have taken to move back by this process from Youngstown to the present site. Frank B. Taylor estimated its age as between 20,000 and 35,000 years.

Waterfalls originate in other ways. The two falls of the Yellowstone River in Yellowstone National Park were formed by vertical bands of hard rock which are transverse to the path of the stream. These bands resulted from the solidification of molten magma which welled up between the softer rocks at some remote date.

One of the highest waterfalls in the world is Yosemite Falls in California. It plunges over a granite cliff into the Merced Valley, making a drop of 1,430 feet, cascades down a steep slope for 625 feet, and then takes a final drop of 320 feet. The Yosemite Falls came into existence as the result of glacial action which scooped out and deepened the river valley.

LAKES

When a river meets some obstruction, such as a hard-rock barrier or a clay or gravel dam, the water spreads out, forming a small lake. But while a river thus brings a small lake into existence, in time the river will destroy it. For as the river continues to flow into the lake it begins to deposit sediment in it. In time it will fill up the lake, leaving only a small channel for itself.

The amount of sediment in a lake is also increased by the remains of the many little creatures that live and die in it. The countless minute organisms add considerably to the sediment.

198 THE EARTH

Plants also aid in the destruction of lakes. This is particularly true in the last stages, when the lake is getting very shallow. At that time a thick swamp vegetation springs up. The decay of these plants results in a dense vegetable sediment known as muck. The shallow lakes of Florida are being slowly filled up by the formation of muck.

A moss known as sphagnum grows very densely in swamps and shallow lakes in some parts of the world. The decay of this moss, and the subsequent rise of new growths, eventually forms a bog several feet deep. This material, as it packs down and solidifies, forms the material known as peat. Ireland and Scotland are noted for their peat bogs, from which the material is cut and dried and used for fuel.

In the western part of the United States and in other parts of the world which have arid climates, many lakes have no outlet. Evaporation is rapid because of the dry climate. The evaporation causes a large accumulation of mineral salts in the water, for these salts, which are brought in by the streams entering the lake, are left behind when the water evaporates. The Great Salt Lake in Utah is this type.

The continuous evaporation from some lakes causes the formation of deposits of ordinary salt, gypsum, and other mineral matter. Frequently such deposits are found in a desert region. These are evidence that lakes once flourished at the spot but have since completely dried up.

South end of the Great Salt Lake Desert, Utah

A U-shaped glaciated valley. Red Mountain Pass in Ouray County, Colorado

UNDERGROUND WATER AND CAVES

A large part of the rain which falls upon the earth soaks into the ground. This underground water is known technically as the ground water. It sinks through cracks in the rocks or between the grains of loose material which form the soil and fills them up to a certain level. This level is known as the water table. Its depth beneath the surface varies with the season and the rainfall. In dry areas it may be several hundred feet below the surface; in wet areas it may only be a few feet.

When the ground has a dip or depression that sinks to the general level of the water table, the result is a spring. When a large portion of ground is down to the level of the water table, the result is a bog. Rivers or lakes exist where the level of the ground sinks below that of the water table. This is an important fact to remember. The greater part of the water in a river is not water which flows into it directly from the runoff of rains. It is rainwater which

V-shaped valley cut by the Yellowstone River in a volcanic plateau, Yellowstone National Park, Wyoming

has soaked into the ground and has come into the river as ground water.

The underground water is one of the forces that wear away the land. As the water soaks through the crevices and bores into the rock, it dissolves much mineral matter out of the rock. The destructive action of ground water is heightened by the fact that the water has become acid by soaking through decaying vegetation on the earth's surface.

The ground water in particular will dissolve limestone, and in regions where the rock is mostly limestone in thick strata these will be much eaten away. Frequently the action of the water will result in the formation of holes or pipelike fissures known as sinkholes. Central Kentucky is known as "the land of 10,000 sinkholes."

The action of ground water in such regions also results in the formation of caves. The most famous is the Mammoth Cave, which consists of a series of passageways and halls, some of them 400 feet wide and 150 feet high.

But while the ground water forms caves, it also tends in time to fill them up. Water dripping in through the roof of the cave is frequently highly charged with calcium carbonate. As the water evaporates it leaves the calcium carbonate behind, so that gradually an iciclelike formation known as a stalactite forms from the ceiling.

In the same way water which drips to the floor of the cave evaporates there and causes a similar growth, a stalagmite, to rise from the floor. In time a stalactite and a stalagmite may meet, forming a solid pillar. And eventually the entire interior of the cave may be filled up.

GLACIERS

Glaciers have played a very important part in certain periods of the earth's history by eroding and sculpturing large portions of the earth's surface. They still play an important role in certain regions today.

A glacier is a huge river of ice. One may form in any region where more snow falls in winter than melts away in summer. As a result each winter's snow falls upon a residue from preceding winters and so in time a great quantity of snow accumulates. It becomes compressed by its own weight into a sort of coarse ice known as neve.

Molas Lake, a glacial lake near Silverton, Colorado

Differential erosion in volcanic sediments created this maze of pinnacles and knifelike ridges. Bryce Canyon National Park, Utah

The snow accumulates in high mountain regions and therefore glaciers are common in these regions. These are known as Alpine or valley glaciers. In Arctic regions such a glacier may extend to the edge of the sea.

Gravity causes a glacier to move downhill. As a result in Arctic regions the end of the glacier gets pushed out over the edge of the land into the sea. Pieces break off because of their own weight. These become icebergs. In warmer regions the glaciers melt as they descend from the high mountaintops to lower levels. Many rivers in mountainous regions have their sources in melting glaciers.

Among the best-known glaciers are those of the Alps. They number about 2,000, and range from 1 to 10 miles in length. Larger glaciers are to be found in the Himalayas, the Caucasus, the southern Andes mountains, and in Alaska. These reach lengths of 50 miles.

The downward movement of glaciers varies, ranging from 300 feet a year in the slowest Alpine glaciers to 70 feet a day in the most rapid Alaskan ones. As a glacier moves down a valley it causes much erosion. Valleys are both widened and deepened by glacial movements.

Rocks and boulders caught in the ice at the bottom of a glacier act like cutting tools as the glacier moves along, making grooves and scratches in the rocks that form the bottom of the valley. Such grooves are signs by which geologists can detect glacial action of past ages. Glaciers carry much rock debris, which is deposited along the sides and at the end where the glacier melts. These deposits of boulders, sand, and ground-up rock are known as glacial moraines. From their study geologists can tell the limits of the areas to which glaciers penetrated in past ages.

Greenland and the Antarctic continent are covered with very thick layers of ice known as icecaps or continental glaciers. The Greenland icecap is 700,000 square miles in area and several thousand feet deep. The Antarctic icecap is several million square miles in area and in places 9,000 feet thick. The huge icebergs of the Antarctic Ocean are pieces of ice which break off from the icecap as it extends into the sea.

WAVES

The shorelines of lakes and seas are the scenes of attacks by another erosive agent, the waves. With the passage of time the waves bite deeply into the shores, wearing away great cliffs in many cases. If there were no interruptions or influences to counteract their work, the waves, with the passing of ages, would wear away the land until the entire surface of each continent was down to the level of the oceans.

fourteen

EARTHQUAKES, VOLCANOES, AND MOUNTAINS

> *... and the mountains shall be thrown down, and the steep places shall fall, and every wall shall fall to the ground.*
> —EZEKIEL XXXVIII: 20

THE eruption of a volcano or the occurrence of an earthquake bursts upon the world with dramatic suddenness. All at once the ground begins to tremble and shake, toppling buildings and destroying life, or the volcano begins to pour out floods of molten lava, swallowing up farms and villages. The geologist recognizes these sudden disasters as the manifestation of forces always at work within the earth. The crust of the earth is restless, responding to forces that arise deep inside it. These forces deform the crust, causing the land to sink in some areas and to rise in others. Geologists call them the forces of diastrophism after the Greek word for "distortion." The famous Blue Grotto of the island of Capri was above sea level in the time of the Emperor Tiberius, nearly two thousand years ago. It has since been flooded as the island has sunk. The coastal region of Holland is sinking today at the rate of 8 inches a century. The Scandinavian peninsula is rising.

In the course of geologic time each continent has experienced a series of great cycles when shallow seas inundated the lowlands and then retreated again. Ten or more such cycles of submergence have taken place in North America in the last 600 million years. The other continents have experienced similar cycles. Geologists are not yet agreed on many aspects of the events. It is not certain whether the cycles were synchronous in the several continents. A second unsettled question is whether the ebb and flow of the shallow seas resulted from the movement of the continents or the movement of the ocean floors.

The manifestation of the forces of diastrophism most difficult to explain is the creation of mountains. Seashells and the fossils of marine creatures are embedded in the lofty peaks of the Alps, the Rockies, the towering Himalayas. These mountain peaks were once the bottom of shallow seas. Where did the force which buckled these ancient sea floors into mountain ranges originate?

Geologists of the nineteenth century sought to explain the deformations of the earth's crust on the theory that the earth was cooling off with consequent contraction of the crust. It was thought that the crust was wrinkling up like the skin of an apple that is drying out.

Today this theory is no longer tenable. Many geologists believe that the earth is growing warmer as the result of the decay of radioactive elements in the crust and mantle. The best explanation of what is happening appears to be the theory of convection currents in the mantle. The crust of the earth echoes the commotion in the mantle. It is dragged down where the convection currents descend. It is buckled up where they ascend.

NEW ROCKS FROM OLD

When the continents were low and shallow seas had crept over much of them, the rivers deposited their burden of sand and mud in them. As the rivers flowed into the seas their current slowed down. The heaviest sediment was deposited first. The coarser bits of rock were deposited near the mouths of the rivers. The finer sediments were carried farther out to sea. In time thick beds of sediment were built up in this way.

A Michigan quarry, showing highly jointed Niagara limestone

As the sediments accumulated a layer grew thicker and thicker. Finally its weight became so great that the sea water was squeezed out of it. As a result certain mineral salts in solution in the water which are cements, like calcium carbonate or silica or oxide of iron, were precipitated into the layer and cemented it into solid rock.

In this way the layer of pebbles and coarse pieces of rock near shore were cemented into the kind of rock known as conglomerate, or pudding stone. Layers of sand were cemented into sandstone, layers of mud into shale, and so on.

As conditions changed, a layer of another type of sediment was deposited over an existing layer. Thus, for example, a layer of mud was deposited on top of one of sandstone. In time it was cemented into shale. Thus, slowly, layers, or strata, of rock formed. These new rocks, made from the cementing together of the remnants of old rocks, are known as sedimentary, or stratified, rocks.

Faulted sandstone near the base of the Green River formation, Rio Blanco County, Colorado, showing a vertical displacement of about 10 feet

BELOW: *Folded strata south of Heavens Peak, Glacier National Park, Montana*

Another type of sedimentary rock, the limestones and chalks, were formed by the accumulation on the sea bottom of the shells and skeletons of sea organisms. These organisms have shells and skeletons of calcium carbonate. They live and die in the seas by the myriads and in the course of time form very thick deposits, such as the chalk beds of England and elsewhere.

Thus the layers of sedimentary rocks were formed on the bottoms of the shallow inland seas. But when the land was elevated again and new mountain ranges created, the seas ran back to the ocean basins and the rock strata were exposed to view and to the processes of erosion. And then another cycle began.

JOINTS AND FAULTS

An examination of the rocks that form the earth's surface furnishes ample evidence of the movements of the earth's crust and the tremendous strains and stresses which the rocks undergo as a result of the movement. We find everywhere that the large areas of rocks are broken up into smaller areas by cracks or fractures. These fractures are most easily observed in quarries or in exposed cliffs along the edge of some stream. They are also very conspicuous in mountain regions. They are known technically as joints. As a rule, a mass of rock will contain two or more systems of joints at various angles to one another. In some cases there are three sets of joints intersecting one another in three directions at approximate right angles.

Often it is found that rock layers are displaced on one side of a fracture, either horizontally or vertically. Fractures where this has occurred are known as faults. In some cases the displacement may be very slight, but in others the displacement may amount to more than 100 miles. Sometimes, when rock layers are tilted at a low angle, the layers on one side of a fault may be pushed up over the layers on the other side. This condition is known as thrust-faulting and is frequently found in mountains.

Faults can be extremely large. Perhaps the best known is the San Andreas fault in California. It is 600 miles long, extending from the Gulf of California to the Pacific coast north of San Francisco.

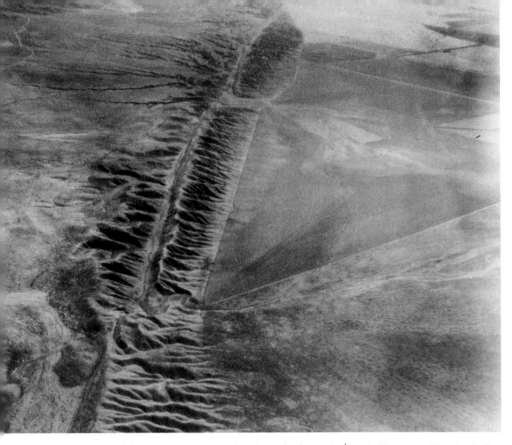

Aerial view of part of the San Andreas fault, California. The fault lies in the rift between the hills, which are several hundred feet high. Farmland lies on either side of the hills.

Faults resembling it exist in Alaska, Canada, Mexico, Venezuela, Chile, Sumatra, Japan, the Philippines, and New Zealand.

EARTHQUAKES

An earthquake occurs when the stresses in the earth's crust reach the point at which rocks suddenly fracture. Geologists speak of this as faulting. The shock sets up vibrations that travel in all directions. The faulting usually starts in the crust at depths of 10 to 20 miles, but it can begin deeper down in the mantle, in some cases at depths ranging from 200 to 400 miles. In big quakes it may extend to the surface, abruptly displacing the ground vertically, horizontally, or both by as much as 20 to 30 feet. It is the vibrations from this displacement, not the displacement itself, that has caused the destruction and loss of life in big earthquakes.

Both earthquakes and volcanoes are most common in the areas of young mountains. In many cases these mountains are still growing, indicating that these are regions in which deformation of the earth's crust is in progress. The rim of the Pacific Ocean is such a region of growing mountains, and because of the many active volcanoes, it has been called a "ring of fire." Japan leads the world in the number and intensity of its earthquakes. But the west coasts of North and South America are not too far behind. Earthquakes appear to be most frequent along active faults, such as the San Andreas fault. The area west of the San Andreas fault is moving to the northwest at the rate of about 2 inches a year. In California, as well as in many other parts of the world, the largest earthquakes have been associated with the longest faults.

Seismographs register about a million earthquakes in the course of a year. Most of these are too small to matter, passing unnoticed except for their seismograph records. The point in the earth where the quake originates is known as its focus. The point on the earth's surface directly over the focus is called the epicenter.

Contact between two rock layers in the Sandia Mountains, New Mexico

ABOVE: *View of Fourth Street area, Anchorage, Alaska, in the wake of the Good Friday earthquake, March 27, 1964*

EARTHQUAKES, VOLCANOES, MOUNTAINS

Seismologists classify earthquakes according to their magnitude on a scale known as the Richter scale, which was devised by the American seismologist Charles F. Richter. The Richter magnitude refers to the physical size, or energy, of the quake and not to the damage it does. A violent quake on the ocean floor or in some uninhabited mountain or desert region does very little damage.

An earthquake with a Richter magnitude of 8 or more is known technically as a great earthquake. The largest known earthquakes, one in Ecuador in 1906 and another in Japan in 1933, had ratings of 8.9. An earthquake with a rating between 7 and 7.9 is known as a major earthquake. An average year has one great earthquake and 18 major ones.

Among the more renowned historic earthquakes is the one which struck Lisbon, Portugal, on All Saints' Day, November 1, 1755. Three major shocks occurred in the course of the morning. Many persons who were in church at the time were killed by the collapse of these medieval structures. A great many people took

BOTTOM LEFT: *Vessels washed ashore in the heart of Kodiak, Alaska, by tsunamis generated by the Good Friday earthquake, March 27, 1964*

BELOW: *Damage to the Government Hill Elementary School, Anchorage, Alaska, in the 1964 Good Friday earthquake*

LEFT: *Ruined houses on Howard Street in the San Francisco earthquake of 1906.* RIGHT: *Looking north along the surface trace of the San Andreas fault in Marin County, California, after the 1906 San Francisco earthquake*

Wrecked Hibernia Bank Building, 1906 San Francisco earthquake

EARTHQUAKES, VOLCANOES, MOUNTAINS 215

refuge from the burning city on a new marble pier that had been constructed on the Tagus River. A great sea wave, or tsunami, set in motion by a submarine shock off the coast of Portugal, swept up the river, smashing the pier and drowning the crowd of refugees huddled on it.

The known history of earthquakes in California begins with one on July 28, 1769. A quake which originated on the San Andreas fault in 1857 is believed to have been equal in magnitude to the better-known quake which struck San Francisco on April 18, 1906. The 1906 quake itself did much damage although greater damage was done by the fire which followed it. Unfortunately, the city's waterlines ran along the San Andreas fault. They were broken by lateral movements along the fault, leaving the city without a water supply with which to fight the fire.

More than 100,000 people were killed in the earthquake and fire that destroyed Tokyo and Yokohama on September 1, 1923. The quake started innumerable fires in Tokyo and more than 40,000 houses were burned down.

The great earthquake which struck Alaska on Good Friday, March 27, 1964, was the largest in the history of North America, and had a Richter magnitude of 8.5. The epicenter was in Prince William Sound. The quake was felt over an area of 700,000 square miles. Rockslides, snow avalanches, and landslides were triggered throughout southern Alaska by the violent ground motion. The city of Anchorage, 80 miles from the epicenter, was severely damaged with a loss of 131 lives. A landslide destroyed much of Turnagain Heights, a suburb of Anchorage. Submarine landslides created destructive sea waves which did severe damage to a number of coastal cities.

VOLCANOES

Active volcanoes are concentrated for the most part in regions where the forces of diastrophism are at work and young mountain ranges are still growing. They form the "ring of fire" around the rim of the Pacific Ocean. They are found along the crest of the Andes in South America and in the mountains of Mexico. Both

White centerline on State Highway 46, one mile east of Cholane, California, shows offset along a fracture in the San Andreas fault in the 1966 Parkfield-Cholane earthquake.

active and dormant volcanoes are found in Japan, the Philippines, the East Indies, and the Aleutian Islands. Extinct volcanoes indicate regions of past crustal deformation. Mt. Vesuvius, near Naples, the only active volcano in continental Europe, is at the end of a long line of volcanoes of which the others are extinct. There are no active volcanoes in the continental United States today, but they exist in both Alaska and Hawaii. There are many volcanoes in the deep oceans.

Because so many volcanoes are within sight of the sea, it was once thought that eruptions were caused by sea water descending into fissures in the earth where it came in contact with molten material. This theory has been abandoned.

There are many pockets of molten rock, or magmas, in the crust of the earth and the upper mantle. These may be due to local concentrations of radioactive material or to local lowering of pressure, or they may be the tops of convection currents. Magmas do not always escape through the surface of the crust. The magmas that solidified below the surface of the earth form the igneous rocks we now see. In many regions there are great granitic intrusions known as batholiths. Some of these are hundreds of miles long. Many geologists believe that great granitic plates form the foundations of the continents.

Volcanoes on the ocean floor pour out lavas rich in magnesium and iron which solidify into basalts. As was said in Chapter 12, the present ocean floors are believed to have been formed by the vast outpourings of such magmas from the rifts in the mid-ocean ridges. The lavas from the continental volcanoes are generally richer in silica and more acidic. It is assumed that they arise from pockets of lava in the upper mantle where the magmas have stood long enough to separate into fractions. The acidic or granitic fraction rises to the surface. As a rule it is highly charged with gases so that the lava is ejected from the volcano with considerable violence. As it explodes into the air, bubbles chill and solidify into volcanic ash while larger fragments form more or less spherical masses known as volcanic bombs. Often such pyrotechnic dis-

This 700-ton block of granite was in the avalanche, triggered by the May 31, 1970, Peruvian earthquake, which slid from the north peak of the Nevados Huascarán Mountain, burying the towns of Yungay and Ranrahirca, killing 20,000 persons.

A sharp bend in railroad tracks records ground movement in the Arvin-Tehachapi, California, earthquake, 1952.

plays are followed by great floods of lava which run down the slopes of the volcano, inundating the surrounding countryside.

Most volcanoes are conical. The cone, however, is merely the solidified lavas of past eruptions. A volcano starts as a fissure or vent, a sort of chimney, reaching down to a pocket of magma. As the lava pours forth the typical mountainlike cone is built up from it. The top of the cone is a huge opening called the crater. In some of the larger volcanoes it is more than a mile in diameter. Frequently there are auxiliary craters and small cones on the sides of the cone.

Many volcanoes of great age have built up cones of great height. Some are more than 12,000 feet high. Where the volcanoes rise from high plateaus, the height of the peak above sea level may be enormous. There are about 470 volcanoes which have been known to be active in historic times. The extinct volcanoes total about 2,500. Most volcanoes are ancient, but a small number have come into existence in the last century. Among the youngest is Paricutín, about 200 miles west of Mexico City.

Paricutín was born on a farm in 1943. The owner, Donisio

Fountains of molten lava and flame rise to height of 1,500 feet during the eruption of Hawaii's Kilauea Volcano in November 1959.

Polido, woke up on the morning of February 20, 1943, to the rumbling sound of thunder. This was strange since the sky was cloudless. After breakfast he saw a column of smoke rising from the far corner of his cornfield. But when he got there, he found no fire. The ground had cracked open and some sort of hot syrupy liquid was bubbling out of the ground. By nightfall things had become a lot worse. The crack had grown larger. Showers of sparks, glowing ash, and huge red-hot rocks were being ejected from it.

By the next morning a cinder cone 30 feet high had formed. Within a month the cone was more than 500 feet high. Clouds of ash and smoke rose to a height of 4 miles above the cone. Then molten lava began to flow from its crater. It engulfed Polido's farm, then the surrounding countryside, finally spreading over an area of 7 square miles and swallowing up the village of Paricutín and the town of San Juan Parangaricutiro. Volcanic ash blanketed the countryside like fallen snow, forming a layer 10 feet deep that destroyed trees and crushed houses. The eruption finally stopped in 1952, when the volcano had grown to a height of 2,000 feet.

The most violent volcanic eruption in history occurred on the island of Krakatoa in the Sundra Straits between Java and Sumatra. The island's central peak, a volcanic mountain, was 2,600 feet high. Suddenly, on August 26, 1883, the entire mountain disappeared in a gigantic explosion. The air was filled with black smoke lit up by continuous flashes of lighting. Glowing ashes rained down on the sea. There were four more tremendous explosions the next day. Clouds of volcanic ash, propelled upward by rising jets of steam, rose to a height of 50 miles. The ash poured down in a muddy paste on Djakarta 80 miles away.

The sea wave, spreading out from the explosion, inundated the coasts of Java and Sumatra, destroying coastal villages and causing the deaths of 30,000 or 40,000 people. Volcanic dust hung in the earth's atmosphere for 3 years, giving the sun a weird green color and turning the sunsets blood red.

A similar volcanic explosion shook the island of Martinique on May 8, 1902. Mt. Pelée, a volcano about 5 miles north of the town of St. Pierre, had become active a few days earlier, ejecting sporadic clouds of smoke, steam, and ashes. On May 8 the top of the mountain vanished in a blinding flash. The city was engulfed

The desolate interior of the vast Haleakala Crater on Maui in the Hawaiian Islands

View of 1959 lava lake in Kilauea-iki Crater, Hawaii, showing drilling rig (lower center)

in a fiery cloud that killed some 30,000 people.

The best-known volcanic eruption in history is that of Vesuvius in A.D. 79. The fall of hot ash destroyed the cities of Pompeii and Herculaneum.

Today's volcanoes of the Hawaiian Islands are far less violent, pouring out their streams of lava in quiet eruptions.

MOUNTAINS

The life cycle of a typical mountain range must be measured on the geologic clock which ticks in millions of years rather than seconds. Strangely enough, it begins with the subsidence of the earth's crust. The Rocky Mountains are a good example of the process, which can be divided into four phases.

The first phase in the history of this range was the formation of the Rocky Mountain geosyncline, a vast trough, or depression, 2,000 miles long and 300 miles wide. As it slowly subsided, waters of the ocean crept into it, turning it into a shallow sea. Rivers from the land on either side deposited their loads of sediment in it. The floor of the geosyncline continued to sink during a period of 100 million years. But the rivers poured sediments into it as fast as it sank so that the sea remained shallow even though the floor of the geosyncline was finally more than 3 miles below sea level. In time the vast accumulation of sediments was consolidated into layers of sedimentary rocks.

The second phase began when the forces of diastrophism uplifted and deformed the geosyncline, buckling up its floor into great arches. These arches, over a period of 15 million years, were elevated above sea level while the overlying layers of sedimentary rocks were folded, crumpled, and faulted, forming towering mountain ranges.

During the third phase, which lasted 30 million years, the crust of the earth was quiet. But the forces of erosion planed down the mountains, depositing the debris in the basins between the ranges. Ultimately the region constituted one vast plain, dotted here and there with solitary peaks of more resistant rock known as monadnocks.

Paricutín Volcano in the state of Michoacán, central Mexico, as it appeared in March 1963. The volcano began as a crack in a cornfield on February 20, 1943.

Cerro Negro Volcano, west central Nicaragua, in action, November 1968

The fourth phase, which began 50 million years ago, was another period of crustal uplift in which the Rockies attained their present contours.

There is evidence that some of the young mountain ranges are still growing and have been elevated more than a mile during the last million years. The evidence for this is the presence at high altitudes of the fossil remains of plants which only grow at lower altitudes. Fossil remains of a tropical fig and a species of laurel which grow in India at an elevation of less than 6,000 feet have been found in the Himalayas at an elevation of 10,000 feet. Similar finds have been made in the Sierra Nevada range and in the Andes.

While the earth's major mountain ranges follow the history just outlined for the Rockies, there are some exceptions, known as block mountains. These seem to be the result of the uplift of a considerable area of the earth's surface between faults without the operation of any lateral pressures which would result in crumpling or folding of the uplifted rock layers. These uplifted blocks were carved into mountains by the forces of erosion. While some areas of the crust were elevated, others were depressed, forming valleys or basins between the ranges of block mountains.

A question puzzling geologists is the possible existence of geosynclines today which will evolve into mountain ranges in some far distant day. There seem to be no extensive geosynclines such as gave rise to the present great mountain ranges. But there are depressions in the crust which can be regarded as partial geosynclines or early stages in the formation of geosynclines. One such partial geosyncline is believed to consist of the area of the Gulf of Mexico bordering the Mississippi Delta. Cores brought up in drilling operations indicate that the layer of sediments is about 40,000 feet thick. This is assumed to be evidence of a slow subsidence of the floor of the gulf.

ISOSTASY

Geologists often speak of the crust of the earth as floating on the mantle. This offers an explanation of why the earth is divided into continents and ocean basins. The ocean floors of heavier basaltic rocks are depressed. The continents, composed of lighter granitic rocks, ride higher on the mantle.

The mountains which rise above the general level of the continents have roots that go deeper into the mantle.

This theory that the crust of the earth floats in equilibrium was first suggested in 1889 by the American geologist Clarence Dutton, who coined the name isostasy for it from two Greek words meaning "equal pressure."

fifteen

THE RECORD IN THE ROCKS

Everything in nature is engaged in writing its own history.
—EMERSON

ONCE the crust of the earth had been divided into continents and oceans, the ceaseless drama of change began. The greater part of the continents is covered by layers, or strata, of sedimentary rocks. Geologists decipher much of the history of the earth in these rocks which were laid down in the shallow seas and later elevated into flat plains and rolling hills or buckled into gigantic mountain ranges, and from the intrusions of igneous rocks into them or their transformation into metamorphic rocks. Except where the rocks have been disturbed by the forces of diastrophism, the lowest layer, or stratum, is the oldest. Nineteenth-century determinations of the relative ages of different strata have been calibrated by the determination of absolute age by radioactive dating. This technique is based upon the determination of the ratio of a radioactive element to its end product in a rock sample, for example, the ratio of uranium to lead or of potassium to argon. Since the rate of radioactive decay is known, the age of the rock can be calculated.

Geologists have called the succession of strata a manuscript in stone. However, it cannot be found in its entirety in any one

Skeleton of a Diplodocus, a swamp-dwelling dinosaur, recovered in Utah

area. Part of the record is found in one locality, other parts in other localities. If all the strata could be piled up in one place, they would form an immense stack 100 miles thick. The enormous span of geologic time is perhaps its most impressive feature. The Geologic Time Scale is summarized in Table III of the Appendix.

THE FOSSIL RECORD

Fossils contained in the stratified rocks reveal the evolution of life on earth as well as geologic history. Geologists have used the fossils to determine the relative age of strata. Rock layers in different parts of the world are assumed to be the same age if they contain the same type of fossils.

Fossils are the remains or evidence of plants or animals that once lived on land or in the sea. They range from the imprint of a leaf or a footprint in sandstone or shale to an entire plant or animal. Usually it is only the hard parts of an organism that have survived. The fossil may be a bone or tooth or shell, in some cases several hundreds of millions of years old. Occasionally an entire skeleton is found.

In most cases chemical action has petrified the fossil, that is, turned it to stone. Sometimes silica or calcium carbonate, precipitated from the ground water, infiltrated the pores of the shell or bone, rendering it harder and more dense. In other cases the original material of the fossil was dissolved and replaced by silica or some other mineral so completely that even the microscopic structure of the original object is preserved. It is possible to count the rings in a petrified tree trunk. Sometimes, however, the entire object was dissolved, leaving only a mold of the original. But in yet others the mold is filled with mineral matter, furnishing a solid cast.

The changing nature of the fossil record reveals the pattern of evolution. It discloses the time of appearance and subsequent development of the major forms of life.

Hipbone of a 70-foot Diplodocus

THE GEOLOGIC TIME SCALE

The oldest known rocks are more than 3 billion years old. Traces of primitive plants are found in rocks more than 2 billion years old. But traces of animal life are extremely rare until 600 million years ago. Then suddenly they are found in great abundance. This is one of the mysteries still awaiting an explanation.

Paleontologists, as students of ancient life are called, divide the last 600 million years into three eras: the Paleozoic era, or Age of Ancient Life; the Mesozoic era, or Age of Reptiles; and the Cenozoic era, or Age of Mammals. The eras get their technical names from the Greek for "ancient, middle, and recent life."

Each of these eras is in turn divided into sections called periods. A new period is said to begin where changes in the nature of the rock layers and their fossils occur. Because the first period of the Paleozoic era is called the Cambrian, the millions of years preceding it are known as the Precambrian era. Geologists feel that so little is known about time relationships in it that there is no point in attempting to divide it into periods. A few widely scattered areas of Precambrian rocks have yielded fossils resembling jellyfish and segmented worms.

THE AGE OF ANCIENT LIFE

The Paleozoic era began 600 million years ago and lasted 370 million years. Its earliest strata are crowded with marine fossils, proving that the seas were already teeming with a varied life. The abundant and varied forms of marine life are ample proof that these creatures must have developed during Precambrian time. But we find no record of this in the rocks. The question of how life may have originated is discussed in Chapter 28.

During much of the Paleozoic era, the continents were very low and covered by wide, shallow seas. Areas ranging from 30 to 50 per cent of North America were flooded by such seas seven different times. These seas were widespread, but from the character of the rock layers which formed in them it is estimated that their average depth was less than 600 feet. The Paleozoic era is

Skeleton of carnivorous dinosaur, Ceratosaurus, found in Colorado

divided into seven periods, known as the Cambrian, Ordovician, Silurian, Devonian, Mississippian, Pennsylvanian, and Permian. The names were adopted from those of regions in which typical rock layers of the various periods occur. The Cambrian is named for Cambria, the ancient Latin name of Wales; the Ordovician and Silurian for ancient Welsh tribes; the Devonian for Devonshire, England; the Mississippian and Pennsylvanian for American states; and the Permian for the ancient kingdom of Permia in Russia. The shallow seas overran the land in the course of each period.

While the earliest seas of the Paleozoic era, the seas of the Cambrian period, were swarming with life, the scene was quite different from that to be found in the ocean today. True fishes and various forms of shellfish are the dominant forms today. But in the Cambrian seas the shellfish were small and few in variety, while true fish were absent. The dominant forms of life were crawling scavengers, the forerunners of the lobsters and crabs of today, known as trilobites. They ranged from 3 inches to 2 feet in length. The trilobite received its name from the fact that the body was divided into three longitudinal sections by two furrows running down the shell—that is, it was tri-lobed. Trilobites are extinct today, the creature most nearly resembling them being the horseshoe crab. The trilobite had a rather heavy shell on its back and a lighter shell on its underside. It had many pairs of double legs and huge compound eyes, as flies have today. In some species each eye consisted of as many as 15,000 facets. With the passage of time a great variety of trilobites developed. Some had legs adapted to swimming; others burrowed in the mud, and as a result in time became sightless, losing the eye structures.

One other form of life was very common in the Cambrian sea, a shellfish known as the brachiopod. The brachiopod has a hinged shell composed of two halves, like the oyster or clam. The brachiopods, unlike the trilobites, have managed to survive the passage of ages and about 200 varieties are found in the ocean today.

Toward the end of the Cambrian period, which lasted 100 million years, the trilobites began to disappear while another form of life became dominant. Mastery of the Cambrian period passed to a cephalopod, which was the forerunner of the present-day octopus, squid, and nautilus, all creatures known as cephalopods. The cephalopod is a shellfish characterized by a circlet of fleshy arms, or tentacles, surrounding the mouth, which is on the front of the head. The first cephalopods which appeared in the Cambrian period had straight cone-shaped shells. Later types had curved shells, and still later ones had shells coiled up like watchsprings and resembling the pearly, or chambered, nautilus, which is found in the South Pacific and Indian Oceans today, or the paper nautilus, or argonaut, of the tropical seas. They grew to immense size, the largest being 15 feet long. It is assumed that these early cephalopods had the aggressive characteristics of present-day squids and octopuses and that the decline of the trilobites is to be ascribed to the fact that the cephalopods fed upon them.

A few types of gastropods, or snails, are also found in the Cambrian seas. It is assumed that there must have been many forms of life which left no fossil traces because of the lack of hard shells or skeletons. It is also thought that there must have been an abundant growth of seaweed and other plant life in the Cambrian seas. There is no trace of any form of life on land during the Cambrian period, but whether the land was actually lifeless or possessed some simple forms which left no fossil traces is impossible to tell.

ABOVE: *Duck-billed dinosaur, Anatosaurus, from Wyoming.* BELOW: *Restoration of the woolly mammoth*

Restoration of the mastodon

The next period of the Paleozoic era, the Ordovician, which began 500 million years ago, saw the appearance in great numbers of the echinoderms. They included starfish, sea urchins, and sea lilies, or crinoids. The echinoderms, of course, are still common in the ocean. The corals also made their appearance in this period. A steady development of oceanic forms of life continued throughout it and throughout the one that followed it, the Silurian.

Some primitive fish occur in the Ordovician and Silurian periods, but true fish first appear in the next period, the Devonian, which began 405 million years ago. The types of animal life found in the ocean prior to this time all come under the heading of invertebrates—that is, creatures without backbones. The fish is the first animal to possess a backbone, the first vertebrate. It is interesting to note that the first fish made their appearance not in the seas but in the fresh-water rivers. These fresh-water fishes were numerous and varied, falling into several major groups. Most of them were armored forms; some of them, however, had armor of bony plates only on the head. These first fish evolved from sluggish creatures living on the river bottoms that lacked true fins and were unable to swim. However, with the passage of time fins were evolved and swimming became possible.

Fossil evidence of life on land is found for the first time in the Devonian period. There is evidence of plants and forests, of air-breathing spiders, scorpions, and thousand-legs, and even of amphibians, the ancestors of the present-day frogs.

Toward the end of the Devonian period upheavals of the earth caused the continents of the world to be particularly high above the oceans. A period of considerable aridity followed. The New England states and portions of Canada, in particular, were elevated into majestic mountains at this time.

The Mississippian period, which followed, is not particularly noteworthy except for the development and subsequent rapid decline of a large group of huge sharks in the seas of the period.

The succeeding period, the Pennsylvanian, is marked by many shifts in the earth's crust. These resulted in the periodic flooding of the land by very shallow seas and the subsequent formation of great swamps. These swamps eventually filled up with decaying vegetation, which was buried under later layers of sediment and transformed by pressure into the great coal beds. Both amphibians and insects flourished in the Pennsylvania swamps, some sluggish amphibians attaining a length of 15 feet, while certain dragonfly types had a wingspread of 29 inches. There were 800 kinds of cockroaches during this period, some of them 4 inches long.

European geologists regard the Mississippian and the Pennsylvanian as a single period which they call the Carboniferous because of the vast deposits of coal.

The Permian period, which brings the Paleozoic Era to a close, was a period of tremendous aridity. Many important mountain ranges came into existence during it, among them the original Rocky Mountains of Colorado and New Mexico, the Appalachians, and the original Alps whose eroded stumps are to be found in Germany, France, Belgium, England, and Ireland. The Per-

Mural by artist Jay Matternes, showing animals of Miocene period including giant hogs and small horses

mian period was marked by a glacial climate in some localities. It is a strange fact that the ice fields were largest in areas presently between 5 and 40 degrees south of the equator in Africa, Australia, and South America. The Permian period closed with great upheavals of the earth's crust, elevating the continents and sending the seas back from them. A biological crisis occurred at this time and the dominant forms of life began to change. The Paleozoic era, sometimes called the Age of Fish, gave way to the Mesozoic era, or the Age of Reptiles.

The first reptiles, however, had made their appearance during the Pennsylvanian Period, about 300 million years ago. They multiplied rapidly during the Permian. One interesting form was the Dimetrodon, which flourished in what is now Texas and had a huge saillike spine running the length of its back.

THE AGE OF REPTILES

The Mesozoic era, or Age of Reptiles, was shorter than the Paleozoic era. It began 230 million years ago and lasted 167 million years. There were only three periods in it, each marked by the advance and retreat of the shallow seas. The three are known as the Triassic, Jurassic, and Cretaceous periods. The Triassic got its name from the fact that in Germany the rocks of this period are divided into three distinctive layers. Jurassic is named for the Jura Mountains. Cretaceous comes from the Latin "*creta*," meaning "chalk."

The climate gradually grew less severe in Triassic time and erosion wore down the mountains that had risen in the Permian period. Later, faulting occurred along a stretch of 1,000 miles from Nova Scotia to the Carolinas, resulting in a series of block mountains, long since eroded away. The western half of North America was arid at the start of the Jurassic period but later the land was depressed and as shallow seas developed the climate became warm and damp. Swamps developed in many places. These swamps disappeared when the western part of the continent began to rise. The great Cordilleran arch was bowed up from Alaska to southern Mexico.

Flooding of North America was particularly widespread during the Cretaceous period. Waters from the Arctic Ocean and the

Skeletons of two giant ground sloths, or Megatheres, which flourished in the Pleistocene period

Restoration of giant ground sloths, or Megatheres, by sculptor Vernon Rickman

Gulf of Mexico created a shallow sea 1,000 miles wide, that bisected the continent. The close of the period saw the rise of the Rocky Mountains and the Andes.

With regard to types of life, the Mesozoic Era is a sort of Middle Ages, bridging the gap between the ancient forms of life of the Paleozoic and the more modern forms of the Cenozoic. The dominant forms of ocean life of the Paleozoic are no longer found in the Mesozoic. Lobsters and crabs made their appearance during this era, as also did many new varieties of corals and echinoids. The most important development was among the mollusks. Many new kinds of shellfish appeared, particularly the oysters. However, the mollusks which showed the most expansion are the cephalopods. Thousands of new kinds developed, including one form, the ammonite, which attained a diameter of 8 feet, and whose coiled shell, if straightened out, would have been 30 feet long.

On land a major development took place among the insects. Flies, butterflies, wasps, and ants made their appearance. Cicadas, grasshoppers, locusts, and cockroaches were also present at this time.

The most important development in the world of life in this era, however, was the rise and fall of the great reptiles. Eighteen great reptilian stocks developed during the Mesozoic era, the most important of which was that of the dinosaurs. These great reptilian monsters spread all over the world. Their fossil remains are particularly common in North America, Africa, China, and Argentina. Some of the first dinosaurs of the Triassic period were no larger than chickens. But as the Mesozoic progressed dinosaurs of greater and greater size arose, reaching their climax in huge creatures weighing 150 tons. But the brain of one of these 150-ton creatures must have weighed less than a pound. It is interesting to note by comparison that the brain of a 150-pound man weighs 3 pounds.

One group of dinosaurs were beasts of prey, feeding upon other reptiles. They had birdlike feet with great claws. Their front legs were small, but their hind legs were large and powerful. They ran on their hind legs, somewhat after the fashion of kangaroos. The largest of this group, known as Tyrannosaurus rex—that is, the king tyrant lizard—attained a length of 50 feet.

Another group of dinosaurs lived in the swamps. These were sluggish creatures with webbed feet, duck-billed muzzles, and long powerful tails which they used in swimming.

The largest dinosaurs were the sauropods, such as Diplodocus, which lived on vegetation. They walked on all fours on short pillar-like legs, but they had very long necks. The largest was the Gigantosaurus, which lived in South Africa. It was 80 feet long, including 36 feet of neck.

There were also a number of armored forms of dinosaurs, such as the Stegosaurus, grotesque creatures with bony plates and spikes on their backs. Near the close of the Mesozoic era a type of horned dinosaur developed. The climax was reached in the development of a type with three horns, the Triceratops.

Two types of dinosaurs forsook the land and went back into the ocean. Some of them attained a length of 50 feet. They included the dolphinlike Ichthyosaurus, or "fish lizard," and the long-necked Plesiosaurus. Other forms took to the air, becoming veritable dragons. Some of these immense carnivorous flying reptiles attained a wingspread of 25 feet. Unlike the dinosaurs on the ground, these

240   THE EARTH

flying reptiles were extremely light, having bodies which weighed less than 30 pounds.

While the reptiles dominated the Mesozoic Era, tiny mammals appeared during this age, about 100 million years ago. The mammal is a warm-blooded animal, whereas the reptile is a cold-blooded one. Birds also made their appearance during this era. For some unknown reason the Mesozoic era came to an end. The great dinosaurs, the flying dragons, the fish lizards all died out. It was time for the tiny mammals and the birds to assume dominance in the world.

THE AGE OF MAMMALS

The Cenozoic era, sometimes called the Age of Mammals, began 63 million years ago and ended some 15,000 years ago when our own era, or Recent Time, began. The Cenozoic was the time of modernization. It was during this period that the types of animals with which we are familiar made their appearance. It is interesting to

Mural of a Devonian forest by Charles R. Knight. The forests of the

note that many types of modern plants had appeared in the Cretaceous.

The era was one in which the continents were very high and only slightly flooded by the oceans. The average amount of North America under water during the era was about 3 per cent. The maximum was only 6 per cent. Shallow seas flooded the margins of the Atlantic coast and the Gulf of Mexico. A small inland sea existed on the Pacific side. As a result most of the sedimentary rocks of this period were formed by sediment carried eastward from erosion of the Rocky Mountains and deposited by the rivers in great flood-plains which formed in the Western states. The era is divided into six periods, known as the Paleocene, Eocene, Oligocene, Miocene, Pliocene, and Pleistocene. These are frequently called epochs rather than periods because they were so much shorter than the periods of the preceding eras. Epochs of this era were first named by Sir Charles Lyell from the Greek word "*kainos*," meaning "recent." To this, other words from the Greek were added to establish their order; thus "Pleistocene" means "most recent."

Local movements of the earth's crust caused mountains again

Devonian period were the earth's first forests.

to rise, beginning in the Miocene. California, Oregon, and Washington were the scenes of mountain-building in this period. In the next epoch, the Pliocene, there was another major readjustment of the earth's crust. This again bowed up the Cordilleran arch, undoing the work which erosion had accomplished since the arch was originally bowed up. The whole of western North America was elevated, in some places as much as 7,000 feet. The Mississippi Valley was also elevated and in South America the Andes were pushed up to new heights. The Alps and the Himalayas were also elevated at this time.

The result of the mountain-making of the Pliocene was to usher in the cold climate of the Pleistocene, which reached its climax in the formation of huge glaciers. The Pleistocene is frequently referred to as the Great Ice Age.

As the Cenozoic era began the climate of much of the earth was warm and humid. Giant forests of subtropical trees, including palms, magnolias, and figs, covered much of North America. As the era progressed the climate grew more like that of today. Forests of beech, chestnut, maple, and oak, and such evergreens as fir, spruce, and pine appeared in what are now the temperate areas of North America and Europe. Great areas of grasslands developed in the Miocene.

Most of the mammals were still very small in the Paleocene and at the start of the Eocene. They included primitive shrews,

Restoration of Devonian sea bottom

A landscape of the Jurassic period. Archeopteryx, primitive toothed bird in the center. Pterosaurus, flying reptile, on tree at right and in flight. Small carnivorous dinosaurs in left foreground. Mural by Charles R. Knight

opossums, and a variety of other insect-eaters. As the epoch progressed the first horse, a small creature known as Eohippus, appeared. There were also rodents and the ancestors of today's cat and dog families.

However, as the Eocene progressed, mammals grew in size and huge archaic forms appeared. These included the titanotheres, ungainly creatures with short thick legs, a body like that of the rhinoceros, and a low-browed head surmounted by two huge, divergent horns. The largest was 8 feet tall. Great flightless birds also made their appearance. Some were 7 feet tall with fearsome beaks more than a foot long.

During the Oligocene and Miocene, starting 36 million years ago, mammals and birds became increasingly modern and the

Landscape of the Permian period. Dimetrodon, large blunt-headed finbacked carnivorous reptile. In the center, Edaphosaurus, herbivorous finback with smaller, more pointed nose. Left foreground, Casea, smaller

archaic forms died out. Grazing animals flourished in the grassy plains of the Miocene. These plains were filled with a wide variety of hoofed animals, including camels, horses, bison, and deer. Many varieties of beavers, squirrels, and mice also appeared. Forerunners of today's carnivores, the ancestors of today's lions, tigers, bears, wolves, and coyotes, emerged and preyed on the grazing animals. The modernization of the mammals continued during the Miocene

Tyrannosaurus, giant carnivorous dinosaur, at-

lizardlike reptile. Left background, primitive conifers. Extreme right, clump of horsetails. Murals above and below by Charles R. Knight

and many of today's fully modern forms emerged during the Pliocene, some 10 million years ago.

Two major families of elephants arose in the Cenozoic era and reached their climax in the Pleistocene, which began a million years ago. They were the mastodons and the mammoths. The mastodon was the smaller of the two. In the icy areas of the Pleistocene both mastodons and mammoths developed heavy coats of coarse

tacking the three-horned dinosaur Triceratops.

Stegosaurus, a large armored dinosaur of the late Jurassic of North America. Mural by Charles R. Knight

hair. Vast herds flourished in Siberia. For centuries in modern times their huge tusks dug out of the tundra have constituted one of the world's important supplies of ivory.

The woolly mammoth was well known to the caveman, and numerous paintings of the curious beast with a strange dome atop

Brontosaurus, a large sauropod dinosaur of the late Jurassic of North America. Crocodiles in left foreground. Mural by Charles R. Knight

his head and a hump on his back are found on the walls of many caves in Europe. The intact frozen remains of these creatures have been dug up in Alaska and Siberia. Small portions of prehistoric mammoth steaks were once served at a scientific banquet in Russia.

The imperial elephant, 14 feet high and with large curving tusks, flourished in the southwestern United States during the Pleistocene.

One of the most prolific sources of Pleistocene fossils has been the La Brea tar pits in Los Angeles, now a small park a few blocks west of La Brea Avenue. Enormous numbers of bones were found when the tar was mined to cover the dusty streets of early Los Angeles. Apparently the animals were trapped in the tar when they sought to drink the thin layer of water which collected on it. Among the most impressive fossils are those of the sabretooth tigers.

The levels of the oceans rose and fell as the waves of glaciation advanced and retreated during the Pleistocene. The levels were lowest when a glacial period was at its maximum, locking up vast quantities of water in the ice sheets. When the levels were lowest, many land bridges emerged, making possible the migration of many animals from one area to another. One of the most important connected Alaska and Siberia across what is now the Bering Strait. The zebra, camel, and horse crossed from Alaska to Siberia, while the elephant, bison, mountain goat, sheep, moose, elk, and musk ox crossed from Siberia to Alaska. This land bridge in the last glaciation enabled man to emigrate from the Old World to the New.

PART 3

✳✳✳✳✳✳✳✳✳✳✳✳✳✳✳✳✳✳✳✳✳

The Atom

sixteen

THE NATURE OF MATTER

Only the atoms and the void are real.
—Democritus

It is striking that the most remarkable researches of the twentieth century have substantiated two ideas about the nature of matter which the Greek philosophers fashioned 25 centuries ago, when the magnificent and miraculous flower of Greek civilization was beginning to unfold on the sunny shores of the Aegean Sea. One is the theory that the many thousands of substances which exist in the world are formed from a smaller number of simpler substances or elements. The other is the theory that matter is constructed of invisible particles or units, the so-called atoms of matter.

Impelled by a passion for generalization, the ancient Greeks sought to devise theories that would unite the diverse phenomena of the physical world. The first of the Greek philosophers, Thales of Miletus, taught in about 580 B.C. that water was the essence or the fundamental substance of the universe. Impressed by the role of moisture in the production and maintenance of life, he imagined that all things were various transformations of water. Subsequent philosophers offered various other notions. Anaximander thought that the primary stuff of the universe was "a continuous infinite medium which filled all space," an idea that reappeared centuries later in the concept of the ether of space. But Anaximenes took a more matter-of-fact view, holding that air was the basic substance. When rarefied, it became fire; when condensed, it turned into water, then earth. Heraclites, however, taught that fire was the basic substance of the universe.

THE FOUR ELEMENTS

These ideas were finally fused into the doctrine of four basic elements held by the school of Pythagoras and subsequently elaborated by Empedocles of Agrigentum. According to this theory, the four were fire, air, earth, and water. Empedocles imagined that all the substances in the universe were combinations of varying amounts of these four fundamental entities.

In all probability this notion of four basic elements arose from a misinterpretation of the phenomenon of combustion. It was assumed that the substance consumed was resolved into its fundamental components. Thus, when a fresh log was burned, fire appeared, the smoke vanished into air, moisture boiled out of the green wood, and earthlike ashes remained.

The theory of four basic elements met with Aristotle's approval. These entities, however, must not be thought of in our present literal sense, but as combinations of the four basic qualities of warmth, cold, wetness, and dryness. Earth was the combination of dryness and cold; water was the combination of wetness and cold; fire, of dryness and warmth; air, of wetness and warmth. This is shown in Figure 21. Aristotle imagined that all substances were composed of some sort of primordial material which he called "*hyle*," the Greek word for "stuff," mixed with various amounts of the four elementary properties. This doctrine of Aristotle's flourished until the seventeenth century.

It was only natural that such theories should eventually give rise to the idea that one substance might be transformed into another and so the pseudoscience of alchemy came into existence. It probably originated in Alexandria in the third century A.D. and continued to flourish until well into the seventeenth century. The chief aim of the alchemists was to transform iron and other base metals into gold. The spirit of mysticism thrived in the atmos-

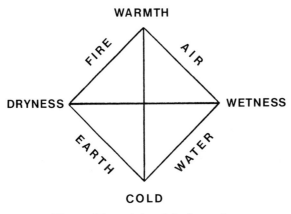

Figure 21.   Aristotle's four elements

phere of alchemy, and soon its devotees were seeking a mysterious "philosopher's stone," which would not only turn iron into gold, but would also insure perfect health and perpetual youth. But though futile, alchemy was not fruitless, for the modern science of chemistry had much of its origin in the mystical vaporings of alchemy.

While many of the alchemists were hard workers, others, of course, were mere rogues, trading upon the credulity of their fellow men. Most of the kings of Europe financed the work of alchemists in the hopes that they might discover the secret of turning iron into gold. A visitor to Hardshin, the beautiful castle at Prague, can see a group of small houses with huge fireplaces which the Emperor Rudolph II built for his alchemists. But frequently a king would lose faith in an alchemist and order him hanged on a gallows which, with grim humor, had been gilded for the occasion.

The downfall of alchemy was heralded with the publication in London in 1661 of *The Sceptical Chymist* by the Irish-born physicist and chemist Robert Boyle who helped organize the Royal Society. Boyle launched a bitter attack upon the alchemists and their obscure manner of writing and urged that chemistry be studied for its own sake. He urged that experimenters discard the old notion of four elements. In its place he suggested that all substances be divided into two classes, compounds and elements. A compound was a substance which could be resolved into simpler substances by chemical means. An element was a substance that could not be split chemically into simpler substances.

This distinction was clearly understood before the end of the eighteenth century. In 1787 the brilliant French chemist Antoine Lavoisier formulated the law of the conservation of matter, showing that in every chemical reaction the weight of the products was exactly equal to the weight of the substances which entered into the reaction. He made a list of 33 chemical elements. This was increased to 50 by the Swedish chemist Jöns Jakob Berzelius in 1819. Today's scientists are aware of 92 chemical elements in nature plus a number of elements heavier than uranium which have been created in the laboratory. The 103 chemical elements known today are listed in Table IV of the Appendix.

THE ATOMIC THEORY

It is possible that the idea of atoms originated in Babylon or Egypt, or even in India, but we find it stated first in Greece in the fifth century B.C. by Leucippus of Abdera and his pupil Democritus, perhaps the greatest of the Greek physical philosophers. In many ways it represents a simplification and extension of the doctrine of the four basic elements. Democritus taught that the universe was composed of an infinite number of invisible particles and empty space. All atoms were composed of the same basic substance although there was a vast array of atoms of differing characteristics, sizes, and shapes. But all atoms were invisible, without pores, very hard, incompressible, and indestructible. They were endowed with a never-ending power of movement. Everything in the universe was the result of their motions, their union into visible objects, and their subsequent dissolution.

It will be seen that the atomic theory was an attempt to find an all-inclusive explanation for the complex phenomena of daily life, to explain how water boiled away as steam, how salt dissolved in water, how fire extracted metals from ores, how seeds grew into plants and trees. It was based on general considerations without the benefit of the meticulous observations and controlled experiments which are the foundations of modern science. Thus Democritus speaks of atoms of iron, atoms of water, and atoms of sand.

But it foreshadowed many of the developments of succeeding centuries.

Unfortunately for the progress of physical science, the atomic theory was opposed by both Plato and Aristotle. The weight of Aristotle's opinion not only influenced the judgment of his own time, but had a similar effect centuries later when his writings became the authority of the Middle Ages in scientific matters.

But the atomic theory always had some adherents. Epicurus, a century after Democritus, embraced the theory and made it part of his philosophy. It was championed in Rome in 55 B.C., two centuries after Epicurus, by the famous Latin poet Lucretius in his immortal scientific poem, *De rerum natura* (Concerning the Nature of Things). It likewise had its followers, though few in number, during the Middle Ages. Nicholas of Autrucia, who taught that all natural phenomena were due to the union or disunion of atoms, was forced to retract this heresy in 1348.

The atomic theory, so greatly in disfavor in the Middle Ages, came to the fore once again with the revival of learning in the Renaissance in Europe. During the sixteenth and seventeenth centuries it was championed by Galileo Galilei, René Descartes, Francis Bacon, Robert Boyle, and Sir Isaac Newton.

Some of the views put forward bordered on the fantastic. Thus the French philosopher Pierre Gassendi suggested that iron owed its strength to the fact that its atoms had rough surfaces which clung together while the atoms of water were smooth and slippery. But the Italian physicist Ruggiero Boscovich proposed an idea far ahead of his time. Influenced by Newton's theory of gravitation, he pictured the atoms as centers of attractive and repulsive forces and rejected the belief that there was any physical contact between the atoms composing a body.

In view of all this, it may seem surprising that the world credits the English chemist John Dalton with having originated the modern atomic theory. Dalton, a Quaker schoolmaster in Manchester, published his *New System of Chemical Philosophy* in 1808. The importance of Dalton arises from the fact that he transformed the atomic theory from a vague philosophical speculation into a specific and workable tool of science.

A contemporary of Dalton, Joseph Proust, had established

the principle now known as the law of definite proportions, showing that any chemical compound always contained the same chemical elements combined in exactly the same proportions by weight. Dalton repeated Proust's experiments and enunciated a second law which Proust had not stated but for which he had laid the groundwork. This was the law of multiple proportions. Certain chemical elements united with each other to form a variety of chemical compounds. The law of multiple proportions states that when this is the case, the different amounts of one element, by weight, which will unite with a given weight of another element, will be exact multiples of each other.

From these considerations Dalton formulated his atomic theory. He showed that the law of definite proportions could be explained by assuming that each chemical element was composed of atoms of definite weight. He showed further that the law of multiple proportions could be explained by assuming that an atom of one element might combine with one, two, or more atoms of another element under various conditions. Perhaps Dalton's greatest contribution was his attempt to determine the relative weights of the atoms of the various chemical elements. He chose hydrogen, the lightest of the chemical elements, for his standard and considered the weight of its atom as unity. He sought, for example, to determine the relative weight of the oxygen atom by determining the proportions by weight in which hydrogen and oxygen combined to form water.

Also in 1808 the French chemist Joseph Louis Gay-Lussac formulated the important law of gaseous volumes. He showed that when chemical action takes place between gases, the volume of the gaseous products bears a simple ratio to the volumes of the reacting gases. However, this seemed to contradict the theory of atoms. How was one to explain, for example, why two volumes of hydrogen combined with one of oxygen to form two volumes of water vapor?

The solution was offered by the Italian physicist Amadeo Avogadro in 1811. Dalton had assumed that all chemical elements consisted of single atoms. Avogadro made the distinction between atoms and molecules. The difficulty disappears once we realize that each molecule of hydrogen consists of two atoms of hydrogen, each molecule of oxygen of two atoms of oxygen, and each molecule of water of two atoms of hydrogen and one of oxygen. Strangely enough, Avogadro's suggestion was not well received and Dalton himself rejected it.

The confusion existed until 1858 when the Italian chemist Stanislao Cannizzaro made clear the distinction between atoms and molecules. The smallest unit of a chemical element is an atom. Atoms combine to form the molecules of a chemical compound. The molecule is thus the smallest unit of a compound. But many gaseous elements also exist in the form of molecules. Thus a molecule of oxygen consists of two atoms of oxygen. There are, however, some gases like helium, for example, which exist normally in the form of single atoms.

As early as 1816 the English physician William Prout suggested that all atoms were composed of hydrogen atoms, suggesting that hydrogen was truly the primordial substance, or *hyle* of Aristotle. This idea was abandoned when it was found that the atomic weights of a number of elements were not exact multiples of the hydrogen atom. But Prout's hypothesis was destined to return, though in much modified form, in the twentieth century.

THE PERIODIC TABLE

As early as 1829 various chemists attempted to find relationships between the properties of chemical elements and their atomic weights. The matter was finally resolved by the Russian chemist Dmitri Ivanovich Mendeleeff who published his famous "periodic classification" in 1869. He showed that when all the chemical elements were arranged in the ascending order of their atomic weights, there were periodic recurrences of elements which resembled each other. Thus, for example, if you started with lithium and counted eight elements down the list, you came to sodium. Counting another eight brought you to potassium. These three elements have many properties in common. They are all soft whitish metals which react chemically with water with considerable violence. Mendeleeff showed his genius by boldly leaving gaps in his table where no known chemical element fitted. He predicted that the missing elements would be eventually found and prophesied from his table what the characteristics of the elements, when found, would be. The discovery of the elements gallium in 1875, scandium in 1879, and germanium in 1886 were brilliant confirmations of his predictions. Although modified from Mendeleeff's day, the periodic table is still one of the foundations of modern chemistry and physics. As we shall see later, the periodic table is the consequence of the internal structure of the atom.

seventeen

X-RAYS, RADIUM, AND THE ELECTRON

Nature does not allow us to explore her sanctuaries all at once. We think we are initiated, but we are still only on the threshold.
—SENECA

THE history of civilization discloses few contrasts greater than that furnished by the difference in viewpoint of the nineteenth-century physicists and their successors in the twentieth century. As the nineteenth century drew near its close physicists felt that they had completed their tasks. One eminent scientist of the time, making an address in 1893, said that it was probable that all the great discoveries in the field of physics had been made. He sketched the history and development of the science, finally summarizing the well-knit, and as he thought, all-sufficient theories of the nineteenth century. The physicist of the future, he said sadly, would have nothing to do but repeat and refine the experiments of the past, determining some atomic weight or constant of nature to an additional decimal place or two.

THE DISCOVERY OF X-RAYS

And then, two years later, on December 28, 1895, Wilhelm Conrad Roentgen presented the secretary of the Physical Medical

Society of Wurzburg with his first written report of his discovery of X-rays. On the first day of 1896 he mailed copies of the printed article to scientific friends in Berlin and elsewhere. With them he sent some prints of the first X-ray photographs he had taken. Some of these were exhibited in Berlin on January 4, 1896, at the fiftieth-anniversary meeting of the Berlin Physical Society. One of these showed a compass needle through its case. Another revealed a set of weights within a closed box. But the most spectacular of all showed the bones of the human hand. Here was exactly what the speaker of 1893 had said could not happen: a new discovery had been made. Here was proof that the nineteenth-century scientists had not completed the story of physics. Roentgen had found some mysterious rays which penetrated opaque objects as easily as sunlight poured through windowglass. There was nothing in nineteenth-century physics to explain this startling phenomenon. Within a few days after the Berlin meeting the newspapers of every capital in the world were printing reports of his sensational discovery. Not only the scientists but people everywhere were excited by the news. Roentgen found himself world famous overnight.

The apparatus with which Roentgen made his discovery represented the results of a long series of developments in the field of electricity. As early as 1705 the English experimenter Francis Hauksbee found that amber enclosed in a glass vessel from which most of the air had been extracted glowed with a faint light when electrified by rubbing. It was subsequently found by William Watson in 1752 that the passage of a charge of static electricity through a partially evacuated glass tube was accompanied by the emission of light.

A significant advance was made in 1854 with the perfection of the Geissler tube. Heinrich Geissler, a glassblower and instrument-maker at the University of Bonn, built a superior mercury vacuum pump. He made vacuum tubes of various shapes with an electrode sealed in either end. Slight traces of various gases were introduced into these tubes. As a result, they lit up with various enchanting colors when an electric current was passed through them. These Geissler tubes will be immediately recognized as the forerunners of the present-day neon tubes used in electric signs on storefronts and theaters in every part of the world. Geissler's chief,

Julius Plücker, noted a greenish glow in the glass of the Geissler tube near the negative electrode, or cathode, as it was called. This was further investigated by Johann Wilhelm Hittorf and Eugen Goldstein who concluded that this glow was due to rays coming from the cathode. Goldstein coined the name "cathode rays."

Further advances were made by Sir William Crookes, the English physicist, who designed improved tubes and better pumps for evacuating them. The highly evacuated tube became known as the Crookes tube. The glow of the Geissler tube was absent. Instead, there were only feeble rays issuing from the cathode. These caused a golden-green luminescence where they struck the glass wall of the tube. The rays traveled in straight lines but could be deflected by a magnet. They cast shadows on the wall of the tube of small objects placed in their path. Philipp Lenard, the Hungarian physicist, built a Crookes tube with a small plate of very thin aluminum in it. He found that the rays came through this "window" into the air for a small fraction of an inch. A Crookes tube is shown in Figure 22.

Crookes became convinced that the cathode rays consisted of a stream of particles bearing a negative electric charge. But he

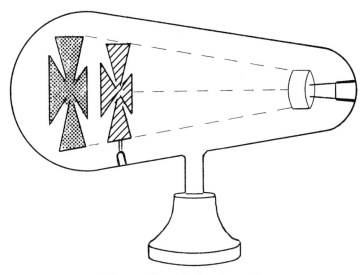

Figure 22. A Crookes tube

made the mistake of considering them a fourth state of matter, suggesting that to the gaseous, liquid, and solid states, there be added the ultragaseous, or radiant, state.

It was in October of 1895 that Roentgen, then the director of the Physical Institute of the University of Wurzburg, began to experiment with cathode rays. Following Lenard's technique he enclosed a Lenard tube in a tightly fitting cardboard jacket covered with tinfoil to protect the tube and exclude the light of the tube which increased the difficulty of detecting the cathode rays in air. This was done by bringing to the aluminum window a cardboard screen coated with barium platinocyanide, a phosphorescent paint. The cathode rays caused the paint to glow.

Having repeated Lenard's experiment to his own satisfaction, it occurred to Roentgen that cathode rays might escape through the glass walls of a more powerful Crookes tube. He determined to test the notion and completely covered such a tube with a jacket of black cardboard.

It was on November 8, 1895, that Roentgen made his epochal discovery with this tube. Darkening the laboratory, he turned on the electric current to test the circuit. To his amazement, he saw a ripple of green light in the dark about 3 feet from the tube. Striking a match, he perceived that the ripple had come from the barium platinocyanide screen which was lying on the bench. He again turned on the current and again there was the same ripple of light. He repeated the experiment, moving the screen farther from the tube each time. But each time the screen became luminous.

Cathode rays were effective only a fraction of an inch from the Lenard tube. Roentgen realized that he was dealing with a brand-new phenomenon. He worked feverishly during the next seven weeks, taking no one into his confidence and hardly stopping for meals. The result was the report of December 28, a veritable model of scientific thoroughness.

Within a few weeks every physicist who possessed a Crookes tube verified Roentgen's discovery. In fact the discovery might have been made anywhere in the previous 20 years. Several physicists had complained of the fogging of photographic plates stored in their laboratories, but none had thought to blame it on the operation of the Crookes tube. Indeed, A. W. Goodspeed in 1890 had produced shadow photographs with a Lenard tube but attributed them to cathode rays. Lenard himself had done the same thing. Lenard grew bitter about the situation when Roentgen received the first Nobel Prize in 1901 and sought thereafter to give the impression that he had noted the X-rays before Roentgen but had not wanted to interrupt his other researches. His bitterness grew greater with the passage of the years.

RADIOACTIVITY

The next major discovery in the realm of atomic physics was that of radioactivity, made a few weeks after Roentgen's announcement, by Antoine Henri Becquerel in Paris. Becquerel's father, also a physicist, had investigated fluorescence, the fact that many substances when exposed to sunlight subsequently glowed in the dark. Becquerel recalled his father's work and wondered if there was any similarity between fluorescence and X-rays. Accordingly, he wrapped a photographic plate in black paper and placed upon it a crystal of a uranium salt which his father had used. He exposed the arrangement to sunlight. Upon developing the plate he found it fogged or darkened, proving that some ray had indeed penetrated the black paper. He supposed that the action of the sunlight had caused the uranium to give off X-rays. He got ready to repeat the experiment on February 26, 1896, but the day was cloudy. He put the paper-covered plate with the crystal on top of it in a desk drawer. Finding it there on March 1, he developed it, thinking to find a very faint silhouette of the crystal on it. To his amazement, the shadow was dark and clear. Further experiments convinced him that uranium gave off rays like X-rays at all times. The sunlight played no part in it. Today, the phenomenon discovered by Becquerel is known as radioactivity.

Working in Becquerel's laboratory at the time were Pierre Curie, a young instructor in physics, and his wife, Marie Curie. Romance and tragedy are combined in the story of the Curies. Mme. Curie was born in Warsaw, Poland, her maiden name having been Marie Sklodowska. Her father was professor of physics at the University of Warsaw. She studied physics at the university and then went to Paris to continue her research work. There she lived in a garret room, up six flights of stairs. In 1894 she met Pierre Curie, a tall young man with auburn hair and sympathetic eyes. Mme. Curie later wrote of this first meeting, "I noticed the grave and gentle expression of his face, as well as a certain abandon in his attitude, suggesting the dreamer absorbed in his reflections." Their mutual interest in science soon expanded into a more personal interest and in 1895 the two were married.

It was a year later that Becquerel made his startling discovery of the mysterious rays given off by uranium salts. Mme. Curie asked his permission to go on with the experiments. Her desire was to find if any substances beside the salts of uranium gave off these rays. After many experiments she found that only one, the salts of thorium, did so. But the most amazing discovery she made was the fact that pitchblende, the ore from which uranium is obtained, gave off rays four times as strong as those of pure uranium.

It was apparent to Becquerel and the Curies that this could mean only one thing. The pitchblende must contain some unknown chemical element which was far richer in these mysterious rays than was uranium. Pierre Curie decided to drop his researches, and he and his wife began the fascinating task of finding this unknown element. The Austrian government presented them with a ton of pitchblende from the mines at Joachimsthal. It was necessary to remove one known substance after another from the pitchblende, carefully conserving the residue for further analysis. The first result of the work was the discovery of a substance giving off Becquerel rays, as the world of science began to call the mysterious rays. Mme. Curie named it polonium in honor of her native Poland. But polonium was not rich enough in the rays to be the end of the search.

In 1898 the search came to a triumphant conclusion. From the ton of pitchblende the Curies had obtained a fraction of a grain of a new element which was 2.5 million times as rich in Becquerel rays as was uranium. They named this new substance radium. It possessed many interesting-provoking properties: it liberated heat, electrified the air in its immediate neighborhood, caused many substances to become phosphorescent when brought near it, and possessed the power of killing bacteria and other minute organisms.

The world of physics was in a whirl of astonishment and excitement. Three great discoveries in three years, each one more astounding than the other—X-rays in 1895, the Becquerel rays in 1896, and radium in 1898. Those were exciting days.

The 1903 Nobel Prize for physics was divided between the Curies and Becquerel. But tragedy entered the lives of the Curies on April 19, 1906, when Pierre Curie, while crossing a Paris street, was run over by a truck and killed.

DISCOVERY OF THE ELECTRON

From a quantitative study of the action of electric currents in decomposing water and various chemical compounds, Michael Faraday in about 1831 came to the conclusion that electricity might exist in discrete units, "atoms of electricity," so to speak. Half a century later the idea was revived by various physicists including Hermann von Helmholtz, Sir Oliver Lodge, and the Irish physicist G. Johnstone Stoney. In 1891 Stoney proposed the name "electron" for the elementary unit of electricity.

By this time it had been generally agreed by physicists and chemists that atoms could not be the simple spherical pellets they were once imagined to be, but that, on the contrary, they must be complex structures. This belief arose from the necessity of explaining electrical and optical phenomena. It was necessary to explain why some substances were good conductors of electricity while others were not, why some substances were transparent while others were opaque. But the most impressive argument for the complex nature of atoms came from the study of spectra, the little rainbow of bright lines revealed by the spectroscope when a chemical element is heated to incandescence (see Chapter 5).

It was assumed in the 1890s by a number of physicists, including Stoney, George F. Fitzgerald, Joseph Larmor, and Hendrik A. Lorentz, that vibrating electronic charges in the atom must be responsible for the genesis of spectra and the other electrical and optical phenomena. The mathematical theory was developed at length by the great Dutch physicist Lorentz. The Lorentz theory

266   THE ATOM

of electrons is one of the monumental accomplishments of nineteenth-century science.

Lorentz held that in electric conductors such as metals the electrons were easily dislodged from the atoms and that an electric current was a stream of electrons in motion. In non-conductors the electrons were fixed in position. Light waves, according to Lorentz, were generated by the vibration of electrons within the atom. The period of vibration accounted for the wavelength of light emitted. Other optical phenomena were also explained by this theory.

No one, however, had as yet isolated or identified the electron. It remained for the English physicist Sir Joseph John Thomson to do this in 1897. It will be recalled that Crookes thought the cathode rays were a fourth state of matter. Thomson undertook to find out what they really were. Extending and perfecting techniques others had used a few years earlier, he developed what was the forerunner of today's oscillograph, or television image, tube. (see Figure 23).

This consisted of a Crookes tube so designed that the cathode ray passed between two charged electric plates. These deflected the rays and the deflection could be measured by the shifting of the luminous spot where the rays struck the end of the tube. A scale was provided to measure the actual deflection. It was possible, however, to restore the beam to its original position by a magnetic field created by a magnet outside the tube. The relative action of the magnetic and electrical fields depends upon the speed of the particles in the rays. Knowing the values of the two fields which just neutralized each other, Thomson was able to calculate the speed of the cathode ray particles. He found they possessed an enormous speed, equal to about 1/10 the speed of light. From the speed the

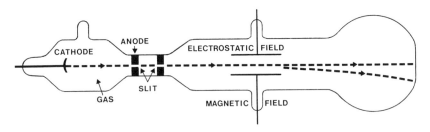

Figure 23. Thomson's cathode ray tube

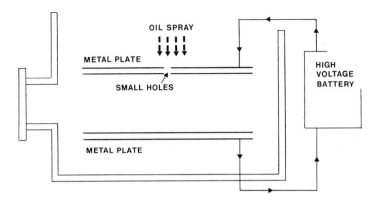

Figure 24. Millikan's oil-drop experiment

ratio of the electric charge of the particle to its mass could be calculated. He found the results to be the same no matter what substance was used for the cathode. From this he concluded that the cathode ray particles are the basic constituents of the atom. At first he called them corpuscles but later adopted the name electron, which Stoney had originally proposed. By subsequent experiments of a related nature Thomson demonstrated that the mass of the electron must be less than a thousandth of the mass of the hydrogen atom. In 1898 Thomson put forward the first definite notion concerning the internal structure of the atom. He suggested that the atom consisted of a configuration of negative electrons distributed in a diffuse sphere of positive electricity.

A more exact determination of the electric charge in the electron was made in 1909 by the American physicist Robert A. Millikan. This was the famous "oil-drop experiment." It consisted in measuring the rate at which an electrified oil drop fell under the influence of gravity between two electrified metal plates (see Figure 24).

It was apparent, as the nineteenth century closed, that a major revolution had occurred in the realm of physics. Four significant discoveries—X-rays, radioactivity, radium, and the electron—convinced scientists that their task was only beginning, not ending as they had imagined only a few years earlier. The time had come to invade the interior of the atom. It is doubtful, however, if anyone foresaw, at the dawn of the twentieth century, the major advances that would be made in theoretical understanding or the spectacular applications that would arise from this new knowledge.

eighteen

THE STRUCTURE OF THE ATOM

Nature is an endless combination and repetition of a very few laws.

—EMERSON

OUR knowledge of the internal structure of the atom is a twentieth-century product. The various notions advanced at the close of the nineteenth century were tentative and vague. An understanding of the true state of affairs emerged from the brilliant researches of Lord Rutherford, one of the giants of twentieth-century physics.

Ernest Rutherford was born in New Zealand but studied under Sir Joseph Thomson at the famous Cavendish Laboratory of Cambridge University in England. In 1898 he became professor of physics at McGill University in Montreal, and it was there that he began his famous researches into the behavior of the radioactive elements and the rays they emitted.

A number of experimenters, including Pierre Curie and Becquerel, had shown by 1900 that the rays given off by radium included charged or electrified particles which were probably electrons. But Rutherford, by placing thin sheets of aluminum in the path of the rays, demonstrated that there were at least two types of particles involved. One type, which he called alpha rays, were

THE STRUCTURE OF THE ATOM 269

stopped by extremely thin sheets. The other, which penetrated thicker sheets, he called beta rays.

Following baffling observations made in various laboratories, including both his own and that of the Curies, in which unexpected radioactive effects appeared, Rutherford proved that they resulted from a radioactive gas which was given off by uranium, thorium, and radium. He called this gas emanation. In 1902 he and his colleague Frederick Soddy announced their revolutionary theory of radioactive disintegration. Nineteenth-century physicists, like all earlier physicists, had assumed that atoms were unchanging and eternal. Rutherford and Soddy, on the contrary, held that the radioactive elements were spontaneously disintegrating, their atoms changing from one chemical element to another by the emission of charged particles.

The theory met with a storm of disapproval and resistance but gradually it became apparent that it was correct. Rutherford and other experimenters gradually worked out the series of transformations for the various radioactive elements. Thus uranium, by a series of such transformations, becomes radium, and radium after more transformations finally becomes lead. The emanation noted by Rutherford is the first product of the radioactive decay of radium. It is a gaseous element now known as radon. In turn, radon decays into a solid called radium A, which is a form, or isotope, of the element polonium.

During the first decade of the twentieth century Rutherford, Becquerel, and other experimenters clarified the nature of the rays given off by radioactive elements, establishing that there were three types. Rutherford collected the alpha rays, or particles, in a glass tube and by spectroscopic analysis proved that they were positively charged helium atoms. (It later became evident that the alpha particles were the nuclei of helium atoms.) The beta rays were shown to be electrons, while the gamma rays were radiations resembling X-rays.

It is possible to separate the three types of rays with the aid of a strong magnetic field. The positive alpha rays are deflected in one direction, the negative beta rays in the other, and the gamma rays are unaffected (see Figure 25).

It is difficult today to realize the impact of Rutherford's

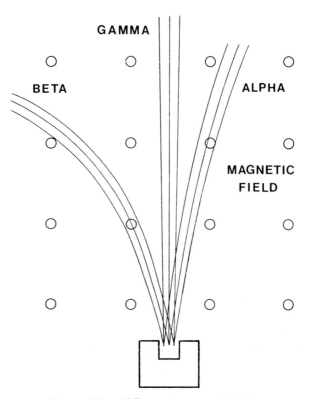

Figure 25. Effect of magnetic field

work on the progress of science. The ancients had dreamed of turning one chemical element into another. All through the Middle Ages the alchemists had worked to change iron and other base metals into gold. With the establishment of the atomic theory, it was assumed that the old dream of alchemy was forever impossible. But now Rutherford had shown that nature herself was an alchemist, turning uranium into radium and radium into lead.

THE RUTHERFORD ATOM

But Rutherford had only begun his revolution of atomic physics with the theory of radioactive disintegration. In 1906 he undertook some new experiments to study the behavior of the alpha particles emitted by radium. He noted that their direction was changed slightly when they passed through a thin sheet of mica. A

THE STRUCTURE OF THE ATOM

few years later one of his colleagues, Hans Geiger, continuing the experiment, found that alpha particles were bent even more by thin films of gold. Rutherford was excited by this and asked Geiger and E. Marsden to repeat the experiments. They found that in some cases alpha particles bounced back from the gold film. The experiment is shown in Figure 26.

Commenting upon this 20 years later, Rutherford said, "It was quite the most incredible event that has ever happened to me in my life. It was almost as incredible as if you fired a 15-inch shell at a piece of tissue paper and it came back and hit you. On consideration I realized that this scattering backward must be the result of a single concussion, and when I made the calculations, I saw that it was impossible to get anything of that order of magnitude unless you took a system in which the greater part of the mass of the atom was concentrated in a minute nucleus." Rutherford published his calculations in 1911 but did not introduce the use of the word "nucleus" until the following year.

Thus was born the idea which is at the center of all modern atomic research, the idea that the atom is organized like the solar system. It will be recalled from Chapter 17 that Sir Joseph Thomson had pictured the atom as a configuration of electrons scattered

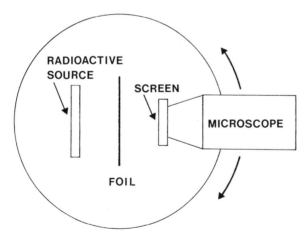

Figure 26. Rutherford's experiment. The radioactive source and gold film remained stationary while the screen and microscope were rotated around the film.

through a uniform sphere of positive electricity. Rutherford had now shown that not only the positive electricity but most of the mass of the atom as well was concentrated in a central nucleus like the sun at the center of the solar system. Electrons, he believed, revolved around this nucleus as planets circle the sun.

Three problems, however, remained to be solved. One was the arrangement of the "cloud" of electrons around the nucleus. The second was the nature and possible structure of the nucleus. The third was the question of whether or not there existed a positive particle corresponding to the electron.

THE CLOUD OF ELECTRONS

Between 1911 and 1913 the key to the situation was found in the analysis of X-ray spectra by a number of experimenters including the British physicist Charles G. Barkla, the Dutch physicist A. van den Broek, and the brilliant young British physicist Henry G. J. Moseley who was killed in 1915 at the Battle of Gallipoli in World War I. Their work showed an orderly increase in the positive charge on the atomic nucleus, corresponding to the place of the element in the periodic classification of the elements. Thus the hydrogen nucleus, the lightest known, was found to have a positive charge equal to the negative charge of the electron. The helium nucleus had double this charge, the lithium nucleus three times, and so on. The size of this positive charge became known as the atomic number of the element. Since the atom is normally neutral, it became apparent that the atomic number also revealed the number of outer electrons in each element. Thus, it was one for hydrogen, two for helium, three for lithium, and so on.

A theory to account for the distribution of the negative electrons in the atom was advanced independently in 1916 by two American chemists, Gilbert N. Lewis and Irving Langmuir. According to their theory, the electrons were arranged around the nucleus in concentric shells. Their conception of the atom has been compared to the Chinese toys which consist of a series of hollow balls, one within the other. The one at the center would represent the nucleus, the others the shells in which the electrons occur. Of course, Langmuir and Lewis did not think that there were any actual shells, and so we must think of an imaginary series of shells whose function is to determine the locations of the electrons.

Further studies by a number of investigators led to the conclusion that there were seven shells around the nucleus. These were designated as the K, L, M, N, O, P, and Q shells. It was established that the K shell could hold two electrons, the L shell eight, the M shell 18, the N shell 32, the O shell 32, the P shell 18, and the Q shell two. This conclusion was reached from the analysis of various chemical reactions and particularly the arrangement of the elements in the periodic table of Mendeleeff.

Hydrogen, according to this theory, consists of an atom with a nucleus and one electron in the K shell. Helium, the next in the series, has an atom consisting of a nucleus and two electrons in the K shell. Lithium, the third in the series, has a nucleus, two electrons in the K shell, and one in the L. Each succeeding element adds an electron to the L shell until neon is reached when it is full. The neon atom has a total of 10 electrons, two in the K shell and eight in the L. The next atom, sodium, has two electrons in the K shell, eight in the L, and one in the M. And so it goes until uranium with its 92 electrons is reached—two in the K shell, eight in the L, 18 in the M, 32 in the N, 21 in the O, nine in the P, and two in the Q. It should be pointed out that the outermost shell of any atom never contains more than eight electrons and that, as in the case of uranium, some of the inner shells are frequently only partially filled.

The actual positions of the electrons in each zone or shell was imagined to be some sort of geometric configuration. The electrons were thought of as adhering pretty well to these locations, although capable of a certain amount of movement or vibration. Many things are explained beautifully on the basis of the Lewis-Langmuir theory. Thus, for example, this theory explains the periodic recurrence of characteristics which led Mendeleeff to formulate the periodic table. Three similar elements are lithium, sodium, and potassium. Lithium has three electrons—two in the K shell and one in the L. Sodium has 11—two in the K shell, eight in the L, and one in the M. Potassium has 19—two in the K shell, eight in the L, eight in the M, and one in the N. Each of these three atoms is characterized by the fact that its outermost shell contains only one electron. This, according to Lewis and Langmuir, accounted for their similarity. On the basis of their theory, therefore, chemical properties are determined by the number of electrons in the outermost shell.

The Lewis-Langmuir theory was adequate to explain most of the problems which confronted chemists at the time. Their model of the atom came to be known, therefore, as the "chemist's atom." But physicists were unwilling to accept it. This was because it was not adequate to explain the way in which light waves are generated. These, according to the principle laid down by Lorentz, arise from vibrations of the electrons.

In 1913 the Danish physicist Niels Bohr developed a theory of atomic structure which took into account the origin of spectra. While he retained the notion of shells, he returned to Rutherford's idea that the electrons were revolving around the nucleus. However, he introduced a number of ideas contrary to the laws of classical mechanics. These will be discussed in Chapter 21 in connection with the quantum theory, which they involve.

THE PROTON

The search for a positive particle corresponding to the electron began in the closing years of the nineteenth century as soon as Thomson had identified the electron. It preceded Rutherford's discovery of the atomic nucleus. Even before the electron had been isolated, positive rays, similar in some respects to the cathode rays, had been noted.

As early as 1886 Eugen Goldstein in Germany had constructed a tube in which the cathode consisted of a perforated metallic disc in the middle of the tube (see Figure 27).

Luminous rays came through these holes, traveling in the opposite direction from the cathode rays. These rays were called canal rays, since they passed through the holes, or channels, in the cathode. A decade later they were shown to be electrically positive but soon after they were proved to consist of electrified atoms, or ions, of whatever gas was in the tube.

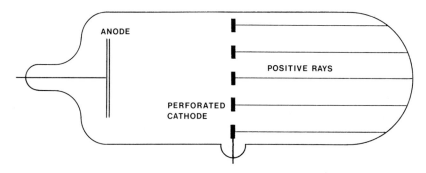

Figure 27. Goldstein's positive ray tube

Because the lightest ion observed in these experiments was hydrogen, Rutherford concluded that the basic positive particle is the hydrogen nucleus. Further researches verified this remarkable intuitive conclusion, and the name proton was given this particle by Rutherford.

The proton was found to be 1,836 times heavier than the electron. This was a surprise to physicists, who assumed that the proton would have the same mass as the electron. Much later it was shown that the proton, while heavier than the electron, was slightly smaller. Apparently it is a much more concentrated and compact particle.

With the establishment of the existence of the proton, physicists undertook to develop a theory of nuclear structure employing protons and electrons for this purpose. The second atom in the atomic table is helium. It has an atomic weight of 4 but an atomic number of 2. An atomic weight of 4 is easily obtained by imagining four protons in the nucleus. Four protons, however, have a total electric charge of 4, whereas the position of helium in the atomic table indicates that it has a charge of only 2.

But this could be explained by imagining two electrons in the nucleus. Two electons would balance two of the protons, leaving an excess positive charge of 2. Since the weight of the electron is negligible in comparison with that of the proton, this seemed reasonable.

It was assumed, therefore, that the nuclei of all atoms heavier than hydrogen contained both protons and electrons. The number of protons was assumed to be sufficient to account for the atomic weight. Associated with these protons there were thought to be enough electrons to neutralize all the protons except a number sufficient to supply the positive charge equal to the atomic number. For example, oxygen has an atomic weight of 16 and an atomic number of 8. It was assumed, therefore, that its nucleus consisted of 16 protons and eight electons.

Further support to this theory of the structure of the nucleus was given by the discovery of isotopes. It was pointed out by Theodore W. Richards of Harvard that while the lead obtained from the disintegration of radium possessed all the characteristics of ordinary lead, it differed from it in atomic weight. Frederick Soddy of England showed that a number of the products formed in the series of transformations from radium to lead possessed similar atomic numbers but different atomic weights. Since these atoms of different weights all occupied the same place in the periodic table, Soddy named them isotopes, a word derived from the Greek meaning "same place." It is obvious at once that isotopes could be accounted for by an increase or decrease in the number of both the protons and electrons so that the excess of protons, and hence the charge on the nucleus, remained the same.

As we shall see in Chapter 22 this notion of the structure of the nucleus was abandoned with the discovery of the neutron and superseded by a more satisfactory theory.

THE SEPARATION OF ISOTOPES

The theory of nuclear structure as outlined makes the atomic weight of an atom the measure of the number of protons in the nucleus. How, then, could the fact be explained that certain atoms had fractional atomic weights? The mystery was solved by Francis W. Aston, working in Sir Joseph Thomson's laboratory at Cambridge.

Thomson, it will be recalled, had identified the electron by using magnetic and electric fields to deflect the cathode rays. He tried similar experiments with positive rays and in one of them, in 1912, noted a strange behavior of neon gas. The experiment seemed to indicate that associated with the neon, whose atomic weight is 20.183, there was a slight trace of a gas with atomic

THE STRUCTURE OF THE ATOM

weight 22. Aston, who was then Thomson's assistant, undertook to separate the two components by diffusing neon through a porous clay barrier but with only moderate success. He returned to the problem after World War I and decided tht the separation could be made most easily by means of an improved positive ray tube. The result was the mass spectrograph in which isotopes are separated by the action of electric and magnetic fields. The separation results from the fact that the amount of deflection in each case depends upon the weight of the isotope. Aston established that neon consisted of two isotopes with atomic weights close to 20 and 22.

Next he showed that chlorine consisted of two isotopes, one of which possessed an atomic weight of approximately 37, the other a weight of approximately 35, the two being mixed in such proportions as to give an ordinary sample of chlorine the atomic weight of 35.46. The two cannot be separated by chemical means, for chemically they are both chlorine, having an atomic number of 17. By the end of 1920 Aston had tried 19 elements and found that nine of them consisted of isotopes.

Meanwhile in the United States, Arthur J. Dempster had in-

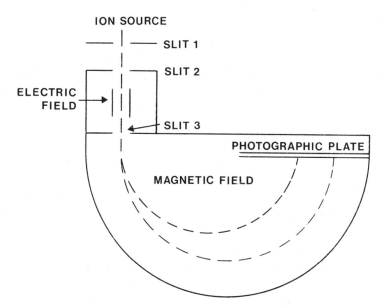

Figure 28. The mass spectrograph

dependently designed an improved mass spectrograph and by 1920 had demonstrated the existence of isotopes of lithium, magnesium, potassium, calcium, and zinc. The mass spectrograph is shown in Figure 28.

It is now known that more than 320 isotopes exist in nature. These include 280 stable isotopes and 40 radioactive ones. In addition, physicists in recent years have created several hundred isotopes in the laboratory which do not exist in nature.

THE SIZE OF ATOMS

The fact must be kept in mind that molecules, atoms, and electrons are all so small that the most powerful optical microscope is inadequate to reveal them. The largest molecule is still smaller than the smallest wavelength of visible light. If a drop of water were magnified to the size of the earth, the molecules composing it would be about the size of oranges. A cubic inch of air contains 800,000 trillion molecules.

Molecules vary in size, depending upon the number and arrangement of the atoms composing them. Inorganic molecules, in general, vary from simple ones consisting of two atoms, such as the hydrogen molecule, to complex ones, which may contain a dozen or more atoms. They range in size from a few hundred-millionths of a centimeter to a ten-millionth of a centimeter. (A centimeter is equal to 0.3937 of an inch.) An exception is the very large molecules of organic compounds, such as carbohydrates, fats, and proteins. The very largest of these contain several hundred thousand atoms and in some cases millions. A few are large enough to be visible in the electron microscope.

Atoms average a few hundred-millionths of a centimeter in diameter. The size varies with the size of the cloud of electrons around the nucleus.

The electron has a diameter of about 4 ten-trillionths of a centimeter. The proton is slightly smaller. Atomic nuclei range in diameter from about 3 ten-trillionths of a centimeter for hydrogen to about 2 trillionths of a centimeter for uranium.

The atom of hydrogen, the lightest atom, weighs 1.67 tril-

THE STRUCTURE OF THE ATOM

lionths of a trillionth of a gram. (A gram is approximately 0.035 of an ounce.) The uranium atom weighs about 4 ten-billionths of a trillionth of a gram.

As the reader may know, scientists have a shorthand method of writing extremely large or small numbers as powers of 10. Thus, 1,000 is written 10^3, a million as 10^6, a billion as 10^9, and so on. The exponent tells the number of 0s needed in ordinary notation. Negative exponents are used for fractional numbers. Thus 0.1 becomes 10^{-1}, 0.01 is 10^{-2}, 0.001 is 10^{-3} and so on. A hundred-millionth is 10^{-8}, a trillionth is 10^{-12}, a trillionth of a trillionth is 10^{-24}.

In this system the diameter of atoms is of the order of 10^{-8} centimeters, that of atomic nuclei of the order of 10^{-13} to 10^{-12} centimeters.

It is startling to realize that atoms contain more empty space than matter. The ratio of empty space to particles in the atom is 10,000 times greater than the ratio of empty space to planets in the solar system.

nineteen

THE FORMATION OF MOLECULES

Everything flows and nothing stands still.
—HERACLITUS

ONCE Rutherford had established that the atom consisted of a nucleus surrounded by a cloud of electrons, it became possible to explain how atoms united to form molecules. Nineteenth-century chemists had no adequate explanation of the mechanisms involved. The Swedish chemist Jöns Jakob Berzelius came closer to an explanation than his contemporaries realized when he suggested in the early part of the nineteenth century that the force which held molecules together was electrical.

It was well understood by the middle of the nineteenth century that atoms differed in their power to combine with other atoms. Thus, for example, a fluorine atom will combine with one hydrogen atom, an oxygen atom with two hydrogen atoms, a nitrogen atom with three, and a carbon atom with four. This combining power of the atom was given the name valence in 1868. Thus hydrogen was said to have a valence of 1, oxygen a valence of 2, and so on.

It is well understood today that chemical activity can be explained by recourse to the behavior of the electrons in the outer

THE FORMATION OF MOLECULES

shells of atoms. Three types of chemical binding are recognized. These are known as ionic binding, covalent binding, and metallic binding.

IONIC BINDING

It has long been known that certain gases, the so-called inert gases, do not ordinarily enter into chemical reactions. They are also called monatomic because they exist as individual atoms and are not combined into molecules. The configuration of the electrons around their nuclei reveals an interesting situation.

Helium, the first of the inert gases, is No. 2 in the atomic table. Its nucleus is surrounded by two electrons, the maximum number which the first, or K, shell can hold.

The next one, neon, is No. 10 in the atomic table. It has two electrons in the K shell and eight in the second, or L, shell. This completes the second shell. In the case of the other inert gases, argon, krypton, xenon, and radon, there are always eight electrons in the outer shell. This configuration of eight electrons has a firm stability and is known as the stable octet.

Other atoms exhibit an attempt to achieve this stable octet either by gaining or losing electrons. Sodium, for example, has one electron in the M shell. It has a tendency to lose this electron. Its outer shell is then the L shell of eight electrons. But when the soldium atom loses an electron, the positive electric charge of the nucleus is no longer balanced by the negative charges of the cloud of electrons. The atom now exhibits a positive charge and is known as a positive ion.

On the other hand, chlorine has seven electrons in its outer shell. It acquires the configuration of the stable octet if it gains an electron. But it then has an excess of negative charges and thus becomes a negative ion.

If a sodium atom and a chlorine atom now come together closely enough, the chlorine atom will capture an electron from the sodium atom. The electrical attraction between the positive sodium ion and the negative chlorine ion binds them into a molecule of sodium chloride.

It is the metallic atoms which have few electrons in their outer shells, while the nonmetallic atoms have fuller outer shells. Consequently, ionic binding accounts for the union of metallic and nonmetallic atoms into molecules.

COVALENT BINDING

It is obvious that ionic binding cannot explain how two atoms of hydrogen combine to form a molecule of hydrogen, or how two atoms of oxygen form a molecule of oxygen. This type of union is the result of covalent binding in which the two atoms share one or more electrons.

Consider the case of hydrogen, which is the simplest. Each hydrogen atom possesses one electron. Two hydrogen atoms, by sharing their electrons, can attain the stable configuration of the helium atom.

It will be recalled that electrons are spinning on their axes. This spin creates a magnetic field so that the electron in essence is a little magnet. The electron can orient itself in one of two directions. When two hydrogen atoms approach sufficiently close, the orbits of their electrons intersect. If the electrons are spinning in the same direction, their magnetic fields increase the normal repulsion between electrons. However, if they are spinning in opposite directions, the magnetic fields create a force of attraction between the two electrons. As a result the two atoms are bound into a molecule by the sharing of two electrons with opposite spins.

The covalent bond explains the formation of molecules between nonmetallic atoms. Carbon, for example, has four electrons in its outer shell. It needs four electrons to attain a stable octet. Consequently, it can share electrons with four hydrogen atoms to form a molecule of methane, or marsh gas.

Hydrogen is unusual in that it can enter into molecules either as a result of covalent binding or ionic binding. When a hydrogen atom gives up its single electron, it becomes a positive ion.

THE FORMATION OF MOLECULES

METALLIC BINDING

The atoms in metals are bound together by a variation of covalent binding known as metallic binding. Because a metallic atom has only one, two, or three electrons in its outer shell, it is not possible for two atoms to form a stable octet. A number of atoms contribute their outer electrons to form pairs which are then shared among them. These bonds are not confined to specific pairs of atoms, but circulate among the atoms, making up for the shortage of electrons by their mobility. The fact that the binding electrons can move freely explains why metals are such good conductors of electricity.

GASES

The electrical forces which cause atoms to form molecules, also operate between molecules. One might ask, therefore, why many substances ordinarily exist in the form of gases. The explanation was supplied in 1738 by the Swiss mathematician Daniel Bernoulli 70 years before Dalton laid the groundwork for the modern atomic theory. Bernoulli advanced the notion that the particles of a gas were in motion, an explanation still accepted today and known as the kinetic theory of gases. The theory explains why heating a gas that is not closely confined causes it to expand. Heating causes the particles to move with greater velocity. As a result they occupy more space. For the same reason, heating a closely confined gas or compressing it into smaller volume causes it to exert more pressure because the collisions of the particles against the walls of the container become more violent and more frequent.

We know now that there is a battle between the motion of the molecules and the forces of cohesion. If the cohesive forces are weak, the substance exists as a gas at ordinary temperatures. But such a gas can be reduced to a liquid or even a solid by cooling it. As the motion of the molecules is slowed down, the forces of cohesion gain the upper hand. But any substance that is ordinarily a liquid or a solid can be turned into a gas by heating it sufficiently.

LIQUIDS

To understand why many compounds exist as liquids at ordinary temperatures, we must explore the forces which exist between molecules and the ways in which they arise. Let us look at the molecule of water which will serve as a typical example.

The oxygen atom, it will be recalled, has six electrons in its outer shell. The two vacant orbits of this shell are at right angles to each other. These, of course, are the orbits shared by the electrons of the hydrogen atoms. One might expect, therefore, that the two hydrogen atoms would be bonded to the oxygen atom so that the angle between them would by 90°. However, the two hydrogen nuclei repel each other and as a result the orbits of the shared electrons are spread apart to form an angle of 105°. If we now turn to the nuclei of the three atoms involved, we find the oxygen nuclei at the vertex of the angle made by lines drawn to the two hydrogen nuclei. This is an angle of 105°. The molecule is not spherical but has two protuberances, or "ears," where the hydrogen nuclei are located. This is shown in Figure 29.

Now the positive charge of the nucleus of the oxygen atom is

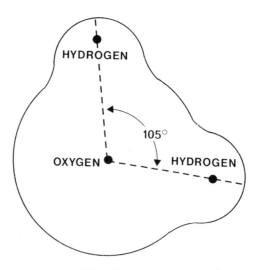

Figure 29. The water molecule

larger than that of the hydrogen atom. As a result each shared electron in the water molecule is drawn closer to the oxygen atom. This gives the oxygen atom an excess of negative charge and leaves the hydrogen atoms with a deficit. As a result of this unequal distribution of electric charge, the water molecule exhibits two positive ends and one negative end. For many purposes one can think of the water molecule as having merely a positive end and a negative end. Such a molecule is called a polar molecule. The distribution of electric charges is known as a dipole. All unsymmetrical molecules are polar to a greater or lesser extent.

It will be seen at once that this polar effect explains why molecules are attracted to each other. The positive end of one molecule is attracted to the negative end of another. In addition there is another force between molecules which arises in the atoms composing them. The electric field created by the cloud of electrons in an atom will affect the similar field of an atom that comes near it. The resulting distortion of the shape of the atom causes it to become a dipole. The positive end of one dipole attracts the negative end of another and this helps to bind molecules into a liquid or a solid.

The forces of attraction between molecules are called Van der Waals forces after Johannes van der Waals, the Dutch physicist who was among the first to study them.

In a gas, as we have seen, the motion of the molecules is so great that the cohesive forces between molecules have no opportunity to come into play. In liquids the motion is less and the cohesive forces greater. The molecules are free to move about but in general they cannot get very far away from one another. The result is that a liquid maintains a fairly constant volume. Occasionally, a molecule does break away from the surface of a liquid and flies off into space. That is the explanation of evaporation.

It is difficult at first to associate the idea of constant motion with a liquid. Yet a little thought will show that all the well-known and familiar properties of a liquid are most easily explained on the assumption that the liquid is composed of tiny particles in constant motion.

Robert Hooke, British physicist of the seventeenth century, recognized this fact. He wrote:

First, what is the cause of fluidness? This I conceive to be nothing else but a very brisk and vehement agitation of the parts of a body; the parts of a body are thereby made so loose from one another that they easily move any way, and become fluid. That I may explain this a little by a gross similitude, let us suppose a dish of sand set upon some body that is very much agitated, and shaken with some quick and strong vibrating motion, as on a millstone turned round upon the under stone very violently whilst it is empty; or on a very stiff drumhead, which is vehemently or very nimbly beaten with the drumsticks. By this means the sand in the dish, which before lay like a dull and unactive body, becomes a perfect fluid; and ye can no sooner make a hole in it with your finger, but it is immediately filled up again and the upper surface of it leveled. Nor can ye bury a light body, as a piece of cork, under it, but it presently emerges or swims, as 'twere on the top; nor can ye lay a heavier on the top of it, as a piece of lead, but it is immediately buried in sand, and (as 'twere) sinks to the bottom. Nor can ye make a hole in the side of the dish, but the sand shall run out of it to a level. Not an obvious property of a fluid body, as such, but this does imitate and all this merely caused by the vehement agitation of the containing vessel; for by this means, each sand becomes to have a vibration or dancing motion, so as no other heavier body can rest on it, unless sustained by some other on either side; nor will it suffer any body to be beneath it, unless it be a heavier than itself.

Liquids boil when heated. Boiling is similar to evaporation. Heating the liquid has simply raised the molecular motion to such a rate that the molecules on the surface of the liquid are able to push back the molecules of air above them and fly off into space en masse.

It is the attractive force between the molecules of a liquid which causes the liquid to form into round drops. This same force also causes the surface of a liquid to form into a sort of dimple when some small object is floated on it. In general, the attractive

THE FORMATION OF MOLECULES

force makes a liquid behave just as though there actually were a thin film over the surface. This phenomenon is known as surface tension.

It is possible, however, to form real films upon the surface of a liquid. The most familiar example is generally regarded as a child's pastime—soap bubbles. But scientists have found it worthwhile to study bubbles and have found out many interesting things about the structure and behavior of molecules from them.

Everyone knows that if a little soap is dissolved in water, it is possible to blow the soapy water into bubbles. The scientist has found that the explanation for this is extremely interesting. It lies in the structure of the soap molecules. Soap is a complex substance consisting of carbon, hydrogen, oxygen, and sodium. Each molecule of soap consists of a chain of carbon atoms fringed on either side with hydrogen atoms. One end of the chain is rounded off with three hydrogen atoms. The other end of the chain consists of a little bunch of sodium and oxygen atoms. The end of the chain consisting of the hydrogen atoms seems to be extremely unsociable. No attractive force manifests itself at this end. But the other end has a great attraction, particularly for the molecules of water. That is why soap dissolves in water. But because only one end of the soap molecule has an attraction for water, the soap molecules have a tendency to stay on the surface of the water, forming a film over it. When this happens, the soap molecules stand on end, the end which has the attraction for the water being rooted in the water, while the other end is exposed to the air. As a result the molecules of soap stand side by side like the bristles of a pig. The soap bubble, therefore, is a thin-walled hollow sphere, its inner and outer surface consisting of films of soap enclosing a thin layer of water.

Many scientists have studied the behavior of such molecules as those of soap. They are known as the long-chain molecules. Oils consist of such molecules, and when a tiny drop of oil is placed upon a surface of water, it arranges itself in a thin film of molecules, each molecule standing on end. It is thought that this explains why ocean waves can be stilled by pouring oil upon them. The oil forms a smooth film upon which the wind has no effect.

SOLIDS

In a solid the attractive forces have the upper hand. In many solids there are no molecules. Instead the atoms are bound to each other to form crystal structures. When molecules do occur, they are likewise bound into crystal patterns. Much of what we know about the structure of solids is the result of X-ray analysis. This technique was developed in the early years of the present century. The pioneers were the German physicist Max von Laue and the British physicists Sir William Bragg and his son W. L. Bragg. When a beam of X-rays passes through a crystal, the rays are refracted by the regular arrangement of the atoms in the crystal. As a result, a pattern of spots is formed on a photographic plate from which the arrangement of the atoms can be determined.

In some cases the atoms are held together by covalent bonds. This is the case with carbon. The carbon atom has four valence electrons, which it can share with four other carbon atoms. This accounts for the structure of the diamond. Each carbon atom is equidistant from four others. These five can be thought of as the basic unit of the diamond. There is one atom at the center of the unit. The other four are located at the corners of a four-cornered pyramid, or tetrahedron.

Metals form crystals in which the atoms are bonded to each other. In the simplest arrangement the atoms are closely packed, like marbles in a box. Each atom has 12 neighbors with which it forms transient metallic bonds. In other arrangements the atoms are less closely packed. The atoms may be located at the corners of cubes. Sometimes there is an additional atom at the center of each cube.

Ionic bonds account for the crystal structure of many chemical compounds. Sodium chloride, ordinary table salt, is a good example. In this crystal the atoms of sodium alternate with atoms of chlorine. The atoms are located at the corners of cubes. Thus each atom of sodium is surrounded by six atoms of chlorine, and each atom of chlorine is surrounded by six of sodium (see Figure 30).

Large molecules, like those of the hydrocarbons, which consist of many atoms bonded into complex structures, form crystals that

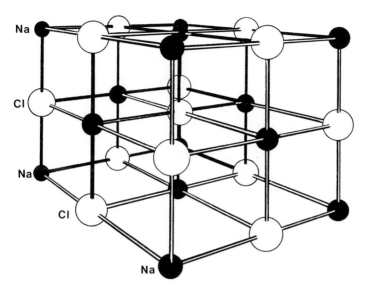

Figure 30. Crystal of sodium chloride

can be compared to lacelike patterns in three dimensions. The location of positive and negative poles on these molecules explain how they are bonded into crystals. Most of these crystals have low density. As a result most organic substances are lighter than water.

It is common to think of crystal structure as something special. The X-ray study of solids has shown just the opposite to be true. Practically all solids have a crystal structure. The crystals, however, are so small and are packed together in such helter-skelter arrangement that it is not possible to recognize the presence of crystals without the aid of the X-ray.

The structure of the crystal and the strength of the bonds holding the atoms or molecules together explain the characteristics of solids. The cohesive forces vary greatly. Such a solid as butter begins to melt on a warm day. Tungsten, to cite a quite different example, can be melted only at a temperature of several thousand degrees.

The twentieth-century understanding of how atoms are put together into molecules has led to the branch of chemistry known as synthetic chemistry in which the chemist synthesizes compounds that were previously known to exist only in living organisms and even to create molecules which do not exist in nature.

twenty

THE NATURE OF ENERGY

To give us the science of motion, God and Nature have joined hands to create the intellect of Galileo.

—Fra Paolo Sarpi

THE modern concept of energy grew out of Galileo's studies of motion. These represent a greater achievement than the astronomical discoveries for which the world best remembers him. Almost from the start of his career he challenged the notions of Aristotle which had dominated scientific thinking for 19 centuries. In his old age, after his trial by the Inquisition, a lonely and blind prisoner at Arcetri, he summed up his views in a book, *Two New Sciences*, which he dictated to faithful disciples. This was a study of the effect of forces on bodies and their resulting motions. Today these branches of physics are known as mechanics and dynamics.

Aristotle believed that the earth was the center of the universe. He taught that bodies fell to earth because every body sought its proper place in the universe. Heavy bodies fell faster than lighter ones. The speed of a falling body increased as it neared the surface of the earth because of a sort of joy at arriving at its proper place. Aristotle regarded this motion as natural motion. All other motion was unnatural motion and required the continuous application of force. It ceased when the force ceased to operate. It is obvious that this notion grew out of consideration of what happened when a horse pulled a cart along a road or sailors tugged on the tow rope of a boat on a canal.

Galileo was not the first to challenge Aristotle's views, but he did so with the aid of physical experiments to which he applied mathematical analysis. In so doing, he inaugurated the modern era of experimental physics.

By a series of experiments in which he timed the speed of balls rolling down an inclined plane, Galileo arrived at two important conclusions. One was that all objects, irrespective of weight or density, fall at the same rate under the influence of gravity. The other was that the application of a force results in a change in the speed or direction of motion of a body. Today we call this change acceleration, and speak of the increase in the velocity of a falling body as the acceleration of gravity.

Galileo saw, contrary to Aristotle's notion, that if left alone, a body will continue in its state of rest or motion forever. A body set in motion on earth eventually comes to a stop because of friction against the air or the surface of the ground. A stone rolled along the ground comes to rest much quicker than one on a sheet of ice. If it were not for friction, it would keep going forever.

The tendency of a body to resist change in its state of rest or motion is known as inertia. It is a property of matter for which no explanation is known. One suggestion is that it represents the sum total of the gravitational effect of every body in the universe.

Galileo died in 1642. Sir Isaac Newton, who was born in that same year, carried Galileo's work to its logical conclusion. He formulated the law of universal gravitation and perfected Galileo's study of motion in his three laws of motion.

Newton's law of gravitation states that every body in the universe attracts every other body with a force that is proportional to the product of their masses and inversely proportional to the square of the distance between them.

NEWTON'S LAWS OF MOTION

Newton's first law of motion states that every body will continue in a state of rest or motion in a straight line with uniform velocity unless some force acts upon it. The second law states that when a force acts, the motion changes, either in speed, direction, or both, and the change depends upon the force and the length of time it acts. The third law states that to every action there is an equal and opposite reaction.

The first law introduces the concept of inertia, the concept which had already been recognized by Galileo. You become aware of inertia when the driver of the automobile in which you are riding starts up too suddenly, goes around a curve too rapidly, or applies the brakes too violently.

In ordinary conversation we usually make no distinction between weight and mass. The physicist defines weight as a measure of the force with which an object is pulled toward the center of the earth by gravity. Newton defined mass as the amount of matter in a body. This definition is good enough for ordinary purposes. But it has been known since the start of the present century, as a result of atomic experiments, that the mass of a subatomic particle increases with its velocity. At ordinary velocities this increase in mass is so tiny that it can be ignored.

The mass of a body determines its inertia. A railroad locomotive exhibits a resistance to a change in its state of rest or motion far greater than does a tennis ball. If a body was taken from the earth to the moon, it would weigh less on the moon because the moon's force of gravity is only one-sixth that of the earth. However, its inertia would be exactly the same on the moon as on the earth because its mass would still be the same.

Newton's second law introduces the concept of acceleration, a familiar idea in this day of the automobile. Acceleration is measured as change in velocity per second. Since velocity is distance per second—for example, 10 feet per second—an increase of this velocity by 1 foot per second is spoken of as an acceleration of 1 foot per second per second or 1 foot per second squared.

Newton introduced the idea of momentum, which he defined as mass times velocity. His second law of motion can be stated, therefore, as saying that the effect of a force acting upon a body is to change its momentum. From this we arrive at the conclusion that the acceleration produced by a force acting on a body is proportional to the force and inversely proportional to the mass of the body as long as the mass is constant. This can be written algebraically as $A=\frac{F}{M}$ where A is the acceleration, F the force, and M the mass. Transposing these quantities, we obtain the equation $F=MA$. This has been called the fundamental law of mechanics.

In the statement of these laws the word "force" has been used. Everyone knows what we mean by force, though it is extremely difficult to put our meaning into words. Our ideas of force are derived primarily from muscular effort. It requires an effort to lift a weight or throw a ball. We therefore say that anything exerts a force when it serves to accomplish something that would require muscular exertion on our part. A force can be defined as any cause that produces, destroys, or changes motion. It can be measured by an application of the equation $F=MA$.

A force which causes an acceleration of 1 foot per second per second in a mass of 1 pound is known as a poundal. In the metric system, preferred by scientists, a force which produces an acceleration of 1 centimeter per second per second in a mass of 1 gram is known as a dyne.

The physicist uses the term "work" to measure the accomplishment of a force. This is measured in foot-pounds. If a weight of 1 pound is lifted vertically 1 foot against the force of gravity, the physicist says that 1 foot-pound of work has been done. If a weight of 50 pounds is lifted 4 feet, 200 foot-pounds of work have been accomplished. The unit of work in the metric system is the dyne-centimeter, which is known as the erg.

The term "power" is used by physicists to denote the rate of doing work. Thus, 1 horsepower is 550 foot-pounds per second. If an engine can lift a weight of 550 pounds 1 foot in a second, it has a rating of 1 horsepower.

POTENTIAL AND KINETIC ENERGY

We come now to energy. Energy is commonly defined as the ability to do work or the capacity for doing work. A bent spring possesses energy, for it is capable of doing work in returning to its natural form. Gunpowder possesses energy, because it is capable of doing work when it explodes. A moving body possesses energy, because it can be made, in coming to a stop, to do work.

Since energy is the capacity for doing work, it is possible to explain physical phenomena in terms of energy instead of the manifestation of force. Mechanical energy is classified as potential and kinetic. Potential energy is the energy which any physical system possesses by reason of its configuration. Thus a weight raised to a certain height above the ground is said to possess potential energy because it can do work in falling. The energy which a body in motion possesses is known as kinetic energy.

It is possible to transform potential energy into kinetic energy and vice versa. Let us say that you are holding a tennis ball in your hand. It possesses potential energy by reason of its position above the ground. If you drop it, its potential energy is transformed into kinetic energy as it moves downward. As it strikes the ground it is deformed by the impact and its kinetic energy is transformed into the elastic energy of its deformation. In a split second the elastic energy is changed back to kinetic energy and the ball rises to your hand. However, if you let it continue to bounce by itself, it finally comes to rest on the floor. The friction of the series of impacts with the floor has turned its energy into heat.

There are many forms of energy and energy manifests itself in many ways. It is possible to transform energy from one form to another. The automobile, for example, transforms the chemical energy of the explosion of gasoline vapor into the kinetic energy of the moving auto. The driver applies the brakes suddenly, and in stopping the car transforms this energy into the heat of friction in the brake bands and the tires. As Sir James Jeans observed, all the life of the universe may be regarded as manifestations of energy, masquerading in various forms and changing from one form to another.

CHEMICAL ENERGY

Chemical reactions can be divided into a number of types. In some atoms unite to form molecules. In others molecules are separated into their component atoms. In more complex reactions one atom or a group of atoms replaces another atom or group of atoms in a molecule.

Molecules and atoms, it will be remembered, are in constant motion. In other words, they possess kinetic energy. When chemical bonds are formed (see Chapter 19), some of this kinetic energy is transformed into potential energy in the bonds. When chemical bonds are broken, this energy is released once more as kinetic energy.

All chemical reactions involve an exchange of energy. They can be divided on the basis of energy into two types, those which absorb energy and those which release energy. A chemical reaction cannot take place unless the atoms are moving fast enough to overcome the natural repulsion of their electron clouds. These clouds must overlap before a reaction takes place. Whether a reaction absorbs or releases energy depends upon a number of factors. One is the amount of kinetic energy needed to make the reaction go. If chemical bonds are formed, another is the amount of energy stored in these bonds as potential energy. If bonds are broken, the amount of energy released from them becomes a factor.

Almost all chemical reactions are reversible. If, for example, energy is absorbed when two atoms unite to form a molecule, the same amount of energy will be released when the bond is broken and the molecule is dissociated into its component atoms.

THE NATURE OF HEAT

An important step forward in an understanding of physical phenomena was made with the comprehension of the true nature of heat. This was first hinted at by Daniel Bernoulli in his kinetic theory of gases (see Chapter 19). He suggested in 1738 that temperature was merely a measure of the vibration of the particles composing a body. The matter was revived sixty years later by Count Rumford, who noted that brass cannon became hot during the process of boring. Later Sir Humphry Davy showed that two pieces of ice could be melted in a vacuum by being rubbed together even though the temperature of the surroundings was kept at zero. The matter was clinched by James Prescott Joule of Manchester who demonstrated by his experiments the mechanical equivalent of heat.

The total heat in a body is the sum of the kinetic and potential energy of the atoms and molecules composing it.

Subsequent experiments have shown that there is a general tendency for all energy to change into heat because of the friction of rough surfaces, the resistance of electrical conductors, and so on. This leads to a lessening of the availability of energy, a fact which has an important bearing upon modern physical theories.

THERMODYNAMICS

While all forms of energy are eventually transformed into heat, it is possible, with the aid of a suitable machine, to transform heat into other forms of energy. This, of course, is what a steam engine does. It is not possible to convert 100 per cent of a given amount of heat into mechanical energy. Heat always flows from a hot body to a cooler one and it does mechanical work only during this time. An engine must have a "cooler" to which the heat can flow. If this were not the case, one could use the heat in the surface water of the ocean to drive a steamship or use the heat of the atmosphere to operate an airplane. We cannot use the potential energy of the surface waters of the ocean because there is no lower level to which they can flow. For the same reason we cannot use the heat of the surface waters because they have no colder place to which they can go.

We can summarize this discussion of energy by saying that it is the nature of mechanical energy to transform itself into heat and for heat to flow from hotter regions to colder ones.

The relationship of mechanical energy to heat constitutes the branch of physics known as thermodynamics. The preceding paragraph constitutes what is known as the second law of thermodynamics. (The first law is the conservation of energy: the concept that energy can be transformed from one form to another, but not created or destroyed.)

The second law of thermodynamics means that energy is constantly running from an available form to a less available form.

In theory all the energy of the universe will eventually be distributed equally throughout space and will be completely unavailable. The universe will have run down and died what has been called a "heat-death."

Today astronomers are not so certain that this is necessarily so. If the expanding universe is an oscillating universe (see Chapter 8), the universe will in some billions of years be gathered together once more in a primordial fireball and ready to start again.

ENTROPY

The net effect of the transformation of energy from an available form to a less available form means a steady progression from organization to disorganization. This increase in disorganization, spoken of as an increase in the random element of the universe, is known as entropy.

Let us return to the example of the tennis ball. The transformation of its kinetic energy into heat means the distribution of its energy to the random motion of a vast number of molecules. Entropy has been compared to the shuffling of a deck of cards. A new deck of cards has the cards in perfect order. The more we shuffle the cards, the more random their distribution becomes. It is conceivable that if we shuffled 52 cards long enough, the original order might finally return. But such a reversal of entropy is inconceivable in a system involving trillions of molecules. For this reason entropy has been called "time's arrow." It marks the direction of time.

SOUND WAVES

An important characteristic of many forms of energy is the ability to move through space. Sound is a familiar example of this, or more exactly, sound waves. Let us say that a bell is struck with a hammer. The rim of the bell begins to vibrate. The vibrations are transferred to the air. The air is set in vibration, that is, the vibrational energy of the bell is transferred to the air. These vibrations travel through the air in ever widening circles, like the circles of waves set up on the surface of a pond when a stone is dropped into it. Finally, at some point the waves strike the eardrum of a person. They set it in vibration and, as a result, certain nerves in the ear are stimulated and the person hears a sound.

RADIATION

We now come to those manifestations of energy which are called electrical and magnetic and those transferences of energy which are electromagnetic in nature and grouped under the heading of radiation.

As everyone knows a magnet will attract a piece of iron. This phenomenon is a manifestation of magnetic force. If we lay a sheet of cardboard over a magnet and sprinkle iron filings upon the card, we find the filings arranging themselves in a series of curves from one pole of the magnet to the other. In some way forces are manifested between the poles of the magnet. We call this a magnetic field and say that the space is filled with lines of force. This expression, coined by the great Michael Faraday, is descriptive rather than explanatory.

Similarly, we find that under certain other conditions bodies exhibit similar kinds of forces and fields which we call electrical. If a glass rod is rubbed with a silk cloth, it becomes electrified. A different state of electrification is obtained if we rub sealing wax with fur. We call the first kind positive and the second kind negative.

As already stated, electrified objects exhibit forces and fields not unlike those of a magnet. Objects electrified the same way repel each other. Objects electrified differently attract each other. Electric fields of forces can be charted around electrified objects.

A great deal of study has shown that all electrical phenomena can be traced down to the positive protons and the negative electrons. The same thing is also true of magnetic phenomena. In other words, each proton and each electron is surrounded by an electric field and a magnetic field.

Normally an atom is neutral because it contains equal numbers of protons and electrons and they balance each other's effect. Atoms become ions, or electrified atoms, when they gain or lose electrons: a gain of negative electrons results in negative electrification, a loss of negative electrons results in positive electrification. When objects show electric fields, it merely means that a great number of atoms on their surface have become electrified. The glass rod rubbed with silk exhibits a positive electric charge because it loses electrons to the silk. The sealing wax becomes negative because it gains electrons from the fur.

An electric current is known to be a stream of electrons in motion. When a current flows through a wire, it means that electrons are moving through the wire, bumping their way between the atoms, billions of them passing along per second. The electric and magnetic fields of an electrified wire can be shown to be the result of the combined fields of the moving electrons composing the current.

In the early part of the nineteenth century Faraday showed the relations between electric and magnetic phenomena and formulated the laws of electromagnetic induction. He showed that a change in an electric current gave rise to a magnetic field and that a changing magnetic field was capable of producing an electric current in a conductor. Every electric generator, motor, and transformer in the world is based upon the application of Faraday's laws.

Another important area of physical investigation during the nineteenth century was concerned with the nature of light. The ancient Greeks had imagined that light consisted of particles, or corpuscles. Newton favored this same theory in the latter part of the seventeenth century and his view prevailed until the start of the nineteenth century. It is interesting to note, however, that Robert Hooke, a contemporary of Newton, proposed the idea that light consisted of waves. This was developed by the Dutch physicist Christiaan Huygens about 1680. New evidence for the wave theory of light was advanced at the start of the nineteenth century by Thomas Young in England and Augustin Jean Fresnel in France. It was necessary, however, to explain how light waves traveled through empty space, and to meet this necessity nineteenth-century physicists imagined space to be filled with a medium which they called the ether. This notion of an ether was eventually discarded.

In 1873 James Clerk Maxwell, the distinguished British mathematician, startled the world of science with the publication of his *Treatise on Electricity and Magnetism*. In it he advanced the revolutionary electromagnetic theory of light. Maxwell had made a profound mathematical analysis of Faraday's laws of electromagnetic induction. His analysis showed that periodic changes in an electric field should result in electric waves, which would spread out in space like the ripples on a pond when a stone is dropped into it. But since changes in an electric field create a magnetic field, this electric wave would be accompanied by a magnetic wave at right angles to it. The combination constitutes an electromagnetic wave. Maxwell found that the speed of such a wave would be identical with the known speed of light, 186,000 miles per second. The conclusion that light consisted of such electromagnetic waves was inescapable.

This conclusion was received with violent antagonism in many quarters. One of the difficulties was that no one at the time was familiar with any form of electromagnetic wave. But in 1887 Heinrich Hertz turned the tide in Maxwell's favor when he succeeded in producing such waves with the aid of an electric spark. These waves were named Hertzian waves in his honor. We still use them today but know them by the more familiar name radio waves.

In 1895 the great Dutch physicist Hendrik Lorentz extended Maxwell's mathematical work to explain the emission and absorption of light by matter. This was the famous Lorentz electron theory, which proposed that the vibrations of electrons in the atoms of matter were responsible for the creation of electromagnetic waves. This theory was spectacularly confirmed in 1896 when another Dutch physicist, Pieter Zeeman, showed that certain spectrum lines were split into component lines if the substance emitting them was placed in a strong magnetic field.

The known types of electromagnetic radiations in space are divided at present into six categories. Ranging in order from the longest to the shortest waves, these are the Hertzian, or radio waves, the heat, or infrared rays, the waves of visible light, ultraviolet light, X-rays, and gamma rays. The range of wavelengths is so tremendous as to be almost unbelievable. The longest radio waves are several miles in length. The shortest gamma rays are about 20,000 trillion times shorter.

The longest waves of visible light are those of red light with a wavelength a little less than 1/100,000 of a centimeter. The shortest are those of violet light with a wavelength equal to about half of that.

twenty-one

THE QUANTUM THEORY

> *The greatest and noblest pleasure which we can have in this world is to discover new truths; and the next is to shake off old prejudices.*
>
> —Frederick the Great

THE structure of twentieth-century physics rests upon two revolutionary theories at complete variance with the ideas of "classical physics" which prevailed at the close of the nineteenth century. They are the theory of relativity and the quantum theory. As erratic as lightning in its behavior, the public chose to center its attention on relativity and to ignore quantum theory. But it would be difficult to say which has played the greater role in the establishment of the "new physics." And, it is interesting to note, Albert Einstein, who formulated the theory of relativity, played a major role in the development of the quantum theory.

Spectroscopic studies in the latter half of the nineteenth century revealed that an incandescent solid body furnished a continuous spectrum, or continuous rainbow, of colors. This meant that it was radiating light of all wavelengths. However, the distribution of energy was not uniform throughout the spectrum. As the body grew hotter, more and more of the energy shifted to the shorter wavelengths, that is, from the red end of the spectrum to the violet end. This explained why the color of an object, a bar of iron for example, changed from red to orange, then to yellow, and finally to white as it grew hotter and hotter.

In the last decade of the century the British physicist and astronomer Sir James Jeans attempted to explain this shift in wavelength with rise in temperature and came to a startling conclusion. On the basis of the classical laws, the light at any temperature ought to shift to shorter and shorter wavelengths almost instantaneously. This meant that if you lit a log in your fireplace, its radiations ought to change in an instant from welcome infrared rays and a cheerful red glow to violet light, ultraviolet light, X-rays, and finally lethal gamma rays. Obviously this does not happen, and so his study became known as Jeans' paradox. It was also called the ultraviolet catastrophe. However, it was clear to physicists that there was something wrong with the classical laws of radiation.

An explanation was put forward by the German physicist Max Planck at the meeting of the German Physical Society in Berlin during Christmas week in 1899. He showed that the distribution of radiant energy with temperature could be explained by giving up the idea of the continuous emission of energy from a hot body. In its place, he suggested that energy was emitted in the form of little bundles or packets, which he called energy elements, but which subsequently became known as quanta. It is for this reason that the theory became known as the quantum theory.

The size of the quantum differed for every wavelength of energy but in every case was equal to the frequency of the wave multiplied by a fundamental quantity. This quantity is now known as Planck's constant and is one of the most basic quantities in the universe. It is extremely small and is equal to 66/10,000 of a trillionth of a trillionth (6.6×10^{-27}) of an erg second. In classical mechanics the erg second is known as the unit of action. It is equal to an erg, the unit of work (see Chapter 20), multiplied by a second. Planck's constant is denoted in equations by the letter h. The shorter the length of a wave, the higher the frequency. It will be seen, therefore, that the quantum of energy grows larger and larger as the wavelength decreases. A quantum of ultraviolet light contains much more energy than a quantum of red light. A quantum of X-rays is even larger.

Most of Planck's contemporaries were unwilling to accept his theory when he announced it. He realized how revolutionary it was and sought to preserve as much as possible of the classical picture by postulating that while energy was absorbed and emitted as quanta, it existed as continuous waves in the ether of space.

It remained for Einstein to make the complete break with the classical theory. This he did in 1905, the same year in which he upset classical notions of space and time with his theory of relativity. Einstein employed Planck's theory of quanta to explain the photoelectric effect and in so doing boldly suggested that physicists abandon completely the notion of continuous waves. He suggested that energy always existed in the form of discrete packets or bullets, which he named quanta. Thus in Einstein's view energy, like matter, became discontinuous, the quantum, as he had named it, serving as the particle of energy. His epochal paper on the photoelectric effect was as important as his theory of relativity. It is interesting to note that Einstein received the Nobel Prize in 1921 for his contributions to quantum theory and not for the theory of relativity.

It had been known for some years prior to 1905 that when a beam of light, and particularly a beam of ultraviolet rays or X-rays, fell upon a surface of metal, electrons were knocked out of the surface. This phenomenon was called the photoelectric effect. The reader will recognize it as the basis of the "electric eye," or photoelectric tube, in common use today. Many disturbing details about the photoelectric effect were observed in the first few years of the present century. It was noted that the number of electrons hurled out of a metallic surface depended upon the intensity or strength of the beam. However, the speed with which the electrons were ejected was independent of the intensity of the beam and depended only upon the wavelength. Everyone had supposed that a strong beam would eject the electrons with greater speed than would a feeble beam.

Einstein's photoelectric theory offered a complete explanation of the observed facts. According to this view, each electron was ejected by the impact of a single quantum. Obviously, the greater the energy of the quantum, the greater would be the speed of the ejected electron.

But this brought the world of physics back to Newton's idea

that light consisted of particles, or corpuscles. Today they are more usually called photons, from "*photos*," the Greek word for "light."

THE BOHR THEORY OF THE ATOM

The next significant advance in quantum theory came in 1913, when Niels Bohr applied the theory to the structure of the atom. In 1913 Bohr was carrying on research in Rutherford's laboratory at the University of Manchester.

It will be recalled that each chemical element when rendered incandescent in the gaseous state emits a characteristic spectrum of bright lines, each line corresponding to light of a definite wavelength. As early as 1885 a Swiss schoolmaster, Johann J. Balmer, showed that the wavelengths of the lines of the hydrogen spectrum formed a sequence which could be expressed by a simple mathematical formula. As a result these lines of the hydrogen spectrum became known as the Balmer series. Similar formulas were developed subsequently when advances in spectroscopy made it possible to chart the hydrogen spectrum in both the infrared and the ultraviolet regions.

In 1908 W. Ritz developed the principle of combination. Instead of using the wavelength, or frequency, of the spectrum lines, he employed the number of waves per centimeter. This is known as the wave number. He showed that it was possible to write a series of terms so that any given spectrum line represented the difference between two terms.

This was expanded by Johannes Rydberg, who showed that the terms of the hydrogen spectrum could be obtained by starting with a given wave number and dividing it by 4, 9, 16, 25, etc., in other words, by the squares of 2, 3, 4, 5, etc. This fundamental wave number, now known as Rydberg's constant, is about 109,700 waves per centimeter.

Rutherford, it will be recalled, had established the fact that the atom consisted of a central nucleus surrounded by a cloud of electrons. It was necessary to assume that the electrons were circling the nucleus as the planets circled the sun. Otherwise electrical attraction would cause the electrons to fall into the nucleus. But

Maxwell's theory required a rotating electron to emit energy. This would cause the electron to slow up so that eventually it would fall into the nucleus. Moreover, such an electron ought to emit light of a constantly changing wavelength as it slowed up. It was impossible, therefore, to reconcile Rutherford's model of the atom with the known facts of spectroscopy.

It was at this point that Bohr undertook to apply quantum theory to the Rutherford model of the atom. Bohr showed that the dilemma could be resolved by making certain arbitrary assumptions contrary to the laws of classical physics. The first was that the electron could revolve around the nucleus only in certain orbits. Each of these possible orbits corresponded to one of the terms in the Ritz principle of combination. The second assumption was that the electron did not radiate energy while revolving in one of these orbits. The third assumption was that when an atom absorbed a quantum of energy, it caused the electron to jump from a smaller to a larger permitted orbit. The fourth assumption was that when the electron fell back from a larger to a smaller orbit, it emitted a quantum of light.

Bohr in 1913 undertook to develop the mathematics of only the simplest case, namely, that of the hydrogen atom. It has a nucleus consisting of a single proton while one electron circles around it. The problem is more difficult for atoms involving more than one electron. Bohr retained the idea of shells as postulated in the Langmuir-Lewis model of the atom (see Chapter 18).

At first Bohr suggested that all electrons revolved in circular orbits, but the Austrian physicist Arnold Sommerfield showed that the facts of spectroscopy necessitated the assumption that some of the electrons moved in circles and some in ellipses. This is shown in Figure 31.

If we concentrate our attention upon the emitted energy, we may think of each permitted orbit of the electron as an energy level. The energy of the emitted photon will be the difference then between the original and the subsequent energy level of the electron.

In his original theory Bohr showed that the energy levels were related by a mathematical formula involving the squares of the integers 1, 2, 3, 4, etc. Consequently, these integers were used to

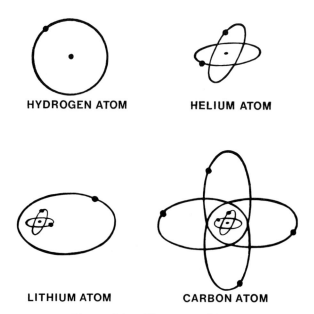

Figure 31. Electron orbits

designate the energy levels and were referred to as the quantum numbers. A second series of quantum numbers were necessary to designate the elliptical orbits. Subsequently, however, it was necessary to introduce two more quantum numbers, one for the orientation of the electron orbit in space and the fourth for the rotation of the electron on its axis, the so-called spin of the electron. Today, therefore, each Bohr electron is designated by four quantum numbers.

In 1925 the Viennese-born physicist Wolfgang Pauli formulated an important law known as the exclusion principle. This states that it is impossible for more than one electron in an atom to have any given set of four quantum numbers. The mathematical analysis of this principle explains why each shell in the atom is restricted to a certain number of electrons.

The Bohr theory of the atom explained many phenomena and brought order into a field that had been greatly confused. It is important to realize that it is essentially a model of the atom. Undoubtedly many laymen obtained the impression that the Bohr model was an established fact. But scientists, including Bohr himself, never regarded it that way. They always felt that it was an extremely ingenious model which explained many of the observed facts but that it would have to be modified with the passage of time. The necessity for such modification grew out of the conflict between the wave theory and the corpuscular theory of light.

WAVES VERSUS PARTICLES

The American physicist Robert A. Millikan, who measured the charge on the electron with great exactness in 1911, used the photoelectric effect to determine the value of Planck's constant in 1916. He employed a photoelectric tube for this purpose. Electrons knocked out of a metallic target inside the tube by a beam of light of known frequency struck a metallic plate. Their motion could be stopped by putting a negative electric voltage on this plate. The speed of the electrons could be determined from the magnitude of this voltage. The energy of the light beam could be calculated from the most energetic electrons. Since the frequency of the light beam was known, dividing the energy by the frequency gave Planck's constant. The fact that this determination agreed closely with Planck's original calculations did much to establish the theory, and Millikan received the Nobel Prize in 1923 for his historic accomplishments of 1911 and 1916.

A decade later two brilliant experiments by the American physicist Arthur H. Compton and the British physicist Charles T. R. Wilson gave added confirmation to the quantum theory. The Nobel Prize in physics was shared by these two scientists in 1927.

Compton showed that when a beam of X-rays was scattered from a block of carbon, the scattered beam possessed a longer wavelength than the incident beam. This meant a lower frequency and therefore a loss of energy. Such a loss cannot be accounted for on the basis of the wave theory of light. It can, however, be explained easily by the quantum theory as the result of the impact of individual photons on electrons, the electrons gaining the energy which the photons lose.

The Compton effect is shown in Figure 32. The photon, striking an electron loosely bound in an atom of the target, loses energy to it. The photon is deflected in a new direction while the electron recoils from the impact.

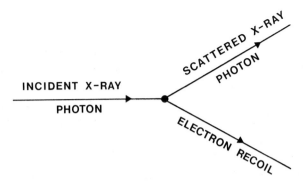

Figure 32. The Compton Effect

Wilson showed that when very feeble beams of X-rays were passed through the cloud chamber apparatus (see Chapter 22), tracks of electrons appeared. These electrons could be explained only on the basis that they had been knocked out of atoms by the impact of X-ray photons.

However, the victory of the quantum theory was by no means complete. For one thing, it must be remembered that in every case the energy of the photon was found to be the product of Planck's constant times the frequency. The frequency cannot be determined directly. The frequency, which is equal to the speed of light divided by the wavelength, must be determined by measuring the wavelength of the light. Here was an obvious contradiction between the wave and corpuscular aspects of X-rays which could not be resolved.

But even more baffling is the fact that the experiments which led to the formulation of the wave theory of light still defy explanation on any other basis. Chief among them is the phenomenon of interference. If light is admitted into a darkened room through two pinholes in a screen and allowed to fall on a second screen, a pattern appears consisting of light and dark bands. This pattern can be explained as the result of the interference of the two trains of light waves with each other. Where the waves meet crest to crest or trough to trough, there is a band of light. Where the waves meet crest to trough, they neutralize each other and there is a band of darkness.

Inability of physicists to explain wave phenomena by quantum theory or quantum phenomena by wave theory led Sir William

Bragg to remark that they were using the wave theory on Mondays, Wednesdays, and Fridays, and the quantum theory on Tuesdays, Thursdays, and Saturdays.

Today physicists have stopped trying to establish one theory at the expense of the other. Instead, an attempt, now thoroughly established as quantum electrodynamics, has been made to reconcile the two views of the nature of light. An analogy will make plain one attempt at reconciling the two views. Sometimes when it is raining, the wind drives the rain in successive sheets so that there is a wave motion running through the rain although the rain is composed of drops. It is possible, therefore, to think of light as composed of photons moving in waves. The wave, therefore, determines the distribution of the photons, and the intensity of the wave at any point is a measure of the number of photons present. More exactly, the intensity of the wave determines the probability of finding a photon at any given point. For this reason the waves may be regarded as waves of probability. But this view breaks down when it is discovered that interference phenomena occur in cases where only a single photon is present.

WAVE MECHANICS

In 1923 Louis de Broglie, the French physicist, while contemplating the dual aspect of the nature of light, asked himself whether it was possible that many of the difficulties concerning atomic structure arose from a similar duality of the nature of material particles. By this time physicists had long recognized that the mass of a particle increases with its speed in accordance with Einstein's theory of relativity. Einstein's equation states that the energy of a particle is equal to its mass multiplied by the square of the velocity of light. De Broglie now combined this with Planck's equation that the energy of a photon is equal to its frequency multiplied by Planck's constant. By assuming that this applied equally to the electron, de Broglie came to the conclusion that every electron was accompanied by a wave whose wavelength was equal to Planck's constant divided by the momentum of the particle.

One might have expected startling results from a mixture of

the quantum theory and relativity. But de Broglie's concept proved immediately useful as a device to explain many things. The wave became known as a de Broglie wave or pilot wave. Electron and wave can be compared to a boat followed by ripples. The picture, however, is too simple. For the pilot wave precedes as well as follows the electron and it is not possible to say exactly where the electron is within the pilot wave.

Bohr had arbitrarily chosen electron orbits to match the terms of the spectrum. He had no explanation of why the electron could revolve in these orbits and no others. This was now explained by the pilot wave. Only those orbits were possible that were multiples of the wavelength of the pilot wave since these were the only orbits into which it would fit.

An even more revolutionary idea than that of de Broglie was advanced in 1925 by the Austrian physicist Erwin Schrödinger. He proposed to go de Broglie one better by eliminating the concept of electron orbits. His proposal was one of three put forward in that same year, all of them starting with this same intention. The second theory was that of the British physicist P. A. M. Dirac. The third was the work of a brilliant group of physicists at the University of Göttingen in Germany, Max Born, Werner Heisenberg, and P. Jordan. Out of these theories has come the new branch of physics known as wave mechanics or quantum mechanics.

De Broglie's theory had received little attention until Einstein gave his support to it in 1925. Attracted by Einstein's praise, Schrödinger turned to it and within a few months had evolved his own theory of wave mechanics which went far beyond it. Bohr's idea of electron orbits was ingenious, but no one had ever seen them. Consequently, Schrödinger proceeded to eliminate them from his theory. He pictured the atom as a fluctuating field, a pulsating sphere of something which he designated by the Greek letter *psi*, a sort of vibrating electrical essence. On the basis of his theory, different atoms possess different types and rates of vibration or pulsation. The emission of a photon, explained in the Bohr model by the sudden jump of an electron from one orbit to another, is explained by Schrödinger as a change from one type of pulsation to another.

From a mathematical point of view, the Schrödinger theory

is more satisfactory than the Bohr theory as an explanation of atomic spectra. But it requires that we picture the electron as spread out into a diffuse sphere of electricity inside the atom. Most physicists have found this unsatisfactory.

A way out of the difficulty was offered by Max Born, who proposed that the de Broglie waves and Schrödinger's pulsations, or psi waves, be regarded as waves of probability. A wave indicated the probability of finding the electron at a given point in the atom at a given time.

In 1925 Born was a professor at Göttingen. Werner Heisenberg, in his early twenties, was a junior member of the faculty. Heisenberg undertook a purely mathematical analysis of the data of atomic spectra, leaving aside all attempts to form any mechanical picture of what was happening. Born and Jordan entered the project, evolving a new form of mathematics known as matrix mechanics. Heisenberg's equations led to a situation similar to Schrödinger's. It was at this point that Born proposed that the wave be regarded as a wave of probability. The wave intensity, therefore, expressed the probability of finding the electron at any given point in the atom at a given time.

Although starting from a different line of mathematical reasoning, Dirac's theory leads to essentially the same conclusion.

But the electron waves first suggested by de Broglie are not so easily dismissed. Although they do not appear to be electromagnetic waves, they nevertheless give evidence of physical reality. The German physicist W. Elsasser suggested that if the de Broglie waves really existed, it ought to be possible to form diffraction patterns with electrons just as is done with X-rays by deflecting them from the surface of a crystal. In 1927 two American scientists, C. J. Davisson and L. H. Germer, succeeded in doing just that. In the same year G. P. Thomson, son of the famed Sir Joseph Thomson, obtained similar patterns by passing a beam of electrons through a thin sheet of metal. Other experiments disclosed the wave nature of electrons leading finally to the development of the electron microscope.

We are faced, therefore, with the same dilemma in the case of the electron as the photon. Like the photon, the electron possesses both wave and particle characteristics.

A reason for our inability to resolve this dilemma was advanced by Heisenberg in 1927 with his uncertainty principle, or principle of indeterminancy. According to this principle, we can never measure both the position and the velocity of an electron with exactness. This is because we must use photons to make our observations of electrons. But the impact of a photon upon an electron disturbs its state and renders the observation meaningless. Similarly, it is impossible to make an observation on a photon without altering its state. According to Heisenberg, what we achieve in exactness in measuring the position of an electron is at the expense of our knowledge of its momentum, and vice versa.

Bohr extended this principle to other pairs of measurements. He called these paired quantities conjugate quantities and the relationship between the two complementarity.

Strangely enough, Planck's constant enters the picture at this point. The product of the uncertainty in the case of two conjugate quantities is never less than Planck's constant. It is "the villain in the piece." Or perhaps we should call it the hero.

With the development of quantum mechanics physicists have abandoned the old view of a mechanistic universe which came to full flower at the close of the nineteenth century. Instead of a strict operation of the law of cause and effect, they speak of the probability of a given event. Bohr became the leading exponent of this point of view, but Einstein opposed it.

twenty-two

THE STRUCTURE
OF THE NUCLEUS

*In Nature's infinite book of secrecy
A little I can read.*
— SHAKESPEARE

SPECTRUM analysis, the study of the fascinating rainbow of bright lines revealed by the spectroscope when a chemical element is turned into an incandescent gas, proved to be the key to the behavior of the cloud of electrons around the atomic nucleus. But it yielded little information about the nucleus itself beyond the strength of the positive charge on it. A more powerful method was needed to explore the nucleus. This was provided by Rutherford, who, you will recall from Chapter 18, had established the existence of the nucleus in 1911. He sensed that it might be possible to explore the nucleus by bombarding it with the alpha particles of radium or with swiftly moving electrons. He pioneered in this attempt and in 1919 reported that when nitrogen nuclei were bombarded with alpha particles, hydrogen nuclei or protons came flying out.

The apparatus used by Rutherford was extremely simple. It consisted of a little metal tube into which the nitrogen was introduced. At its center was a tiny amount of radium salt. At one end

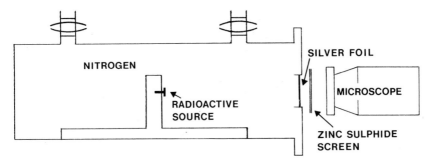

Figure 33. Rutherford's atom-smashing experiment

of the tube there was a thin metal wall. Beyond the wall Rutherford placed a fluorescent screen which could be observed through a microscope. The apparatus is shown in Figure 33. The length of the tube was such that alpha particles would not reach the screen, but high-speed particles could do so, revealing their arrival by causing bright sparks easily visible in the microscope. The scintillations thus observed indicated that high-speed particles were being knocked out of the nitrogen nucleus. Subsequent researches by Rutherford and James Chadwick established that these particles were indeed protons or hydrogen nuclei. Since H is the classical symbol for hydrogen, a scientific wag described the exploit by saying Rutherford "had knocked the H out of nitrogen."

An important question was whether the alpha particle had merely chipped a proton out of the nucleus as a bullet might chip a flake from a rock, or whether some sort of reaction had taken place in which the alpha particle entered the nucleus with the subsequent ejection of a proton. It was eventually established that the latter was the case. The nitrogen nucleus, by absorbing an alpha particle and ejecting a proton, was transformed into an oxygen nucleus. Rutherford had succeeded in realizing the old dream of the alchemists by changing one chemical element into another.

During the 1920s Rutherford and Chadwick found that alpha particles would knock protons out of the nuclei of most of the lighter atoms. They also observed that from certain nuclei the ejected protons possessed more energy than the bombarding alpha particles. It became evident, therefore, that energy was released in such nuclear transformations. This gave rise to the feeling that the atomic nucleus was a storehouse of energy which somehow could be tapped for the good of mankind. The difficulty was that only a few alpha particles in a million made direct hits on nuclei and thus released energy.

It was subsequently predicted by George Gamow that atom-smashing experiments could be carried on with greater success if streams of protons accelerated by strong electric fields were used as the bombarding projectiles. Various methods of obtaining the high voltages necessary for this purpose were devised. The first devices used electric transformers and condensers. These attained voltages in the neighborhood of 1 million volts.

The most important development was the invention of the cyclotron by Ernest O. Lawrence of the University of California in 1929. The heart of the machine is a cylindrical box placed between the poles of a powerful electromagnet. A stream of protons introduced into the box is accelerated by a high high-frequency electric field while the magnetic field bends its path into a circle. As the protons gain speed they move in an ever widening spiral until they emerge from the box ready for atom-smashing (see Figure 34).

The first cyclotron built by Lawrence was a small affair, almost pocket size. Its success led to larger machines, requiring more powerful electromagnets. Improved cyclotrons built after World War II employed magnets weighing more than 2,500 tons. These machines furnished streams of protons with energies of more than 400 Mev (million electron volts).

The most powerful atom-smashers now in use are adaptations of the cyclotron known as proton synchrotrons. One in Geneva, Switzerland, at the laboratory of the European Center for Nuclear Research, a cooperative venture of 14 European nations, accelerates protons to 28 Bev (billion electron volts). Energies of 33 Bev are attained by one at the Brookhaven National Laboratory on Long Island. These machines accelerate the protons in a big ring of oval pipe, nicknamed the "racetrack," more than 800 feet in diameter.

The U.S. Atomic Energy Commission has built a still more powerful proton synchrotron at Weston, Illinois, near Chicago. It will accelerate protons to energies of 200 Bev.

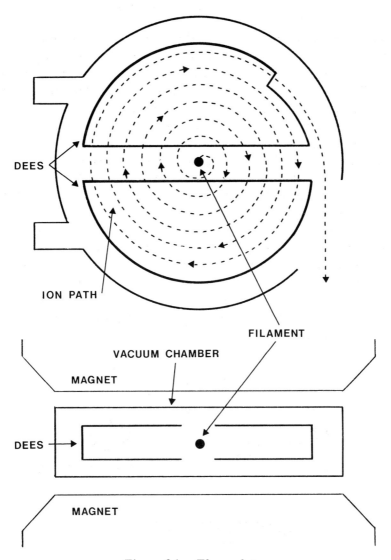

Figure 34. The cyclotron

RADIATION DETECTORS

The interpretation of atom-smashing experiments requires the employment of instruments capable of detecting and measuring the particles and radiations released in them. Sophisticated extensions of Rutherford's method of observing the scintillations caused by particles striking a phosphorescent screen are still in use.

An extremely useful device is the Geiger counter, first invented by Hans Geiger in Rutherford's laboratory in 1908 but subsequently much improved. It is a small gas-filled tube with a wire through the center. The passage of a subatomic particle through it triggers the release of an electrical impulse which activates a recording device or counter.

Of the utmost importance is the Wilson cloud chamber, invented by the Scottish physicist Charles T. R. Wilson in 1911. When a particle bearing an electric charge moves through a gas, it knocks electrons out of the atoms of gas, forming an ionized trail. The device is a glass cylinder filled with moist air. When a piston at the bottom of the cylinder is dropped, expansion of the air causes the air to cool and the water vapor to condense into

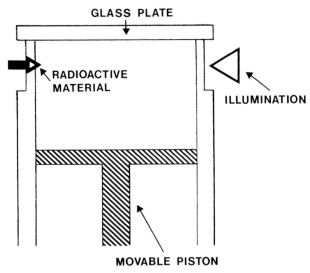

Figure 35. The Wilson cloud chamber

A 1934 photograph of the 27-inch cyclotron at the University of California. At the left of the instrument is Ernest O. Lawrence, inventor of the cyclotron.

droplets. These form along the ionized trails of subatomic particles which have entered the cylinder. A beam of light makes these trails visible, revealing the paths of the particles (see Figure 35).

A superior device along similar lines is the bubble chamber invented by Donald A. Glaser at the University of Michigan in 1953. In this device subatomic particles cause bubbles in a liquid, usually liquid hydrogen, to form along the ionized trail of a subatomic particle. Bubble chambers of large size have been built. One at the University of California is about the size of a bathtub and holds 150 gallons of liquid hydrogen.

THE COCKCROFT-WALTON EXPERIMENT

One of the deductions which Albert Einstein drew from his theory of relativity in 1905 was that matter could be converted into energy and vice versa. He expressed this in the equation $E=mc^2$, where E is the energy, m the mass, and c the velocity of light. It is undoubtedly the world's most famous equation.

If Einstein's equation is right, the energy of particles ejected in nuclear transformations must represent a loss of mass by the nuclei involved. That this is indeed the case was verified in 1932 by the Cockcroft-Walton experiment, 27 years after Einstein had written his famous equation.

J. D. Cockcroft and E. T. S. Walton, working in Rutherford's laboratory at Cambridge, used a stream of protons to bombard a target composed of the element lithium. It was found that a lithium nucleus absorbed a proton and subsequently split into two helium nuclei or alpha particles. The mass of two alpha particles is less than the combined mass of a lithium nucleus and a proton. According to the Einstein equation, this loss of mass should be transformed into energy and exhibited by the energy with which the alpha particles were ejected from the target. Cockcroft and Walton measured the energy of the alpha particles and found that it verified the Einstein equation.

THE DISCOVERY OF THE NEUTRON

With uncanny insight into atomic problems three physicists in as many countries predicted in 1920 the existence of a subatomic particle with the mass of the proton but without an electric charge. The three were Lord Rutherford, William D. Harkins of the University of Chicago, and Orme Masson in Australia. The name neutron was proposed by Harkins for this then hypothetical particle. An unsuccessful search for the neutron was conducted at the Cavendish Laboratory of Cambridge University by Rutherford and Chadwick.

A decade later scientists got on the trail of the neutron without at first realizing what was happening. In 1930 two German physicists, Walter Bothe and H. Becker, noted that when beryllium and certain other light elements were bombarded with alpha particles, they gave off a highly penetrating radiation. It was thought

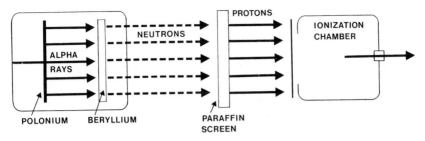

Figure 36. The Curie-Joliot experiment

that this might consist of gamma rays of very great energy. In 1932 these experiments were repeated in Paris by Irène Joliot-Curie, the daughter of Pierre and Marie Curie, the discoverers of radium, and her husband, Frédéric Joliot. It is an interesting coincidence that just as the Curies worked together on scientific research, so did their daughter and her husband. The Joliots set up an arrangement in which a thin sheet of paraffin was placed beyond the sheet of beryllium. The arrangement is shown in Figure 36.

They found that the penetrating radiations from the beryllium caused protons to be ejected from the paraffin at high speeds. It was impossible to reconcile the energy of these protons with the notion that the radiation consisted of gamma rays.

The correct interpretation of this experiment was made that same year by Chadwick with the aid of the Wilson cloud chamber. He showed that the ejection of the protons could be explained by assuming that the radiations consisted of the long-sought neutrons. This has been amply confirmed by many experiments since then.

The neutron was found to be slightly heavier than the proton. Physicists designate the mass of atoms and subatomic particles on a scale which places the oxygen atom at 16. On this scale the proton has a mass of 1.00759, the neutron 1.00897, and the electron 0.000548.

NUCLEAR STRUCTURE

As soon as Chadwick's discovery of the neutron was announced in 1932 Heisenberg suggested a scheme for the composition of the nucleus which has been amply verified by subsequent studies. According to it, two particles occur in the nucleus, namely, the proton and the neutron. These are now referred to as nucleons.

The simplest atom of all is ordinary hydrogen. Its nucleus consists of a single proton. All other nuclei are combinations of protons and neutrons. Since the proton has a positive electric charge, the number of protons in the nucleus determines its electric charge. This is known as the atomic number. It determines the place of the atom in the periodic table of the elements and also determines the number of electrons in the cloud around the nucleus.

The total of both protons and neutrons in the nucleus determines the mass of the nucleus. Physicists had long been aware of the fact that the atomic weight of atoms in general is roughly double the positive charge on the nucleus. The term mass number is used to express the atomic weight rounded off to the nearest whole number. It will be seen, therefore, that the mass number is the total of protons and neutrons in the nucleus. If the atomic number is subtracted from the mass number, the result is the number of neutrons in the nucleus.

Ordinary hydrogen has a nucleus consisting of a single proton. Therefore, it has a mass number of 1 and an atomic number of 1. There are two other isotopes of hydrogen, double-weight hydrogen, or deuterium, and triple-weight hydrogen, or tritium. Deuterium has a mass number of 2, and an atomic number of 1. Its nucleus, therefore, consists of one proton and one neutron. Tritium has a mass number of 3 and an atomic number of 1. Its nucleus consists of one proton and two neutrons. Mass numbers are commonly used to identify isotopes. Thus we speak of Hydrogen 1, Hydrogen 2, and Hydrogen 3.

We come next to helium, which has an atomic number of 2. There are two isotopes of helium with mass numbers of 3 and 4. It will be seen at once that Helium 3 has a nucleus consisting of two protons and one neutron, while Helium 4 has a nucleus of two protons and two neutrons. This latter nucleus is the alpha particle.

The third element in the periodic table is lithium. As a consequence its atomic number is 3, indicating the presence of three protons in the nucleus. There are three isotopes of lithium, their mass numbers being 6, 7, and 8. It is obvious, therefore, that they

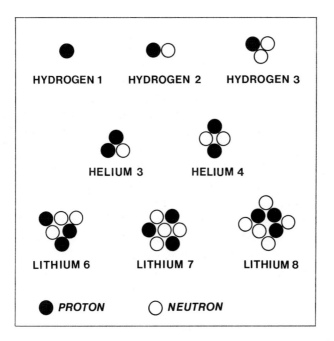

Figure 37. Composition of hydrogen, helium, and lithium nuclei

contain, respectively, three, four, and five neutrons. These and the foregoing nuclei are shown in Figure 37.

And so it goes for all the chemical elements. In every case the atomic number tells the number of protons. The mass number reveals the total of protons and neutrons. Simple arithmetic discloses the number of neutrons.

Let us take element No. 92, uranium, the heaviest element occurring in nature. It has 92 protons in its nucleus. Four isotopes occur in nature, namely, Uranium 234, 235, 237, and 238. They contain, respectively, 142, 143, 145, and 146 neutrons.

When nuclear transformations take place, there must be a complete balance of atomic numbers and mass numbers.

THE BINDING FORCE

A moment's thought will make it obvious that electrical forces cannot hold the nucleus together. The electrical force between two protons is one of repulsion, since they are both positively charged. The neutron, being electrically neutral, is unaffected by electrical forces. Consequently, if only electric forces operated in the nucleus, the neutrons would fall out of it while the protons would fly apart violently, completely disrupting it.

Atom-smashing experiments have made it clear that another type of force operates within the nucleus. At the infinitely small distances within the nucleus this binding force is millions of times more powerful than the electrical forces. But this is true only within these extremely small distances. At greater distances the electrical forces are more powerful.

At first glance it would appear that the atomic weight of the isotope of any element ought to equal the combined weight of the protons and neutrons in its nucleus plus the weight of the electrons surrounding the nucleus. This, however, is not the case. The atomic weight is found to be less. This difference is known as the mass defect.

It is evident from the Einstein equation $E=mc^2$ that if the nucleus of a given isotope was created by bringing together the requisite number of protons and neutrons, the ensuing loss in mass would be released in the form of energy. It would take this same amount of energy to break up the nucleus into its constituent protons and neutrons. Consequently, this amount of energy corresponds to the binding force which holds the nucleus together.

Experiments have shown that the binding force varies from one element to another. It rises rapidly with increase in atomic number to about element No. 40 in the periodic table. It remains at an approximate flat maximum for elements from No. 40 to No. 120 with a value of about 8.4 Mev per nucleon. Then it falls off slowly for heavier elements to a final value of 7.5 Mev for uranium.

Atomic physicists recognized more than four decades ago that there were two ways by which atomic energy might eventually be released for the use of mankind. One was the fusion of light elements into heavier ones. As was related in Chapters 4 and 5, this process is believed to account for the energy of the sun and stars. It is also the source of energy of the hydrogen bomb. The other process is the disruption of heavier atoms into lighter ones. This occurs in the uranium bomb.

Since the nucleus contains both protons and neutrons, it is

An electron microscope

apparent that there are three types of attractions at work in the nucleus. These may be called proton-neutron, proton-proton, and neutron-neutron forces of attraction. The latter two forces appear to be approximately equal when the electrical repulsion between protons is subtracted, while the proton-neutron force is slightly greater. This is borne out by the fact that almost all of the stable isotopes of the elements up to No. 20 in the periodic table have nuclei consisting of equal numbers of protons and neutrons. In general, isotopes with unequal numbers are radioactive, emitting radiations and transforming themselves into stable atoms.

When the nucleus contains more than 20 particles, the atom is not stable unless there is an excess of neutrons over protons. This increases as the atomic table is ascended. Thus Uranium 238 contains 92 protons and 146 neutrons. This balances the greater electrical proton repulsion by separating protons by neutrons.

Unusually stable nuclei are those of Helium 4, Carbon 12, and Oxygen 16. The latter two, it will be noted, represent multiples of 4. In this connection it is interesting to note that six isotopes

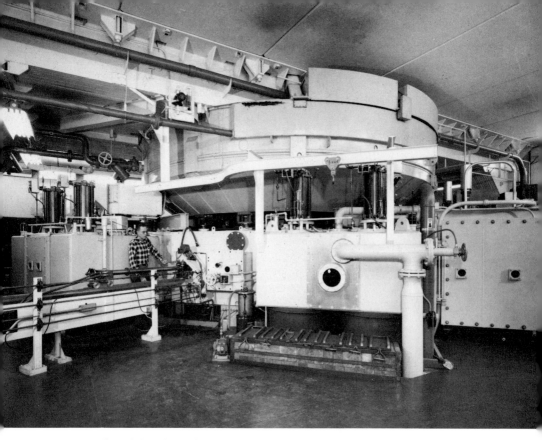

The 184-inch cyclotron which produced the first manmade mesons

constitute about 80 per cent of the earth's crust. They are Oxygen 16, Magnesium 24, Silicon 28, Calcium 40, Titanium 48, and Iron 56, all divisible by 4.

An attempt to explain these facts has been made by assuming that the nuclear particles are arranged in shells, just as are the electrons in the cloud around the nucleus, and that the principles of quantum mechanics and Pauli's exclusion principle (see Chapter 21) apply to them also.

Just as electrons possess "spin," so protons and neutrons are spinning, or rotating, on their axes. The protons, therefore, can occupy the same shell, or energy level, only if they are spinning in opposite directions. The same thing applies to neutrons. It will be seen, therefore, that a closed, or filled, shell is created by four particles, namely, two protons spinning in opposite directions, and two neutrons doing likewise. Such a "foursome" is the alpha particle, or Helium 4. It is possible to think of Carbon 12 as an aggregation of three such closed shells and Oxygen 16 as four of them.

It is interesting to speculate upon why atoms heavier than No. 20 in the periodic table require more neutrons than protons for stability. The binding force between protons is a short-range force whereas the electric force between them is a long-range force. As the number of protons in the nucleus increases, the electric forces, which are forces of repulsion, become important and render the nucleus unstable unless they are counterbalanced by proton-neutron forces. In addition, the neutrons increase the distance between protons and thus reduce the electrical repulsion between protons.

Next to the closed shell of two protons and two electrons, the most stable configuration within the nucleus is that of two neutrons with opposite spins. It follows, therefore, that nuclei containing even numbers of protons and even numbers of neutrons will be the most stable.

However, there are stable nuclei containing an even number of protons and an odd number of neutrons, or vice versa. But every nucleus containing an odd number of each, with a few exceptions, are radioactive. These exceptions are Hydrogen 2, Lithium 6, Boron 10, and Nitrogen 14.

Certain heavier nuclei are more stable than others. These are nuclei containing 20, 50, or 82 protons, or 20, 50, 82, or 126 neutrons. These numbers are sometimes called "magic numbers."

NUCLEAR TRANSFORMATIONS

The nuclear transformations which occur in the naturally radioactive elements or in atom-smashing experiments are easily understood if we remember that the addition or subtraction of a proton changes both the atomic number and the atomic mass, while the addition or subtraction of a neutron changes only the atomic mass.

Students of nuclear physics use a system to express reactions copied from that long used by chemists. The same symbols are used for the chemical elements, but a subscript is added to signify the atomic number and a superscript to express the mass number.

It was found by the Joliot-Curies in 1933 that many atoms formed in atom-smashing experiments were not stable but behaved like the natural radioactive elements, ejecting particles after a lapse of time and transforming themselves into other elements. This phenomenon became known as artificial radioactivity.

It will be recalled from Chapter 18 that one of the particles emitted by natural radioactive elements is the electron. In artificial radioactivity a positron, a particle like the electron but possessing a positive charge, is frequently emitted.

If physicists agreed that nuclei contained only protons and neutrons, it was difficult to explain where these electrons and positrons came from. Study of the transformations involved showed that when an electron was emitted, a neutron in the nucleus was changed into a proton. A positron was emitted when a proton was changed into a neutron. These changes will be discussed further in the next chapter.

NEUTRON BOMBARDMENT

It occurred to the Italian physicist Enrico Fermi that neutrons could be used for atom-smashing experiments. His work inaugurated a new chapter in nuclear physics, earning the Nobel Prize for him in 1938. Alpha particles and protons, possessing positive electric charges, had to fight their way through the barrier posed by the positive charge on atomic nuclei. Since the neutron was electrically neutral, this barrier had no effect on the neutron. It was soon realized that swiftly moving neutrons easily passed through a nucleus without disturbing its structure. This led Fermi to the invention of the slow neutron technique. A stream of neutrons was permitted to pass through water or a block of paraffin before striking the target. Collision with the hydrogen nuclei greatly slowed up the neutrons. Entering an atomic nucleus at slow speed, the neutron now stayed in the nucleus long enough to be captured by it with the result that a transformation then took place. This neutron capture, as it was called, proved the most efficient method of producing artificial radioactivity.

NUCLEAR FISSION

It occurred to Fermi that the technique of neutron capture could be used to create chemical elements beyond uranium, No. 92 in the periodic table, so-called transuranium elements. He tried the experiment in 1934 and reported his belief that Uranium 238 was transformed into Uranium 239 by the capture of a neutron. However, the subsequent ejection of an electron indicated that a neutron in the nucleus had been changed into a proton. Thus the atom changed from No. 92 in the periodic table to a new element, No. 93. The subsequent emission of a second electron indicated that No. 93 had been transformed into No. 94.

A number of eminent scientists, among them Otto Hahn and Lisa Meitner in Berlin, immediately sought to isolate these transuranium elements. The task of separating by chemical means the microscopic amounts of these elements which would be created in the average experiment posed a problem that challenged all the skill of the best chemist. Hahn and Meitner, with the aid of a colleague, Fritz Strassmann, soon found that the changes in Fermi's experiment were more complicated than he had supposed. Hitler's attitude made it necessary for Fräulein Meitner, a Jewish scientist, to flee Germany and she continued her researches in Copenhagen with Otto R. Frisch, her nephew.

Hahn and Strassmann found that one of the radioactive products of the experiment was apparently barium. They were at a loss to explain this. The solution was offered by Meitner and Frisch in January 1939. Barium, No. 56 in the periodic table, has an atomic weight about half that of uranium. They suggested that the uranium atom had split in half and proposed the name "fission" for this process. They pointed out that this would release a very large amount of energy because of the transition from the far end of the periodic table to the very center. This meant a transition from a binding energy of 7.5 Mev to one of 8.4 Mev.

Ordinary uranium is a mixture of three isotopes. Approximately 99 per cent of it is Uranium 238. There are small amounts of Uranium 234, 235, and 237 present. Subsequent studies by the American physicists Edwin M. McMillan and Glenn T. Seaborg showed that Uranium 235 underwent fission. Uranium 238 was transformed into No. 93 and No. 94 as Fermi suspected. These were named neptunium and plutonium.

When Uranium 235 undergoes fission, it releases from one to three neutrons, making possible a chain reaction. This led to the creation of both the uranium bomb and the uranium reactor for producing atomic energy for peaceful purposes.

twenty-three

ELEMENTARY PARTICLE PHYSICS

Nature never breaks her own laws.
—LEONARDO DA VINCI

THE newest branch of physics is known as elementary particle physics. It is concerned with the nature and behavior of the particles which compose the atoms of matter or make their appearance in the course of atomic phenomena. It is not possible to say what the eventual achievements may be in this field. In addition to providing new knowledge of the structure and behavior of the universe, it may yield new methods of releasing atomic energy.

The great difficulty at present is that physicists are aware of the existence of more elementary particles than can be fitted into an orderly picture. Several hundred are now known as a result of experiments with the giant particle accelerators, or atom-smashing machines, particularly the proton synchrotrons. Apparently the discovery of new particles is not yet ended.

An even more astonishing discovery has been that one particle can be transformed into another. Most of them are also radioactive, spontaneously disintegrating, or decaying, to use the technical term, into other particles in brief periods of time measured in millionths, and in some cases billionths, of a second. A cynic has

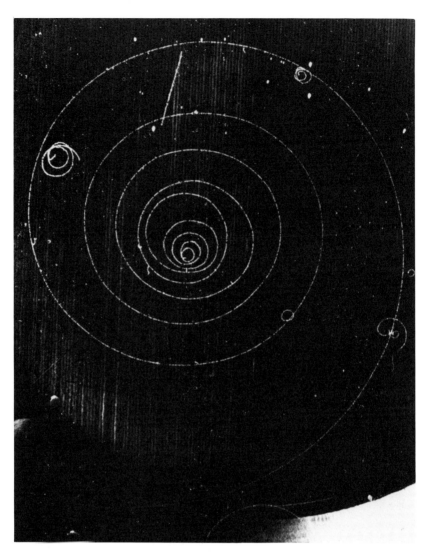

The spiral track made by an electron as it loses energy in the magnetic field of a bubble chamber

said that there are only two things certain about elementary particles today: they are not elementary and they are not particles. The latter part of this remark is based upon the fact that under proper circumstances these particles exhibit the characteristics of waves.

The atom, as the reader already knows, appears to be constructed of three elementary particles. Protons and neutrons, known together as nucleons, occur in the nucleus, and electrons form the cloud of particles around the nucleus. Both the proton and the neutron are heavy particles, having approximately 1836 times the weight or mass of the electron. The neutron is slightly heavier than the proton. To be exact, we must speak of the rest-mass of a particle, since its mass increases with motion in accordance with an equation written by Lorentz before 1905. If we call the rest-mass of the electron 1, that of the proton is 1836.1 and that of the neutron 1838.6.

Mass and electric charge are the most basic properties of these three particles. The proton exhibits a positive electric charge. The electron has a negative electric charge equal in intensity to that of the proton. The neutron shows no electric charge.

All three particles are spinning on their axes just as the earth turns on its axis. This spin, or angular mechanical moment, is found to be the same for all three. This is measured in units based on Planck's constant and is found to be equal to one-half unit in each case. Because of their spin, each of the three particles possesses a magnetic field. (Internal electric currents are responsible for the magnetic field of the neutron.)

The proton and the electron are stable particles in free space. But the neutron is unstable outside the nucleus and disintegrates in about 12 minutes, decaying into a proton, an electron, and a neutrino. (Neutrinos will be discussed shortly.)

Radiant energy, as we have seen, exists in the form of particles known as photons. The energy of a photon is always equal to its frequency multiplied by Planck's constant (see Chapter 21). The photon is the unit of electromagnetic radiation. However, it is not possible to assign a definite size to a photon because of its wave characteristics. Photons have a continuous range of sizes determined by the frequency of their radiation. The amount of energy carried by a given photon is also determined by the frequency. Radio photons possess very little energy. A photon of visible light contains more energy, a photon of X-rays still more. A photon is sometimes regarded as a particle of zero rest-mass immersed in a packet of waves. It is not possible to say where the particle is in the packet at any given moment. The photon has no electric charge. It always moves with the speed of light. It possesses momentum, and therefore mass, as the photoelectric effect demonstrates. It has a spin of one unit.

All particles with spin of one-half unit obey Pauli's exclusion principle (see Chapter 21). This was fully developed by Fermi. Consequently, such particles are said to obey Fermi-Dirac statistics and are known as fermions. Particles with spin of zero, one, or two units do not obey the exclusion principle. Any number can be packed in a given space. Their behavior was formulated by the Indian physicist Satyendra Nath Bose and Einstein. They are said to follow Bose-Einstein statistics and are known as bosons.

THE POSITRON

At the start of the present century physicists expected to find a particle with the mass of the electron but with a positive instead of a negative charge. They gave up the search for it when it was realized that the positive particle in the nucleus of the atom was the much heavier proton.

However, the idea of a positive counterpart of the electron was revived in 1930 by the British physicist P. A. M. Dirac in connection with his development of quantum mechanics. According to his theory, electrons should be able to exist in negative as well as positive energy states. These energy states refer to kinetic energy and must not be confused with the electric charge of the electron which is always negative. It is not possible to observe these negative energy states, which according to Dirac's theory are always completely filled with electrons. If, however, a photon collides with one of these electrons within an atom's electric field, it will give it sufficient energy to lift it to a positive energy state. This leaves a "hole," which manifests itself as a positive electron. It will disappear, however, as soon as the hole is filled by another negative electron.

Needless to say, this Dirac theory was received with considerable skepticism. But in 1932 Carl D. Anderson in Millikan's laboratory at California Institute of Technology discovered the existence of positive electrons in the course of cosmic ray studies.

Anderson was using a Wilson cloud chamber to study the tracks of particles ejected by atoms which had been hit by cosmic ray particles. He used a magnet to deflect the particles, thus obtaining a means for determining their energies. He observed some tracks exactly like those of electrons except that the magnetic field deflected them in the opposite direction. This indicated that they possessed a positive charge instead of a negative one. Anderson proposed the name positron for these positive electrons. The discovery won him the Nobel Prize.

Further experiments in the United States, England, and France showed that the positron is stable. However, it disappears in about a billionth of a second as the result of a collision with an electron. Both disappear, being transformed into a pair of photons. It was subsequently found that a photon of sufficiently high energy, on passing near a heavy nucleus, was transformed into a positron and an electron. This phenomenon became known as pair production. Here, then, was added proof of Einstein's principle of mass-energy equivalence.

Richard P. Feynman of California Institute of Technology has advanced an alternative theory for the positron which the layman will find as odd as that of Dirac. According to Feynman, the electron, instead of moving between a negative and positive energy state, moves forward and backward in time. When it is moving backward, we observe it as a positron.

THE NEUTRINO

It will be recalled that the beta rays ejected by the naturally radioactive elements consisted of ordinary negative electrons. Nuclear transformations achieved with atom-smashing machines created numerous isotopes which were then found to be radioactive. These ejected electrons in some cases and positrons in other. Since the atomic nucleus consists of protons and neutrons, the obvious question is that of where the positrons and electrons come from.

An examination of these transformations shows that in some a neutron has been transformed into a proton with the emission of an electron. For example, radioactive sodium, which has an atomic number of 11 and a mass number of 24, is transformed into magnesium with an atomic number of 12 and a mass number of 24. It is obvious that a neutron has been transformed into a proton, thus increasing the positive charge on the nucleus by one. In other cases a proton is changed into a neutron with the emission of a positron. Both these processes are known as beta decay.

However, this is not the whole story. It was found in every case that the ejected positron or electron possessed less energy than it should. This appeared to be a contradiction of the law of the conservation of energy. There was also a loss of spin or angular momentum.

To save the laws of conservation, Pauli suggested in 1930 that there must be another particle ejected in each transformation which carried away the missing energy and spin. Fermi gave it the name neutrino, the "little neutral one." Its discovery posed difficult problems since it possessed no electric charge and virtually no mass. It was compared to the Cheshire cat in *Alice in Wonderland* which disappeared, leaving only a disembodied grin. The neutrino was called a disembodied spin. Twenty-six years elapsed before its existence was detected by Frederick Reines and Clyde L. Cowan of the Los Alamos Scientific Laboratory with the powerful nuclear reactor at the Savannah River Laboratory of the U.S. Atomic Energy Commission.

MESONS

In 1932 Heisenberg advanced the idea of exchange forces to account for the bonds in the nucleus. According to this theory, a proton and a neutron in close proximity in the nucleus did not maintain their identity as such. Instead the positive charge was tossed back and forth by the two particles so that at a given instant of time one of the two was the proton and at the next instant the other became the proton. Thus the two nuclear particles carried on a sort of ping-pong game in which the electric charge was batted back and forth millions of times per second. Attempts to picture the "ball" in this ping-pong game as a positron ran into mathematical difficulties. In 1935 Hideki Yukawa suggested that the basis of the exchange between the two nuclear particles was a new particle with a mass between 200 and 300 times that of the electron. Stated another way, Yukawa suggested that a new kind of field existed within the atomic nucleus, consisting of quanta of energy which took the form of these particles midway in mass between a proton or neutron and an electron. It was for this reason that he called them mesons. Because it was necessary to account for the bond between proton and proton, proton and neutron, and neutron and neutron, he postulated the existence of positive, negative, and neutral mesons.

The existence of mesons was established in the next two years by Carl D. Anderson and Seth Neddermeyer at California Institute of Technology and J. C. Street and E. C. Stevenson at Harvard University. As in the case of the positron, the mesons were first found among the fragments of atoms shattered by cosmic rays. The mesons first discovered were found to be about 200 times heavier than the electron. However, these mesons presented one difficulty: they showed almost no absorption by atomic nuclei. This contradicted the notion that the meson was responsible for the binding force in the nucleus.

Ten years later, in 1947, Robert E. Marshak of the University of Rochester suggested that the mesons which had been observed were not the ones which existed in the nucleus but a lighter variety which represented a decay product from the heavier sort. According to this "two meson" theory, the heavier mesons would be found at high altitudes in the earth's atmosphere where cosmic rays had knocked them out of nuclei. The decay products, the lighter mesons, would be found at lower altitudes.

A few weeks after Marshak had made his suggestions, Cecil F. Powell and his colleagues in England sent photographic plates to the United States on which these heavier mesons had been recorded in the Bolivian Andes. Subsequent experiments showed that these would react with protons and neutrons as Yukawa's theory required. The heavier mesons are known as pi mesons while the lighter, discovered first, are called mu mesons. (*Pi* and *mu* are letters in the Greek alphabet.) Positive, negative, and neutral pi mesons are now known. They are 273 times heavier than the electron. Positive and negative mu mesons have been found. They are 210 times as heavy as the electron. No neutral mu meson has yet been found. The names of these particles are frequently shortened to pion and muon.

In open space a positive or negative pi meson decays into a positive or negative mu meson and a neutrino in about one two-hundred-and-fifty-millionths of a second. But the mu meson, in another one two-millionths of a second, decays into a positron or electron plus two more neutrinos. The neutral pi meson decays in about a trillionth of a second into two photons.

Continuing studies of the collisions of cosmic rays with atoms of air resulted in the discovery of additional elementary particles. These were subsequently produced in large numbers by the newer and more powerful accelerators. They include other types of mesons, among them the positive and negative kappa (or K) mesons. Particles heavier than the nucleons have also been found, and these are called hyperons. Because of their complex behavior, they became known as strange particles. Physicists have catalogued their behavior into various degrees of strangeness.

Since 1960, the combination of powerful accelerators and improved methods of detection have revealed a vast number of new particles known as resonance particles. They are extremely short-lived, existing for something like a thousandth of a trillionth (10^{-15}) of a second. Physicists are now aware of a hundred or so particles which are heavier than the proton. About 60 mesons, all heavier than the pi meson, are now known.

COSMIC RAYS

Prior to the construction of the proton synchrotrons physicists were largely dependent on cosmic ray phenomena for knowledge of particle transformations. Both the positron and the charged mesons were first identified in the course of cosmic ray studies.

The story of cosmic rays begins shortly after 1910 when the Austrian physicist Victor F. Hess and the German physicist W. Kolhörster established that penetrating radiations entered the earth's atmosphere from outer space. Investigations of these cos-

Two antiprotons, entering the bubble chamber, give rise to mesons on striking protons. One meson decays into electrons, which spiral in the magnetic field.

mic rays, as they came to be called, were resumed after World War I. Among the chief investigators were Robert A. Millikan and Arthur Compton in the United States. For some time a debate raged over whether the rays consisted of radiations like gamma rays or subatomic particles.

It was eventually established that the rays from outer space, the so-called primary rays, consisted chiefly of protons, but contained some nuclei of atoms heavier than hydrogen. These primary rays possess extremely high energies, ranging upward to many billions of electron-volts and even trillions of electron-volts. Their origin is an unsettled astronomical problem. They may arise in the explosion of supernovae. Fermi thought they were accelerated to high energy by passage through the magnetic fields of our galaxy.

Striking the atoms of air at the top of the atmosphere, these primary waves shatter many atoms, gradually losing energy after a series of collisions. Protons and neutrons, knocked out of atomic nuclei at high speed, continue the process of atom-smashing. A great many pi mesons are created which subsequently decay into mu mesons. Finally the so-called secondary cosmic rays found in the lower atmosphere consist chiefly of mu mesons, positrons, electrons, low-energy neutrons, and neutrinos.

ANTIPARTICLES

The positron, which resembles the electron in everything but electric charge, can be thought of as an antielectron. Dirac's theory of the electron, which predicted the existence of the positron, also predicted the antiproton, which would be the negative counterpart of the proton, and the antineutron, which would differ from the neutron in the polarity of its magnetic field. While neutrons exhibit no electric charge, they contain internal electric currents which create magnetic fields.

In the case of mesons, each charged meson can be regarded as the antiparticle of the meson of opposite electric charge.

Physicists have succeeded in creating both antiprotons and antineutrons with powerful accelerators. It has been found, however, that both a proton and an antiproton or a neutron and an antineutron are created simultaneously. The energy of the bombarding particle is split between the two new particles. This is similar to the transformation of a photon into a positron and an electron. In similar fashion a proton and an antiproton annihilate each other when they collide.

Frederick Reines and Clyde L. Cowan showed that antineutrinos as well as neutrinos existed. These differ in the direction of their spin. Still later it was found that there were two types of neutrinos, each of which was matched by an antineutrino. The neutrino or antineutrino which appears in beta decay differs from those which appear in the decay of a pi meson.

THE FOUR INTERACTIONS

Physicists now believe that everything that takes place in the physical universe can be regarded as the manifestation of one of four forces: the force of gravity, electromagnetic force, the strong nuclear force, and the weak interaction force. The physicists prefer to speak of four interactions rather than forces. This is because events in the realm of elementary particles go beyond the effects of forces of attraction and repulsion. Subatomic particles are created, transformed into other particles, and annihilated.

The layman thinks of gravitation as the most powerful of these forces. It may startle him to discover that it is by far the weakest. It is the size of the earth that creates the strength of terrestrial gravity. Gravity holds the solar system together because of the immense mass of the sun. However, the gravitational attraction between ordinary objects can be measured only with the most sensitive instruments.

The electromagnetic force, or interaction, is responsible for the formation of atoms and molecules. It is 10 trillion times a trillion times another trillion (10^{36}) times stronger than the force of gravity. Unlike gravity, which is always a force of attraction, the electromagnetic force can be one of attraction or repulsion. As the reader knows, some particles possess a positive electric charge, others a negative electric charge.

According to the newest theory, so-called quantum electrodynamics, the electric repulsion between two electrons results from the exchange of photons between them. The electric field is created by the constant emission and absorption of photons. Since these photons cannot be observed directly, they are known as virtual photons.

The strong nuclear force, or strong interaction, holds the atomic nucleus together and accounts for the interactions between nucleons. It will be remembered that it is the exchange of mesons between nucleons that holds the nucleus together. The strong nuclear force is 100 times stronger than the electromagnetic force and 10^{38} times stronger than the force of gravity.

The weak interaction force manifests itself in beta decay or other interactions between nucleons with light particles—mu mesons, positrons, electrons, and both kinds of neutrinos and antineutrinos. It is weaker than the electromagnetic force but 10 trillion times a trillion (10^{25}) times stronger than the force of gravity. It operates only within distances smaller than the diameter of the atomic nucleus.

THE CONSERVATION LAWS

The four types of interaction between elementary particles are governed by the conservation laws. These include the conservation of energy, momentum, angular momentum, electric charge, and nucleonic charge. In any interaction the total amount of energy, momentum, and angular momentum, or spin, remains constant. If a positive charge is regarded as $+1$ and a negative charge as -1, the total of electric charges remains constant. To understand the reactions between nucleons, we must assume the existence of a property called nucleonic charge. This is $+1$ for the proton, 0 for the neutron, and -1 for the antiproton. Only those interactions are possible in which nucleonic charge is conserved.

THE REPEAL OF PARITY

Other conservation laws include the symmetry laws. One of these is known as time reversal. This states that interactions do not depend upon the direction of time.

Physicists have found it necessary to modify one of the conservation laws. This is the law of parity which deals with mirror symmetry. It states that there is no absolute distinction between a real event and its mirror image. In other words, the laws of atomic behavior are the same whether the physicist employs a right-handed description of space or its mirror image, a left-handed description. A consequence of this law is that there ought to be no correlation between the direction in which a particle is moving and the direction of its spin. As many electrons moving in a given direction should show a right-handed spin as show a left-handed spin. This symmetry is found in electromagnetic interactions and in strong nuclear reactions.

Two Chinese-born physicists in this country, C. N. Yang and T. D. Lee, concluded in 1956 that parity was not conserved in weak nuclear reactions and proposed several tests for their opinion. In an experiment performed at Columbia University the physicist Chien-shiung Wu and her colleagues found that in the beta decay of radioactive cobalt, the emitted electrons showed a preferential spin. This was contrary to the principal of mirror symmetry.

Physicists believe that the neutrino is responsible for the lack of symmetry in weak nuclear reactions. When a positron or electron is emitted in beta decay, a neutrino or antineutrino is shot out in the opposite direction. Unlike other particles, the neutrino always spins counterclockwise, looking along its direction of motion, while the antineutrino spins clockwise.

THE EIGHTFOLD WAY

Attempts are now being made to bring some order out of the maze of elementary particles by grouping them into families. A beginning can be made on the basis of mass. This results in four groups. One, known as the baryons, from the Greek for "heavy," consists of the proton and all heavier particles, and, of course, their antiparticles. A second group consists of the pi mesons and all heavier mesons. The third group, known as the leptons, from the Greek for "small," are the mu mesons, the electron, the positron, the two neutrinos, and the two antineutrinos. The fourth group has only one member, the photon with zero rest-mass.

It was noted further that many particles formed triplets that differed only in electric charge, for example, the positive, neutral, and negative pi mesons. Some of the confusion is reduced by regarding such a triplet as different states of a single particle.

A scheme for grouping these multiplets into families was suggested independently about 1961 by Murray Gell-Mann of California Institute of Technology and Yuval Ne'eman in Israel. This arranges particles in groups according to eight characteristics which are now known as their quantum numbers. Since for the most part the particles fall into groups of eight, Gell-Mann has named the scheme the Eightfold Way, from a passage in the teachings of Buddha.

Gell-Mann showed mathematically that all the elementary particles could be formed from combinations of three more primitive, or fundamental, particles which he named quarks. Gell-Mann, who appears to have a considerable sense of humor, derived this name from a line in James Joyce's *Finnegans Wake* which reads "Three quarks for Muster Mark." Whether quarks are a mathematical invention or something real is not yet known. Each quark would have to possess fractional electric charges, something that is unknown today. It is thought that the 200 Bev proton synchroton may tell us whether quarks are fact or fiction.

twenty-four

THE THEORY OF RELATIVITY

For a thousand years in thy sight are but as yesterday when it is past, and as a watch in the night.
—Psalms xc: 4

ALBERT EINSTEIN

ALBERT Einstein was one of the greatest scientists of all times. His work, like that of Galileo and Newton, was a turning point in the history of physics and astronomy. He revolutionized scientific thought by his theory of relativity, creating a new understanding of space, time, motion, mass, gravitation, and the structure of the universe. Despite the fact that his theory was difficult to master, it caught the imagination of the public more than any other physical theory in history. His rotund figure and shock of unruly hair became known to the whole world.

He was born in 1879 in the town of Ulm in Württemburg, Germany, the son of German-Jewish parents. He received his education in mathematics and physics at the Federal Institute of Technology in Zurich. He published the first part of his theory in 1905, the same year in which he presented his paper on the photoelectric effect. It became known as the Special Theory of Relativity. In 1913 he was called to Berlin and made director of the Kaiser Wilhelm Institute for Physics, the highest honor that Germany could bestow upon him. In 1916 he published the second part of his theory, known today as the General Theory of Relativity.

Driven out of Germany by the Hitler regime, he came to the United States in 1933, becoming a professor at the Institute for Advanced Study in Princeton, New Jersey, and in 1940 an American citizen. He died in 1955.

THE GALILEAN-NEWTONIAN RELATIVITY PRINCIPLE

Both Galileo and Newton had recognized the relativity of motion, now referred to as the Galilean-Newtonian Relativity Principle. Newton pointed out that it would be impossible for a sailor in a ship moving in a straight line at uniform velocity on a perfectly calm sea to tell whether the ship was at rest or in motion unless he looked outside the ship. "The motions of bodies included in a given space," he wrote in the *Principia*, "are the same among themselves whether that space is at rest or moves uniformly forward in a straight line." The laws of mechanics, that is, the laws governing motion and all mechanical activities, are the same in either case.

However, Newton was not happy with this state of affairs. He regarded space as a physical reality and thought that in some way it might serve as a frame of reference against which absolute motion could be determined.

THE MICHELSON-MORLEY EXPERIMENT

With the development of the wave theory of light, there came the belief that all space was pervaded by a motionless medium in which the waves were propagated. This medium, which became known as the ether of space, was thought to be the fixed frame of reference against which absolute motion might be determined. It ought to be possible to determine the absolute motion of the earth through the ether by comparing the speed of light in different

An early form of the Michelson-Morley interferometer, the instrument which led to Einstein's theory of relativity

directions. Due to the motion of the earth, a slight variation should exist. But since the speed of light is 186,000 miles a second and the speed of the earth around the sun is 18.5 miles a second, there seemed to be no way of performing the experiment with sufficient accuracy to obtain a significant result.

The situation changed, however, with the invention of the interferometer by the American physicist Albert A. Michelson in the 1880s. This instrument made it possible to compare, with the necessary accuracy, the difference in speed of two beams of light moving at right angles to each other. The experiment was tried in Cleveland in 1887 with an improved form of the interferometer by Michelson, then professor of physics at Case School of Applied Science, and Edward W. Morley, professor of chemistry at adjoining Western Reserve University.

To the surprise of the two scientists and of the whole world of science, the experiment failed to reveal any motion of the earth through the ether. The speed of light was the same in all directions.

An explanation of the negative result of the Michelson-Morley experiment, as it came to be called, was advanced independently by the Irish physicist George F. Fitzgerald and the Dutch physicist Hendrik A. Lorentz. This became known as the Lorentz-Fitzgerald contraction.

According to it, objects moving through the ether were contracted, or shortened, in the direction of their motion. This altered the interferometer just enough to conceal the motion of the earth. Lorentz suggested that the contraction was an electromagnetic effect resulting from the motion of the electrons of matter through the ether. He wrote a series of equations, since known as the Lorentz transformations, to express the contraction.

It was obviously impossible to carry out any physical test of the Lorentz-Fitzgerald contraction since the contraction would affect equally a measuring rod or any other measuring device.

THE SPECIAL THEORY OF RELATIVITY

It is fair to say that by 1900 relativity was in the air. A number of scientists, including Lorentz and the French mathematician Henri Poincaré, saw the need for a new view of physical phenomena. Lorentz suggested that motion through the ether would affect time as well as length, slowing down the clock. In 1904 Poincaré advanced what he called "a principle of relativity." The laws of all physical phenomena, he said, ought to be the same for two observers moving with uniform velocity in a straight line with reference to each other, and it might be impossible to determine absolute motion by any means whatsoever. The next year Einstein showed that this was indeed the case. (Physicists employ the term speed to describe the rate of motion. Velocity is the rate of motion in a designated direction.)

Lorentz had tried to save the old, or classical, view of the universe. He continued to believe in the existence of the ether. His electromagnetic theories led to the conclusion that absolute motion existed, but that Nature was engaged in a conspiracy to prevent scientists from discovering it.

Einstein broke with classical physics. He started from a premise, first advanced by the German physicist Ernst Mach in 1872, that there was no such thing as absolute motion, only the relative motion of two objects with reference to each other. Einstein arrived at the Lorentz-Fitzgerald contraction but without any consideration of the ether of space or of electromagnetic effects. He placed the emphasis on the question of measurement, and contended that there was no way to detect absolute motion by either mechanical or optical means.

He based the special theory of relativity on two postulates:

1. The speed of light in a vacuum is always the same, irrespective of the motion of the source of the light or that of the observer.

2. The laws of physics are always the same in any frame of reference, moving with constant speed in a straight line relative to the observer.

From these he deduced by bold and original thinking the surprising results of the special theory.

According to Einstein, measurements of space and time depend upon the motion of the observer in such a way that the observed speed of light is always the same. He began by a critical analysis of the notion of simultaneity, showing that scientists had taken too much for granted.

In his original paper Einstein asked us to imagine a long straight railroad track. Lightning strikes the track at two points. An observer, midway between the two points, equipped with suitable instruments, sees the two flashes at the same instant. The two events are simultaneous for him. But, asks Einstein, will they be simultaneous for an observer on a train speeding along the track? His answer is no. This second observer is moving toward one flash and away from the other. Consequently, the light waves from one flash will reach him sooner than the waves from the other. In other words, there is no absolute simultaneity. What is "now" for one observer is not necessarily "now" for another.

Once the notion of absolute simultaneity is given up, it is seen that there is no such thing as absolute time or distance. Since observers will disagree about the lapse of time between the same two events, there are only relative times and distances whose measurement depends upon the motion of the observer.

A clock carried by a moving observer will appear to go slower to an observer on the ground than his own clock. Similarly, a measuring rod carried by the moving observer will appear shorter to the observer on the ground. This is not the same as the mechanical contraction imagined by Lorentz. The moving observer will arrive at the opposite conclusion. His measurements will show the clock on the ground to be going slower and the measuring rod there to have been shortened. In both cases the observed effect is the result of the relative motion of the two observers.

The difference in measurements of length and time between the two observers can be calculated with the aid of the Lorentz transformations. It is extremely minute if the difference is small, but becomes extremely large as the difference approaches the speed of light. A body moving at 90 per cent of the speed of light relative to an observer would appear to him to have shrunk to approximately half its length.

THE EQUIVALENCE OF MASS AND ENERGY

Einstein extended this concept of relativity to mass. It is interesting to note that as early as 1900 the German physicist W. Kaufman observed that the mass of an electron increased with its speed. Lorentz sought to explain this as an electromagnetic effect arising from the fact that the electron is an electrically charged particle. Einstein showed on the basis of his theory that this increase in mass, like the changes in length and time, applied to all moving bodies. The increase at ordinary speeds is microscopic but the mass of a subatomic particle moving at 90 per cent of the speed of light is increased about twelve times. This fact has to be taken into account in the design of big cyclotrons and similar atom-smashing equipment.

Carrying his calculations further, Einstein demonstrated that the mass of a body is a measure of its energy content and that any change in the energy of a body will be accompanied by a change in mass. This is Einstein's Principle of the Equivalence of Mass

An etching of Albert Einstein made by the German artist Emil Orlik in the 1920s

and Energy, which he expressed in the equation $E=mc^2$, where E is the energy, m the mass, and c the speed of light.

When you put the coffeepot on the stove, the pot weighs a trifle more as it grows hot. It loses mass as it cools off. Until Einstein enunciated this principle of mass-energy equivalence, scientists had accepted as two of the basic laws of the universe the conservation of matter and the conservation of energy, and had regarded matter and energy as forever separate and distinct entities. According to the law of the conservation of matter, it was possible only to alter the form of matter, not to create or destroy it. The other law maintained the same thing with respect to energy. But now Einstein had shown that matter changed into energy and vice versa. It was necessary, therefore, to merge the two conservation laws into one. This was done by regarding the particles of matter as forms of concentrated energy, "bottled energy," so to speak, a view that had been suggested earlier by Sir Oliver Lodge and others.

One might ask why these changes in mass had not been detected earlier in ordinary chemical reactions which absorbed or released energy, for example, the combustion of coal. Calculations show that the loss of mass in such reactions is less than one part in a billion, an amount too small to be measured by the most sensitive chemical balances.

THE SPACE-TIME CONTINUUM

In 1908 the Polish physicist Herman Minkowski extended Einstein's Special Theory of Relativity by introducing the concept of four-dimensional space-time. According to this view, the universe is a four-dimensional continuum with three dimensions of space and one of time. The different measurements of space and time made by observers moving relative to each other are the result of how they divide the four-dimensional continuum into space and time. Two events in the space-time continuum are separated by a combination of space and time known technically as an interval. This is the same for all observers. The location of an event in space-time is given by four coordinates, three for space and one for time.

THE GENERAL THEORY OF RELATIVITY

Newton understood the relativity of uniform motion. He was convinced, however, that accelerated and rotational motion were absolute. To support his view, he advanced the famous illustration of the bucket of water. If the bucket is rotated rapidly, the surface of the water becomes concave. Imagine a motionless bucket at the center of a room which is rotating rapidly. The surface of the water would again become concave. A person in the room could tell that the room was rotating by looking at the bucket. Consequently, Newton maintained, there must be absolute motion.

A contemporary of Newton, the Irish philosopher-bishop George Berkeley, opposed this view. He held that the water in the bucket was subject to a centrifugal force that was due to the presence of the stars in the heavens. If nothing existed in space but the bucket, it would be impossible to imagine any motion of the bucket at all. Berkeley's view was championed some 200 years later by the Austrian physicist Ernst Mach.

Einstein struggled with the problem of accelerated motion for 10 years after the publication of his Special Theory of Relativity in 1905. Was there, he asked himself, something absolute about accelerated motion? Seated in a railroad train, we cannot tell whether it is standing still or moving smoothly unless we look out the window. But if the train goes around a curve or the engineer jams on the brakes, we are aware of what has happened. The force of inertia makes our bodies resist the change in direction or speed.

In 1915 Einstein published his General Theory of Relativity. This stated that the laws of nature are the same for all reference systems whether in uniform or accelerated motion. There is no experiment that will enable an observer in any sort of motion to determine his absolute motion. This was a direct contradiction of the Newtonian view, which appeared to be supported by everyday experience.

THE EQUIVALENCE OF GRAVITATION AND INERTIA

At the center of the general theory is Einstein's Principle of the Equivalence of Gravitation and Inertia. This states that there is no way to distinguish motion produced by inertial forces, such as acceleration or rotation, from motion produced by gravitational force.

Newton had recognized the fact that inertial forces and gravitational forces produced similar effects. This is implicit in his law of universal gravitation, which states that the force of gravity between two objects is proportional to the product of their masses. Heavier bodies fall to earth no faster than lighter ones because they possess greater inertia. Inertial mass and gravitational mass, or weight, are always equal. It remained for Einstein to declare that they were the same thing. The distinction between them is the result of the frame of reference used by the observer.

Einstein asked us to imagine a man in an elevator in a region of gravitation-free space. Someone is pulling on a cable attached to the elevator so that it is rising with accelerated motion—faster and faster. The occupant will experience all the effects of a gravitational field. His feet will press against the floor of the elevator just as our feet press against the earth. If he drops an object, he will think it has fallen to the floor. He will not know that the floor has risen to meet the object.

This brings us to the most important facet of the general theory, namely, Einstein's view of the nature of gravity. Newton spoke of the force of gravity. He realized that this involved the notion of action at a distance. The earth, according to Newton's theory, exerts a force on the moon 240,000 miles away. He states in the *Principia* that his aim was to describe the action of gravity, but that he had no idea of what it was. Einstein abandons the notion that gravity is a force.

To explain the nature of gravitation, Einstein adopted the notion of curved space, first suggested by the German mathematician Bernhard Riemann in 1854. Gravitation, Einstein concluded, was the distortion or warping of space, more exactly of the four-dimensional space-time continuum, by the presence of matter.

According to Newton, the planets move in their orbits around the sun because of the gravitational pull of the sun. According to Einstein, planets move as they do because their orbits are the straightest path possible for them in curved space.

An analogy will help to explain this. Imagine a flat rubber sheet like a trampoline. A bowling ball placed in the center of the sheet will depress the center, distorting the surface of the sheet. A marble placed on the edge of the sheet will now roll downhill, colliding with the ball. A person looking down on the trampoline from a high building would think that a force had drawn the marble to the ball. He would not realize that it had merely rolled downhill.

The geometry of curved space is not that of Euclid—that is, it is non-Euclidean. The shortest possible path is called a geodesic.

In addition to the local distortion of space in the neighborhood of every body, there is a general overall curvature of space (this was discussed in Chapter 8).

EXPERIMENTAL CONFIRMATION

Let us return to our observer in the elevator rising with constant acceleration in gravitation-free space. A ray of light enters the elevator through an opening in one wall and strikes the opposite wall. Since the elevator is rising, our observer sees the light path as a parabola. It will be remembered that he is unaware of his accelerated motion and believes that he is in a gravitational field. Consequently, he concludes that light is affected by a gravitational field. Since it is impossible to distinguish between the effect of acceleration and the effect of gravity, Einstein concluded that light is affected by a gravitational field. A light ray, he said, should be bent or curved in passing through a gravitational field.

During an eclipse of the sun, the stars become visible at the moment of totality, when the moon blots out the sun. It is then possible to see stars which appear close to the sun. These stars are immensely distant, but the light from them passses close to the sun on its way to us. In 1915 Einstein predicted that the images of these stars would be found to be displaced outward with respect to the edge of the sun due to the bending of their light rays in the sun's gravitational field.

Here was a prediction that lent itself to experimental verification. But World War I was in progress at the time. In 1919, after the war was over, two British expeditions undertook to test the prediction at the solar eclipse of May 29. They found the displacement predicted by Einstein. It was the report of these British expeditions which touched off the furor over the Einstein theory.

It is ironic that subsequent tests have given variable results, throwing doubt on the original British reports. Astronomers today feel that the test presents so many difficulties as to make it not worth trying.

Other tests, however, appear to support the General Theory of Relativity. Einstein's theory of gravity predicts that the frequency of a light beam is decreased by a gravitational field. This so-called gravitational red shift has been found in the spectrum of the sun and more clearly in the spectra of extremely dense stars. Certain irregularities in the motions of the planet Mercury can be explained on the basis of Einstein's theory of the nature of gravity but not on the basis of Newton's law of gravitation.

PART IV

✳✳✳✳✳✳✳✳✳✳✳✳✳✳✳✳✳✳✳✳✳✳

Life

twenty-five

FROM MAGIC TO SCIENCE

> *Let us first understand the facts, and then we may seek the cause.*
>
> —Aristotle

THE beginnings of biology, like those of astronomy, are lost in the mists that veil the dawn of civilization. Long before the start of written history early man had learned a great deal about plants and animals and developed a crude medicine and surgery. The caveman was faced with dangers on every side. He might be clawed by a wild beast or break an arm or leg in a fall. Individuals who were more skillful at binding up a wound or setting a broken bone became the world's first surgeons.

But there were mysterious illnesses for which there was no visible cause. And so medicine became entwined with magic. It is possible that there were medicine men in the Old Stone Age who differed little from those of today's primitive tribes.

Many prehistoric skulls have round holes drilled in them. The edges of the holes show healing, making it evident that the individual survived the operation. We can only guess at the reason for the trepanation, as it is called. Perhaps it was an attempt to cure such diseases as epilepsy or migraine headache by letting out the demons thought responsible.

By 3000 B.C. both Egypt and Babylonia had developed a considerable body of medicine and surgery. Egypt had many physicians and surgeons and medicine was already divided into specialties. There were eye specialists, skin specialists, stomach specialists, and dentists, among others.

The first doctor mentioned in Egyptian records is Imhotep, who lived about 2950 B.C. He was also an astronomer, an architect, and a statesman. By 700 B.C. he was worshipped as the god of medicine and temples built in his honor became centers of healing. Details of Egyptian medicine and surgery are contained in a number of papyri that have come down to us. Much of the surgery is sound. The physicians made use of a wide variety of gargles, salves, and plasters, some made from the internal organs of lions, elephants, camels, and crocodiles.

We know less about Babylonian medicine than we do of Egyptian medicine. Apparently the Babylonian physician leaned more heavily on magical incantations. Disease was caused by the gods and a cure required their appeasement. A disease was frequently diagnosed by divination, which included the examination of the entrails of a sacrificed animal. Surgery is dealt with more rationally in the famed Code of Hammurabi, which sets the fees for successful surgery and the penalties for unsuccessful operations. The surgeon ran the risk of having his right hand cut off if the patient died.

GREEK MEDICINE

The first Greek doctor whose name we know was Asclepius. Homer mentions him in the *Iliad*. He may have lived about 1200 B.C. However, some medical historians believe he was purely a creature of mythology. To begin with, Greek medicine was greatly influenced by that of Egypt. Like Imhotep in Egypt, Asclepius was eventually worshipped as the god of medicine. Temples built in his honor became centers of healing like those of Imhotep in Egypt. The Greek physician priests became known as Asclepiads.

One of these temples stood on the island of Cos and it was here that there appeared the greatest figure in Greek medicine, Hippocrates, sometimes called the Father of Medicine. He lived from 460 to 377 B.C. Many writings credited to him have come down to us, but scholars are no longer certain that he wrote most of them.

With Hippocrates medicine made the transition from magic to science. He brushed aside the notion that disease was caused by an angry or vindictive god. Disease was the result of some failure of bodily function. It was the task of the physician to locate the trouble and do something about it. Hippocrates observed his patients with such meticulous care that modern physicians have no difficulty in recognizing many diseases from his accurate descriptions of them. He noted the connection between certain diseases and impure supplies of water. He prescribed adequate rest and wholesome food for his patients, believing that Nature frequently accomplished the cure.

GREEK BIOLOGY

Side by side with the development of medicine in ancient Greece went the formation of the science of biology. The Greeks seem to have been the first to emphasize the philosophical side of the study of living creatures and to investigate the phenomena of life for their own sake. Investigations of the structure and habits of living creatures were made as early as 500 B.C. Alcmaeon of Croton appears to have been the first to practice dissection of animals. He described the nerves of the eye and studied the development of the chick within the egg.

Greek biology reached its climax with the profound studies of Aristotle. It is curious that while his physical theories dominated scientific thinking until the time of Galileo, his biological researches were largely ignored.

Aristotle was born in Stagira on the northeast coast of Greece in 384 B.C. His father was an Asclepiad who served as physician in the royal court in Macedon. At about 17 years of age Aristotle went to Athens, where he entered the Academy of Plato, remaining there until Plato's death 20 years later. In 342 B.C. Aristotle was called back to Macedon by King Philip to tutor his son Alexander, later Alexander the Great. When Alexander ascended the throne, Aristotle returned to Athens and founded a school in the grove sacred to Apollo Lyceius and hence known as the Lyceum.

Aristotle was a prolific writer. He not only made many original discoveries in the field of biology but also made a broad survey of existing knowledge. He accumulated knowledge concerning some 500 species of animals, a few of them, alas, fictitious. It is believed that he personally made dissections of some 50 species. He suggested a general classification of animals on the basis of their structure. He studied the changes which took place day by day in the chick as it developed in the egg. He seems to have made little study of plants, but his pupil Theophrastus did work of such importance that he is regarded as the Father of Botany.

With the rise of Rome, Greek knowledge passed into the hands of the Romans and many Greek scholars and physicians flocked to Rome. History records the names of two famous Greek physicians, Dioscorides, an army surgeon under Nero, and Galen, physician to Emperor Marcus Aurelius.

Galen wrote on medical subjects and became the greatest anatomist of antiquity. But since human dissection was frowned upon, he had to confine his studies to animals. He also studied the functions of organs and hence can be regarded as the first experimental physiologist. His studies of anatomy and physiology became the standard of the world of medicine for 15 centuries.

Another important Roman figure was Pliny the Elder, whose works on natural history filled 37 volumes. However, he was an uncritical collector of data and put down myth and fable along with fact.

MEDIEVAL BIOLOGY

During the Dark Ages that followed the downfall of Rome progress in biology and medicine was slow. Arabian and Jewish physicians did much to keep Greek medicine alive. Aristotle and Galen were translated into Arabic. Chief among the Moslem physicians was the Persian Avicenna, who wrote numerous treatises based on the medical theories of Hippocrates.

After A.D. 1000 Greek learning began to return to Europe as the continent emerged from the Dark Ages. The first of the great cathedral schools was founded soon after. The Arabic works of the leading Jewish scholars were translated into Latin. The most important of these medieval Jewish sages was Rabbi Moses ben Maimon, better known as Maimonides. He was a physician as well as a philosopher, mathematician, and astronomer, and he left Spain for Egypt, where he became the physician of the Sultan.

In time many hospitals were built near the monasteries of Europe. The doctors were mostly priests or monks. The monasteries also organized schools of medicine. However, medicine gradually passed into the hands of physicians who had no church connections. This was due largely to the founding of medical schools and universities in many cities. The first of the renowned medical schools was at Salerno in Italy, a seaport not far from Naples.

However, the medical schools taught anatomy from the writings of Galen, who had never dissected a human body. He had dissected various animals including pigs and monkeys and taken it for granted the human anatomy was the same.

By the fifteenth century medical schools were permitted to make dissections of the bodies of criminals who had been executed. But the professor of anatomy would lecture to his class from a high desk with a volume of Galen in front of him, completely ignoring the fact that what he read failed to agree with what the dissection, made by an assistant, disclosed.

THE NEW ANATOMY

The revolution in astronomy began in 1543 with the publication of the great work of Copernicus, *De revolutionibus orbium coelestium* (Concerning the Revolutions of the Heavenly Spheres). The same year saw the start of a revolution in biology with the publication of *De corporis humani fabrica* (On the Structure of the Human Body) by Andreas Vesalius.

Vesalius was born in Brussels on December 31, 1514. His ancestors included many physicians and learned men. During his school days he showed keen interest in anatomy, dissecting frogs, mice, rabbits, dogs, and birds. While a student at the University of Louvain he found the skeleton of a robber that had been left hanging on a gallows. The vultures had cleaned the bones. In the dead of night Vesalius cut down the skeleton and carried it off. Before long he could identify every bone when blindfolded. From Louvain he went to the University of Paris. But he was bitterly disappointed with the way anatomy was taught. Later he went to the University of Padua, where he was permitted to make some of the dissections. He quickly gained a reputation for his skill. On December 15, 1537, he received his medical degree and although he was not quite 23 years old, was elected professor of anatomy the next day.

Two years later he was asked by a Venetian publisher to prepare a new edition of Galen's work on anatomy. As he began the work it dawned upon him that Galen was describing the anatomy of pigs and goats and monkeys, not human anatomy. He decided that the time had come to break away from Galen, whose writings had ruled the medical world for 1,400 years. He determined, therefore, to write an anatomy of his own, based on the dissection of the human body.

Fortunately the great artists of the day, Leonardo da Vinci, Michelangelo, Raphael, and others had become interested in the study of anatomy, realizing that they needed to know more about the play of muscles under the skin. Leonardo dissected both men and animals. Vesalius was able to get Stephen Calcar, one of Titian's best pupils, to do the illustrations for his book. Calcar's illustrations were not only correct, but magnificent works of art.

THE NEW PHYSIOLOGY

The revolution begun by Vesalius was completed almost a century later by the English physician William Harvey. Vesalius had disclosed the correct anatomy of the human body. But the medical schools went right on teaching Galen's ideas of how the body functioned. This study of function is known as physiology.

Harvey received his medical degree in 1602 at Padua, where Vesalius had once taught anatomy. Returning to London, Harvey eventually became the physician of King James I and later of Charles I. Harvey's great contribution to biology was the discovery of the circulation of the blood.

Galen had taught that the blood was manufactured in the liver, passing upward into the right side of the heart. From there it made its way to the left side of the heart through invisible pores in the wall between the two sides of the heart. But Vesalius had already shown that no such pores existed. Galen said that the blood moved outward from the heart with a pulsating, or rhythmic, motion in both the arteries and the veins.

Harvey's teacher at Padua, Fabricius, showed that there were tiny valves in the larger veins. Harvey noted something that Fabricius had missed. These delicate cuplike structures were so arranged that they permitted blood to flow toward the heart, but prevented it from flowing in the opposite direction. Blood could not oscillate back and forth in the veins.

Harvey wrestled with the problem for 20 years, studying the beating of the heart and the motion of the blood in fish, frogs, birds, snakes, and dogs, even in wasps and flies. At last he arrived at the proper answer: the blood moves in an endless circle, leaving the heart by way of the arteries and returning by way of the veins.

There was only one gap in his work. He was aware that both arteries and veins branched into finer and finer vessels. But he was unable to find the connection between the finest arteries and the finest veins.

In 1661, four years after Harvey's death, the Italian anatomist Marcello Malpighi, one of the first to use the microscope, found the connecting vessels while studying the structure of the lungs of frogs. These microscopic connecting vessels are known as capillaries.

Just as Vesalius is remembered as the Father of Anatomy, so Harvey has been called the Father of Physiology. Once the circulation of the blood was properly understood, it was possible to make steady progress in understanding the functioning of the human body.

THE CELL THEORY

In 1665 Robert Hooke, a contemporary of Sir Isaac Newton, examining a thin slice of cork with his primitive microscope, saw that it was composed of tiny rectangular compartments. He coined the name cell for them. Actually these were dead cells since their contents had disappeared. By the start of the nineteenth century many scientists had noted the existence of cells in both plant and animal tissues. In 1838 the German botanist Matthius Jakob Schleiden advanced the theory that the cell was the structural and functional unit of all plants. The following year the German physiologist Theodor Schwann advanced the same theory about animals. Also in 1839 the Czech biologist Jan Purkinge proposed the name protoplasm for the living contents of all cells.

By this time it was well understood that living matter consists chiefly of water, carbohydrates, fats, proteins, and mineral salts. Protoplasm became known as the physical basis of life. The term protoplasm is falling into disfavor today. It is now well understood that the contents of the cell are far more than a mere colloidal mixture of various chemical compounds. It is now recognized that the cell is the basic unit of life and that in the last analysis all the phenomena of life are varied aspects of cell behavior.

MODERN BIOLOGY

With the advent of the eighteenth century we find biology splitting up into many subdivisions, paving the way for the specialists of the nineteenth and twentieth centuries. On the basis of organisms, modern biology can be divided into microbiology, the study of bacteria and other microorganisms; botany, the study of plants; and zoology, the study of animals. Each of these has been split into many specialties. One biologist may be interested only in birds, another only in viruses. Anthropology, the study of human characteristics, is an important subdivision of zoology.

It is also possible to divide the field of biology into subdivisions according to approach. One biologist may be interested in taxonomy, the classification of living organisms, another in anatomy, a third in physiology, a fourth in cytology, the structure and functions of cells.

Greater progress has been made in biology since World War II than in the previous 20 centuries. The present era has been called the Golden Age of Biology.

Many of the spectacular contemporary discoveries resulted from the union of biology with chemistry and physics to form biochemistry and biophysics and from the use of the electron microscope.

twenty-six

THE CHEMICAL NATURE OF LIFE

Life is simply a process of combustion.
—MATTHIUS JAKOB SCHLEIDEN

IT is an interesting fact that if we construct a scale with a supergiant star at one end and an electron at the other, a living organism about the size of a man will come near the center of the scale. Man is about as many times larger than the electron as he is smaller than the star.

In general, we have no difficulty in distinguishing between living and non-living things. Living organisms are characterized by a number of typical attributes. These include growth, self-repair, reproduction, movement, response to stimuli, the ability to adjust to environmental changes, and a complex series of chemical reactions grouped under the general term metabolism.

While there are more than a million kinds of plants and animals in the world, exhibiting a vast difference in size, appearance, and behavior, the startling fact is that all living organisms, whether a typhoid bacillus, a tree, a man, or an elephant, are basically alike in their physical and chemical composition and in their metabolism.

Six chemical elements compose the bulk of any living organism, although not always in the same proportions. Your body is

65 per cent oxygen, 18 per cent carbon, 10 per cent hydrogen, 3 per cent sulphur, and 1 per cent phosphorus. About 25 elements make up the other 3 per cent. They include, among others, calcium, chlorine, cobalt, copper, fluorine, iodine, iron, magnesium, potassium, sodium, and zinc. It can be seen, therefore, that living organisms are composed of quite common chemical elements. The important fact is not the elements themselves but their level of organization.

The atoms of these elements are organized into an immense variety of molecules, some of them quite simple, but others extremely complex, containing hundreds of atoms and in many cases thousands and even millions. These molecules, in turn, are organized into complex and dynamic units known as cells. Except for unicellular organisms, the cells are organized into tissues and the tissues into organs.

Water, one of the simplest chemical compounds, composes from 60 to 90 per cent of all living organisms. Your body is about 65 per cent water.

At the start of the nineteenth century the Swedish chemist Jöns Jakob Berzelius proposed that those chemical compounds, other than water and mineral salts, which occur in living organisms be known as organic compounds. His belief was that they could only be elaborated by living organisms and that their manufacture depended upon the presence of a "vital force" in living organisms.

However, this notion was upset in 1828, when the German chemist Friedrich Wöhler succeeded in synthesizing urea. This was followed by the synthesis of numerous other organic compounds.

Before the end of the century it was realized that the chief organic compounds in all living organisms were carbohydrates, fats, and proteins. These are all carbon compounds of varying complexity. The presence of nucleic acids was recognized, but it remained for twentieth-century biologists to discover their importance.

The term organic chemistry is used today to describe the chemistry of carbon compounds. It deals not only with the carbon compounds in living organisms, but such relics of ancient life as coal and petroleum, and with many synthetic compounds that do not exist in nature.

The outer shell of the carbon atom contains four electrons. Since another four are needed to form a stable octet (see Chapter 19), carbon has a valence of 4. Consequently, carbon has the ability to form a vast number of compounds. The framework, or "backbone," of these compounds is a straight or branched chain of carbon atoms or a ring of six carbon atoms. Atoms of other elements, chiefly hydrogen and oxygen, are attached to these frameworks. A number of chains can be linked together to form longchain molecules. Similarly, a number of rings can be linked together to form the heart of a complex compound. Giant molecules, formed by linking together hundreds or even thousands or millions of simpler molecules, are often referred to as macromolecules.

CARBOHYDRATES

The carbohydrates are compounds of carbon, hydrogen, and oxygen. The hydrogen and oxygen atoms are generally present in the same ratio as in water (H_2O), namely, two atoms of hydrogen to one of oxygen. Their importance arises from the fact that they are the chief energy sources of living organisms. They also form some of the structural components of cells.

Sugars are the most abundant carbohydrates. The simplest sugars are known as monosaccharides. These are constructed around chains of carbon atoms containing from three to ten atoms. The five-carbon sugars, known as pentoses, form part of the structure of nucleic aids. The six-carbon sugars, known as hexoses, are the principle source of energy of both plants and animals.

One of the most abundant hexoses is glucose, known also as dextrose, grape sugar, and blood sugar. Other hexoses include fructose, or fruit sugar, and galactose. They all have the chemical formula $C_6H_{12}O_6$ but differ in the arrangement of the hydrogen and oxygen atoms around their "backbone" of six carbon atoms.

Two monosaccharides are frequently linked together to form a disaccharide. When this occurs, two hydrogen atoms and one oxygen atom are lost in the form of a molecule of water. The reaction is known as dehydration synthesis. Thus, for example, two molecules of glucose unite to form a molecule of maltose. The combination of glucose and fructose results in sucrose, or cane sugar. Glucose and galactose produce lactose, or milk sugar.

More complex saccharides are known as polysaccharides. These may consist of the linkage of dozens or even thousands of monosaccharides. Glucose is the chief constituent and in many cases the only one. They include the starches, which are composed wholly of glucose units. Animal starch, which is stored in the liver, is known as glycogen. Another polysaccharide is cellulose, which forms the walls of plant cells.

PHOTOSYNTHESIS

The sun not only supplies the light and heat which make our earth habitable, but furnishes the energy of life itself. The basic process upon which all life depends is photosynthesis, the process by which green plants use the energy of sunlight to synthesize glucose from carbon dioxide (CO_2) and water (H_2O). In the course of this process oxygen is set free. From glucose the plants manufacture more complex carbohydrates, fats, and proteins, utilizing nitrates and other mineral salts in their environment.

It is estimated that plants annually use about 150 billion tons of carbon dioxide and 60 billion tons of water to produce 100 billion tons of organic matter and 110 billion tons of oxygen. About 80 per cent of all photosynthesis is carried on by the microscopic algae in the upper waters of the oceans, the rest by land plants or plants in bodies of fresh water. Plants use the organic matter to build their own structures and carry on their own metabolism, but in the cycle of the seasons it is eventually decomposed back to water and carbon dioxide by the decay of dead plants in water and on land.

A small proportion of the organic matter produced by plants is eaten by the animals of land and sea. Animals are incapable of photosynthesis and can exist only by eating plants or each other. They are parasites on the plant world.

Photosynthesis is possible only in green plants and microorganisms which contain a green pigment known as chlorophyll and some associated pigments. The chlorophyll molecule is a complex one, containing carbon, hydrogen, oxygen, and nitrogen with a single atom of magnesium at its very center. Whole batteries of enzymes play important roles in the mechanism of photosynthesis and all the steps are not yet understood. Studies employing radioactive isotopes as tracers have revealed that photosynthesis consists of two sets of reactions known as the light phase and the dark phase.

In the light phase the energy of sunlight is absorbed by the chlorophyll. Part of this energy is used in reactions not yet completely understood to split the molecules of water into hydrogen and oxygen. The oxygen is set free as molecular oxygen. The hydrogen atoms, however, are linked to a compound in the cell known by the formidable name nicotinamide adenine dinucleotide phosphate. Biochemists call it NADP. The rest of the energy absorbed by the chlorophyll is stored in the chemical bonds of another compound known as adenosine triphosphate, or ATP.

In the dark phase, which gets its name from the fact that it does not require light, the NADP gives up the hydrogen atoms and the energy of the ATP is used to attach the hydrogen to the carbon dioxide, transforming it into glucose. This is accomplished in a series of complex steps in which enzymes play a part. As a result the energy of sunlight is now stored in the high-energy bonds of glucose.

FATS

Fats, like carbohydrates, contain only carbon, hydrogen, and oxygen, but unlike carbohydrates, are not soluble in water. They also differ in that the ratio of hydrogen to oxygen is much greater than two to one. Because of this a given weight of fat can release much more energy than the same weight of carbohydrate. Fat is the most concentrated form of energy in living systems. After a heavy meal the blood is crowded with tiny droplets of fat. The body stores these high-energy droplets in the cells of fatty, or adipose, tissues. When the diet is rich in carbohydrates and contains more calories than are needed, the carbohydrates are converted into fats and stored in adipose tissues.

Fat molecules consist of the union of three molecules of fatty acid with one of glycerine. For this reason they are called triglycerides. The backbone of the glycerine molecule is a chain of three carbon atoms. A molecule of fatty acid is attached to each carbon atom. Fatty acids are built on chains of 12 to 24 carbon atoms. The most common of these acids have chains of 16 to 18 carbon atoms.

Biochemists group fats with a heterogenous class of organic compounds known as lipids, which share the common characteristic of being insoluble in water. The most abundant lipids are the phospholipids, which contain phosphorus and sometimes nitrogen as well in addition to carbon, hydrogen, and oxygen.

A highly important group of lipids are the steroids, which include ergosterol, cholesterol, the bile acids, vitamins A, D, E, and K, adrenal hormones, and sex hormones. Steroids, chiefly cholesterol, and phospholipids are components of cell membranes.

PROTEINS

Proteins are the most abundant organic compounds in living things, constituting about half the dry weight of all organisms. They are part of every structural component of all cells and play a key role in the chemical and physical activities that make up the life of the cell. A cell may contain as many as 2,000 different proteins.

About a third of the proteins in your body are in the muscles. Two proteins, myosin and actin, form the fibers which are the contractile elements in muscle cells. Another 20 per cent of your protein occurs in the bones and cartilages of the body in the form of collagen. The skin has about 10 per cent of the body protein. The oxygen-carrying compound in the red blood cells, hemoglobin, is a protein.

Many bacteria give off toxins which are proteins. These antigens, as they are called, cause the body in defense to form antibodies which are also proteins. Many hormones, such as insulin, are proteins. The enzymes, the catalysts which make possible the thousands of chemical reactions that go on in the living cell, are all proteins. Every protein has a structure peculiar to itself and serves a specific function.

Protein molecules consist primarily of carbon, hydrogen, oxygen, and nitrogen. Many also contain sulphur, and some contain phosphorus and iron. They are large, complex molecules, varying greatly in size and shape. The smallest protein molecules contain several thousand atoms, the largest millions of atoms.

All proteins consist of chains of smaller molecules known as amino acids. There are about 20 commonly occurring amino acids, although all 20 may not occur in a particular protein. Each one has a backbone consisting of two carbon atoms followed by a nitrogen atom. Some hydrogen and oxygen atoms are attached to this backbone. In addition there is a complex arrangement of atoms known as a side-chain attached to the second carbon atom. Each amino acid has its own distinctive side-chain.

Amino acids are linked together to form chains known as polypeptides. The chemical bonds between units of the chain are known as peptide bonds. Some proteins consist of a single polypeptide chain, others of two or more chains linked together by the chemical bonds between atoms of sulphur. The chains are not straight but coiled into a helix. This was established by the American biochemist Linus Pauling, who received the Nobel Prize in 1954 for his achievement. In many cases the helix is folded back and forth on itself to form a ball. The structure is held together by hydrogen bonds between loops of the helix.

The number of amino acid units in a protein can vary from 50 to more than 100,000. An average protein contains about 600 units. They can be linked together in any order. Five different amino acids can be joined to form 120 different combinations. Ten will form more than 3 million combinations, 20 more than 3 billion. A chain of 500 units of all 20 amino acids can form more combinations than there are atoms in the universe (about 10^{600}).

Certain types of proteins are common to all living things but they differ slightly from one species to another. Thus, while all animals contain insulin, there are slight differences among pig insulin, beef insulin, and human insulin. But the insulin of every individual of a given species is alike.

On the other hand there are many proteins which differ from one individual to another in the same species. No two human beings, except identical twins, have exactly the same array of proteins. This causes the difficulty in attempts to transplant organs. The body resists the introduction of foreign proteins and produces antibodies to attack them.

The proteins in our food are broken down into their constituent amino acids by digestive enzymes. They are then used by the cells of the body to create their own proteins. Because the cells need these amino acids, a shortage of proteins in the diet can lead to serious disease.

ENZYMES

The enzymes are proteins which make possible the thousands of chemical reactions which go on in the living cell. The cell is constantly building up complex molecules and tearing them down. The enzymes are biochemical catalysts. They speed up the rate at which these chemical reactions go on in the cell and enable them to occur at ordinary temperatures. They are large globular proteins with molecular weights ranging from 100 to more than a million. (The molecular weight is the sum total of the weights of all the atoms in a molecule. See Chapter 18.) Apparently an enzyme enters momentarily into a chemical reaction and then drops out unaltered, ready to repeat the performance. For example, a particular enzyme can break down 5 million molecules of hydrogen peroxide into water and oxygen in one minute. In general, enzymes are highly specific in their action and one enzyme will promote only one particular reaction. It is believed that the reacting molecules attach themselves to so-called active sites on an enzyme. The combination, however, quickly results in a chemical reaction in which a complex molecule is formed or split, as the case may be, while the unaltered enzyme separates from the products.

RESPIRATION

The main source of energy for many organisms, including man, is the oxidation of carbohydrates. Biochemists call this process respiration. It is just the opposite of photosynthesis in which the energy of sunlight is captured and stored in the high-energy bonds of carbohydrates formed by the union of carbon dioxide and water. In respiration the carbohydrate is broken down, step by step, into carbon dioxide and water and the chemical energy of its hydrogen bonds is set free to do the work of the living cell. The reader will note that this use of the word respiration differs from that of the medical man, who uses it to describe the process of breathing. Respiration takes place in plants as well as in animals. It is also important to note that the biochemist uses the term oxidation to mean the removal of hydrogen as well as the addition of oxygen to a compound.

In animals the decomposition of carbohydrates begins in the digestive tract, where starches are split into glucose by the action of various enzymes. The molecules of glucose are absorbed into the bloodstream and carried by the blood to all the tissues of the body where they are absorbed by the cells. Similarly, the cells absorb oxygen which has been brought to them by the red blood cells.

However, the utilization of glucose in the cell begins with a series of steps which do not require the presence of oxygen. These steps are similar to those by which yeast converts glucose to alcohol and carbon dioxide. The glucose molecule is broken down into two molecules of pyruvic acid. The next series of steps utilize oxygen and result in the breakdown of the pyruvic acid to carbon dioxide and water. The energy released in the process of respiration is trapped in the high-energy chemical bonds of adenosine triphosphate, or ATP, where it is available for all the life processes of the cell.

While glucose is the chief source of cellular energy, the cell can also use fatty acids and amino acids as "fuel" for respiration. Life goes on only as long as the cell receives the nutrients from which it can extract energy.

twenty-seven

THE CELL

A mighty maze, but not without a plan.
—POPE

THE cell is the basic unit of life. Unicellular organisms, as the name implies, consist of single cells. The great majority of organisms are multicellular, being composed of aggregates of cells. In the case of some microscopic organisms there may be no more than 16 cells. But the body of an infant at birth is formed of about 2 trillion cells. Your body is composed of some 50 to 60 trillion cells, depending on your height and weight. Cells form tissues and these in turn form organs. Since many tissues consist of several kinds of cells, your body contains more kinds of cells than organs.

With a few exceptions, cells are too small to be seen without the aid of a microscope. Most cells are a few ten-thousandths of an inch in diameter, although some bacteria, among the smallest cells known, are about 1/250,000 of an inch in diameter. It may startle the reader to realize that the yolk of a bird's egg is a single cell. Most of its bulk, however, consists of foodstuffs stored in it for the developing embryo. The largest known cell is the yolk of the ostrich egg, about 3 inches in diameter.

The single cell of a unicellular organism must carry on all the activities necessary to sustain life. There is, however, a con-

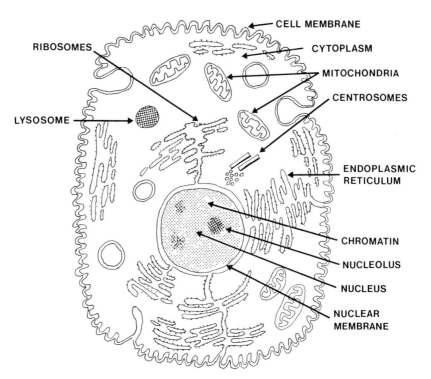

Figure 38. A generalized cell

siderable variety of size and shape among them. Bacteria may take the form of spheres, rods, or spirals.

In multicellular organisms there is a division of labor. Cells carry on specialized functions of importance to the organism as a whole as well as the basic functions needed to keep themselves alive. One of the functions of the cells of the liver, for example, is to secrete bile. Cells in various glands manufacture the powerful secretions known as hormones. The result is that they exhibit a wide variety of size and shape, depending upon their function. Many muscle cells are spindle-shaped. Nerve cells are characterized by extensions, or fibers, branching from the cell body.

The tissues of an animal, including man, can be classified into five fundamental types, namely, epithelial, connective, muscle, nervous, and blood. Each consists of cells designed to facilitate the functions of the particular tissue.

Because of the diversity of plants and animals, it is not possible to select any one cell as a typical cell. However, it is possible to delineate the basic structure which is common to all cells. Such a generalized cell is depicted in Figure 38.

A cell, it will be seen, consists of three major components, the outer skin, or cellular membrane, the nucleus, and the cytoplasm. We owe most of our knowledge of the detailed structure of the cell to the electron microscope. This instrument has revealed a wealth of detail which cannot be seen with the less powerful optical microscope. As a result biologists have been able to gain a new concept of the activities within the cell.

Each cell of an organism is a chemical factory. Like any other factory, it has its board of directors, its assembly lines and other machinery for making its product or performing its service, its source of supplies, and its source of energy.

THE CELLULAR MEMBRANE

The electron microscope has shown that the membrane around the cell is a complex structure consisting of a double layer of fat molecules sandwiched between an outer and inner layer of protein. The membrane is not smooth, but has blisters extending outward and pouches extending inward. It is more than a skin around the cell. It controls what molecules can enter the cell or leave it. Glucose and amino acids pass through it much more easily than many smaller molecules. Water moves easily into and out of the cell.

Plant cells have an additional wall outside the cellular membrane. This consists chiefly of cellulose, manufactured and secreted by the cell. It gives the plant its structural strength. The cellulose wall is non-living.

THE NUCLEUS

A dark spot is easily seen with an optical microscope in a cell which has been properly stained. This is the nucleus, the controlling center of our cellular factory. It contains the board of directors. As befits an executive office, it is closed off from the factory. It is surrounded by the nuclear membrane, a structure of protein and fat similar to the cellular membrane. It controls the molecules that can enter or leave the nucleus.

The structure of the nucleus can be revealed by staining the cell with appropriate dyes. The interior of the nucleus contains a colloidal fluid, the nuclear sap, or nucleoplasm, in which a fine network of threadlike material is distributed. This network is known as the chromatin and consists of twisted, elongated forms of the chromosomes. It is not possible to make out the individual chromosomes. This network occupies most of the nucleus. But there are also a number of small rounded bodies known as nucleoli. These vary in number in different species of plants and animals. Both human beings and onions have four. They appear to play some role, not yet understood, in cell division.

THE CHROMOSOMES

The chromosomes become visible in the optical microscope as distinct entities when the cell undergoes division, a process which will be discussed later in this chapter. It can then be seen that the cells of each species contain a definite number of chromosomes which are grouped into pairs, each member of a pair resembling its partner. The usual number varies from 10 to 50. Human cells contain 46 chromosomes (23 pairs). Chromosomes vary in size and shape. Human chromosomes are about 1/5,000 of an inch in length.

Chromosomes contain two types of nucleic acid and a number of proteins. The nucleic acids are deoxyribonucleic acid, usually referred to as DNA, and ribonucleic acid or RNA. The proteins include a number of proteins of low molecular weight, known as histones, as well as more complex proteins. While a great deal is known about the chemistry of the molecules which form the chromosomes, the physical structure of the chromosome is not yet completely understood.

It is now well understood that the chromosomes are both the carriers of heredity and the controlling directors of the activities of the cell. For a long time biologists thought that these important functions were carried on by the proteins in the chromosomes.

It is now known that this is not so. The controlling constituent of the chromosome is DNA. This was established by a succession of experiments by various biologists initiated in 1944 by the American biochemists Oswald T. Avery, Collin McLeod, and Maclyn McCarty at the Rockefeller Institute.

The DNA molecule is a long threadlike entity of extremely high molecular weight, ranging from a million to tens of millions. It is composed, however, of the repetition of four smaller molecules known as nucleotides. A strand, or molecule, of DNA contains upwards of 10,000 nucleotides.

Each nucleotide consists of a molecule of deoxyribose, a five-carbon sugar, linked to a molecule of phosphate. One of four molecules known as bases is attached at a right angle to the sugar molecule. Two of the bases are pyrimidines, built around a ring of four carbon and two nitrogen atoms: they are thymine and cytosine. The other two are purines, built around a double ring: they are adenine and guanine. Biochemists refer to these bases by their initials, T, C, A, and G.

It is now believed that the DNA molecule is a double helix constructed of two strands of nucleotides held together by their bases. The structure can be compared to a spiral staircase. The sides of the staircase are formed by a seemingly endless succession of sugar and phosphate molecules. The steps are formed by the union of the bases jutting out from the sugar molecules (see Figure 39). The bases are joined together by hydrogen bonds. However, adenine can be linked only with thymine, and guanine with cytosine. This means that only four types of steps are possible in the staircase, A-T, T-A, G-C, and C-G. These are referred to as base pairs.

This model of DNA, known as the Crick-Watson model, was put forward in 1953 by the British biophysicist F. H. C. Crick and the American biochemist James D. Watson. They were aided in arriving at their model by X-ray diffraction studies made by the British biophysicist M. H. F. Wilkins. All three shared the Nobel Prize for the achievement. Their model received major support from the success of the American biochemist Arthur Kornberg in synthesizing DNA from the four nucleotides in test tubes. Kornberg also received the Nobel Prize.

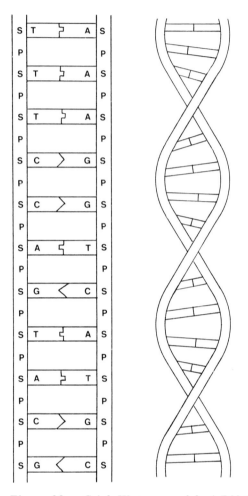

Figure 39. Crick-Watson model of DNA

The sequence of the nucleotides in one strand of the DNA molecule can be any random order. However, the other must be the complement of the first one, containing T where the first contains A, etc., so that the proper base pairs can be linked together. Because a molecule of DNA contains so many nucleotides, no two individuals, with the exception of identical twins, have identical strands of DNA in their chromosomes.

It is thought that each of the 46 chromosomes in the human cell contains one strand, or molecule, of DNA composed of more than 150 million nucleotides. It is known that this strand of DNA is associated with structural proteins, but the exact arrangement is not known. One suggestion is that the DNA strand is coiled tightly around a central post of protein like a rope coiled on the deck of a ship. The strand of DNA is only 18 or 20 atoms in diameter but if uncoiled it would be about 1.5 inches. The total DNA in the 46 human chromosomes contains about 4 billion base pairs. Placed end to end, the strands of DNA would make a string about 5 feet long.

There is also some RNA in the chromosomes. RNA consists of a single helix, not a double one like DNA. Its sugar is also a five-carbon sugar but contains more oxygen than the sugar molecules in DNA. One of its bases is different: in place of thymine it contains another pyrimidine, uracil.

Let us complete the picture of cell structure before we turn to the functions of DNA and RNA.

THE CYTOPLASM

The cytoplasm constitutes the major bulk of the cell. It contains the main assembly lines of the chemical factory. The structure of the cytoplasm was poorly understood until the invention of the electron microscope. It was originally believed that the cytoplasm consisted of a colloidal mixture containing particles of various sizes. It is now known that many of these particles are extremely complex structures and that the cytoplasm itself has a highly organized structure.

The chlorophyll in green plants and certain bacteria is contained in complex structures in the cytoplasm known as chloroplasts. These contain membranous laminations, called grana, which are arranged like loose stacks of coins. They are coated with layers of chlorophyll one molecule thick. The chloroplasts also contain the enzymes which make possible the complex steps of photosynthesis (see Chapter 26).

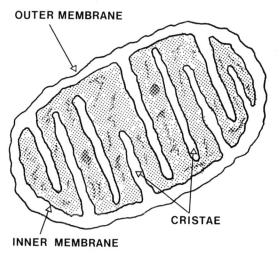

Figure 40. The mitochondrion

Rod-shaped bodies in the cytoplasm, known as mitochondria, are the "powerhouses" of the cell. They are the centers of respiration (see Chapter 26), the process which furnishes the energy that keeps the cell going. A cell may contain as many as 1,000 mitochondria.

Each mitochondrion is contained within a membrane of protein and fat like the cellular and nuclear membranes. The inner protein layer is folded into a great many convolutions, providing a considerable surface area within the mitochondrion (see Figure 40).

Carbohydrates entering the cell are broken down with the aid of enzymes into smaller molecules of pyruvic acid, fats into their constituent fatty acids, and proteins into amino acids. These smaller acid molecules undergo further changes and then enter the mitochondrion, where with the aid of more enzymes, many of which may form part of the inner layer of the membrane, they are degraded step by step by a series of oxidations to carbon dioxide and water. In these oxidations they give up the energy of their chemical bonds which is transferred to the bonds of adenosine triphosphate, or ATP. The ATP is secreted by the mitochondrion into the cytoplasm, where it is available as an energy source for the many reactions that go on in the cell. One can think of the mitochondria as secreting chemical energy for the use of the cell.

Other structures in the cytoplasm include the Golgi complex, the centrioles, vacuoles, plastids, and lysosomes.

The Golgi complex is a membranous structure containing a number of vesicles. Although its existence was first noted in nerve cells by the Italian neurologist Camillo Golgi in 1898, its function is still a mystery. It is possible that in secretory cells it takes part in the manufacture or storage of hormones or other cellular products.

The centrioles are two small bodies at right angles to each other near the nucleus in animal cells. They play a role in cell division.

The vacuoles are liquid-filled cavities in the cytoplasm surrounded by membranes. Small vacuoles occur in animal cells, but in many plant cells a large vacuole may occupy most of the cell, pushing the nucleus and contents of the cytoplasm against the cellular membrane.

Plastids are small bodies in plant cells engaged in manufacture or storage. The most familiar plastids are the chloroplasts. Starch is stored in plant cells in plastids known as leucoplasts.

Lysosomes are small rounded sacs. They contain enzymes which promote the digestion of food materials in the cell including proteins, nucleic acids, some carbohydrates, and some fats. They break down food materials stored in cells. In white blood cells they break down invading bacteria or foreign proteins. They have been called "suicide sacs," because they cause the destruction of aged and worn-out cells.

When viewed in an optical microscope, the various components of the cytoplasm, often referred to as organelles, appear to be suspended in a colloidal fluid. The electron microscope, however, has shown that the ground substance of the cytoplasm is as highly structured as any of the organelles. It consists of a network of meandering double membranes that tend to lie parallel to each other. This network forms a system of canals extending from the cellular membrane to the nuclear membrane and is believed to form the chief assembly lines of our chemical factory. The membranes are dotted with particles rich in RNA and known as ribosomes. They play an important role in the fabrication of proteins.

THE GENES

As early as 1880 a number of German biologists concluded that the chromosomes were the carriers of heredity. By 1920 the American geneticist Thomas Hunt Morgan and his students, A. H. Sturtevant, C. B. Bridges, and Herman J. Muller, by means of breeding experiments with the drosophila, or fruit fly, established that the units of heredity, the genes, were arranged in linear fashion along the chromosomes. In the 1940s the American biologists George Beadle and Edward Tatum showed by a series of brilliant experiments with the common pink bread mold, or neurospora, that genes controlled the metabolism of the cell by controlling the manufacture of the enzymes. By the 1950s it was established that DNA was the component of the chromosome, which was the carrier of heredity, and that the genes were segments of the strand of DNA. It is estimated that the 46 chromosomes of the human cell contain 2.5 million genes. The genes control the production of the structural proteins as well as the enzymes.

The major problem facing biologists was to explain how a gene, a segment of DNA composed of four base-pairs, could specify the sequence of 20 amino acids in a protein. It was obvious that if each base-pair signified an amino acid, only four amino acids could be designated. A sequence of two base-pairs permitted 16 combinations (4×4), and therefore could designate only 16 amino acids. However a sequence of three base-pairs permitted 64 combinations ($4 \times 4 \times 4$). This was more than adequate. Researches since 1960 have established that a sequence of three base-pairs signifies a particular amino acid. Thus, if a protein contains 200 amino acid units, a gene must consist of 200 triplets, or 600 base-pairs. Researches to establish the triplet which specified a particular amino acid became known as "cracking the code."

PROTEIN PRODUCTION

The production of a protein begins when a segment of DNA constituting a gene comes apart, or "unzips." RNA nucleotides in the nucleus are attracted to this gap and form a strand of RNA, using the gene as a template. This strand of RNA, known as messenger RNA, or m-RNA, is now a negative print of the gene, carrying the instructions of the gene in its sequence of nucleotides.

The strand of m-RNA separates from the gene and makes its way from the nucleus to the cytoplasm, where a ribosome attaches itself to one end of the strand and begins to move along the strand. The ribosome has been called a traveling workbench.

Among the particles in the cytoplasm are very short strands of RNA known as transfer RNA or t-RNA. It is a single strand of 70 to 80 nucleotides. However, it is doubled back upon itself to form a double strand. There are 20 different types of t-RNA, one for each amino acid, differing from each other in the sequence of their nucleotides. One end of a strand of t-RNA picks up a specific amino acid. The t-RNA now attaches itself to the strand of m-RNA where the triplet specifies that particular amino acid. As one t-RNA after another "plugs in," enzymes link the amino acids to each other. The t-RNA now separates from both the m-RNA and the amino acid. A protein is thus created as the ribosome moves along the m-RNA. But as the strand of protein is being formed, another ribosome attaches itself to the start of the m-RNA and a second protein is created, then a third, etc. This complicated process is shown in Figure 41.

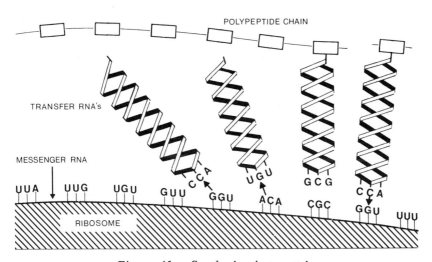

Figure 41. Synthesis of a protein

No attempt has been made in this figure to indicate the amino acid which corresponds to a particular triplet of bases in the m-RNA.

A large number of biologists and biochemists in laboratories all over the world contributed to this picture of how the genes control and direct the production of proteins. It was Crick who first demonstrated that the code consisted of triplets, three base pairs designating an amino acid. Marshall W. Nirenberg and a group at the U. S. National Institutes of Health were the first to identify a specific triplet with a specific amino acid. Other triplets were identified by Severo Ochoa and his colleagues at New York University. Transfer RNA was discovered by M. B. Hoagland of the Harvard Medical School. Messenger RNA, first suggested by Jacques Monod and Francois Jacob of the Pasteur Institute in Paris, was subsequently found by a number of teams in American laboratories.

MITOSIS

The division of a cell into two daughter cells is an amazing process which never ceases to awaken the wonder of the biologist. It has been compared to a ballet or an exquisite minuet in which the chromosomes are the dancers. It is known as mitosis.

The "dance" is preceded by a period during which the nucleus of the cell enlarges and nucleotides and large protein molecules are synthesized in it. During this period, while the chromosomes are still spread out in the elongated, twisted chromatin network, the DNA in each chromosome is duplicated. Experiments with radioactive tracers have enabled biologists to understand what happens. The process is known as replication.

The strand of DNA begins to come apart, or "unzip," as the hydrogen bonds of the base pairs are successively broken. Each of the strands of nucleotides acts as a template on which a complementary strand is formed from the pool of nucleotides in the nucleus. In this fashion the original DNA molecule is replaced by two identical molecules. If we designate the two strands of the original molecule as "old A" and "old B," one of the new molecules consists of old A and new B, while the other consists of new A and

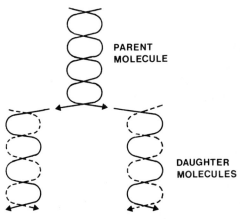

Figure 42.   Replication of DNA

old B. This is shown in Figure 42.

Once replication has been completed, the chromatin network begins to separate into the individual chromosomes which gradually grow shorter and thicker by a process of coiling, while the nucleoli grow smaller and finally disappear. It is now seen that each chromosome is split longitudinally into identical coiled halves, known as chromatids, which are held together at some point along their lengths known as the centromere. Each chromatid consists of proteins and one of the new strands of DNA.

Next the nuclear membrane disappears and in animal cells each centriole divides into two halves which migrate to opposite sides of the cell. Fibers radiating from the centrioles form a three-dimensional spindle, widest at its center and tapering at both ends. A similar spindle forms in plant cells, despite the absence of centrioles. The chromosomes now migrate to the equator of the spindle where they are attached to it by their centromeres. The centromeres divide and the chromatids of each chromosome begin to move in opposite directions until they are at opposite ends of the spindle. Each chromatid is now a full-fledged chromosome and there is now a complete set of chromosomes at either end of the spindle.

A nuclear membrane now forms around each set of chromosomes. In plants the cell is now divided into two cells by the formation of a cellulose wall, or cell plate, at the equator of the spindle. In animals division into two cells is accomplished by a process known as furrowing, which pinches the cell in two.

It is important to realize that the process of mitosis insures that each of the two daughter cells receives a full set of chromosomes, each chromosome containing a molecule of DNA which is a replica of that in the same chromosome of the original cell.

MEIOSIS

Mitosis is sufficient to account for the reproduction of most unicellular organisms. The two daughter calls separate so that there are now two individual organisms in place of the original one. This type of reproduction is known as asexual.

Most multicellular organisms reproduce by sexual processes. The organism produces sex cells known as gametes. The fusion of two gametes produces a single cell known as the zygote. It grows by mitosis into the new organism. In some organisms the gametes appear alike. But in general they differ in size and structure. The female gamete, the larger of the two, is known as the ovum, or egg. The male gamete is the sperm. The zygote is spoken of as a fertilized egg.

A moment's thought will show that if each gamete possessed a full number of chromosomes, the zygote would have twice that number. Since, the human cell has 46 chromosomes (23 pairs), if each ovum and sperm contained this number, the fertilized egg would have 92 chromosomes. Nature does not permit this to happen. The gametes are produced by a process resembling mitosis, but differing from it in important details so that each gamete possesses only half the number of chromosomes. Thus each human ovum or sperm has 23 chromosomes. It is important to understand that this is not any 23, but one of each of the original 23 pairs.

The process of meiosis begins like that of mitosis with the separation of the chromatin network into individual chromosomes. However, the two members of each pair wind around each other, so that the cell appears to contain only half the usual number of chromosomes. Meanwhile a spindle forms as in mitosis and each chromosome is seen to have divided into two chromatids. At this point, therefore, there are four chromatids joined together, forming a so-called tetrad, which takes its place near the equator of the spindle.

The tetrad now separates into two chromosomes. However, the chromatids of each chromosome remain attached to each other. One of each pair of chromosomes (consisting of two chromatids) now migrates to one pole of the spindle, the other chromosome to the other. Two cells now form, each possessing half the usual number of chromosomes. But each chromosome consists of two chromatids. Finally, each of the two cells undergoes a division in which there is no replication of DNA, but merely an equal distribution of the chromatids, which are now full-fledged chromosomes. The end process of meiosis, therefore, is the production of four cells, each with half the usual number of chromosomes. Biologists refer to the full number in any organism as the diploid number. The number possessed by the gametes is called the haploid number.

In the case of the sperm cells the cytoplasmic division is equal so that four functional sperms result. The division in the case of the egg cells is unequal so that one functional egg and three small non-functional cells, known as polar bodies, are produced.

The genetic significance of meiosis is discussed in Chapter 28.

CONTROL OF THE GENES

The development of a multicellular organism begins with the fusion of two gametes into a zygote. The zygote divides by mitosis into two cells, the two into four, the four into eight, and so on. Mitosis insures that every one of the 2 trillion cells present in a newborn infant contains a full set of chromosomes and, therefore, a complete complement of genes. How, then, are we to explain cell differentiation? As the human embryo develops why do some cells become muscle cells, others nerve cells?

Once it was established that genes control the production of structural proteins and enzymes, it became evident that there must be some mechanism which turns the genes on and off so that each cell produces only the proteins it requires. Every cell of the human body possesses the genes for the manufacture of hemoglobin. However, these genes operate only in those cells which give rise to the red blood cells. They are suppressed or turned off in every other cell of the body. Similarly, the genes for making muscle proteins are turned on only in muscle cells.

In 1950 a husband-and-wife team of British biochemists, Edgar and Ellen Stadman, suggested that the histones, the low-weight proteins in the chromosomes, were the gene inhibitors. A decade later this was confirmed by two teams of American biologists, one at California Institute of Technology headed by James Bonner and Ru-chih Huang, the other at Rockefeller University led by Vincent G. Allfrey and Alfred E. Mirsky. Apparently a histone molecule wraps itself around a gene, preventing it from functioning. More exactly, the histone molecule is bonded to the gene by electrostatic attraction.

Bonner and his group established the existence of another RNA in chromosomes which they named chromosomal RNA. This in Bonner's picturesque phrase is the "seeing-eye dog" which leads the histone molecule to the proper gene.

It is estimated that in any of the cells of the human body 99 per cent of the genes are turned off. Since the human cell contains 2.5 million genes, this means that 25,000 genes are still turned on.

Once it was established that histone turns off the genes, the next problem was to find out what turns on a repressed gene. Researches in a number of laboratories here and abroad have established that various small molecules, chief among which are the hormones, have the ability to turn the genes on.

The orderly development of any organism from a zygote proceeds along an established orderly plan as though it had been programmed on a computer. Biologists see this sequential "switching" on and off of the genes as the reason the human embryo does not often develop two heads or three arms. But much research is still needed to reveal the details of how it is done.

twenty-eight

HEREDITY AND EVOLUTION

> *Fierce eagles do not produce timorous doves.*
> —Horace

FROM time immemorial it has been recognized that children usually resemble their parents. All living things, both animals and plants, pass on inherited traits from one generation to the next. This must have been recognized in the Neolithic Age, soon after man changed from a roaming hunter to a settled farmer and began to raise crops and domesticate wild animals. Primitive man improved his crops and his animals by selective breeding, planting the seed of the most desirable crops and mating the animals whose characteristics pleased him most. But while the phenomenon of heredity was recognized before the dawn of written history, early attempts to expalin it were futile. Aristotle, noting the long-known fact that the child inherited some of his characteristics from one parent and some from the other, sought to explain this as "the mixing of bloods." This idea that the blood determined heredity was accepted for centuries and gave rise to such phrases as "blood will tell" and "blue blood." A beginning of the scientific understanding of heredity was not made until the middle of the nineteenth century. Biologists speak of the branch of their science concerned with heredity as genetics.

GREGOR MENDEL

The science of genetics began in 1866 with the publication of a monumental paper by an obscure Austrian monk, Gregor Mendel, in the little-read journal of the Natural History Society of Brünn. Ironically, Mendel's conclusions, which are the foundation of modern genetics, went unnoticed until 16 years after his death, when his paper was discovered in 1900 by three eminent biologists, Hugo De Vries of Holland, Karl Correns of Germany, and Erich Tschermak of Austria.

Mendel, a teacher in one of Brünn's high schools, carried on a series of experiments with garden peas in his little garden at the Augustine monastery where he lived. He had noted that there was a variety of garden peas. Some plants were tall, others were short. Some produced yellow seeds, some green seeds. Mendel noted seven pairs of contrasting characteristics.

The flower is the reproductive structure of seed plants. Pollen grains, produced by the stamen, contain the sperm cells. The egg cells are found at the base of the pistil. Fertilization of the egg cells, known as pollination, occurs when pollen grains are transferred from stamen to pistil. When this occurs in the same flower, or two flowers on the same plant, it is known as self-pollination. Cross-pollination occurs when the wind or insects carry pollen from one plant to another. Mendel allowed several generations of plants to self-pollinate. He found that characteristics were passed on faithfully from one generation to the next. Seeds from tall plants gave rise to tall plants. Yellow seeds produced plants that always developed yellow seeds, and so on.

Next he tried cross-pollination, carefully transferring pollen from hundreds of tall plants to short plants. In every case the resulting seeds from the short plants gave rise to tall plants. He tried transferring pollen from short plants to tall ones. Again, the resulting seeds gave rise to tall plants. He called the plants which he crossed the parental, or P_1, generation and the next generation the first filial, or F_1, generation.

His third step was the most important. He permitted plants of the F_1 generation to self-pollinate, thus producing a second filial, or F_2, generation. Mendel found that 75 per cent of the F_2

generation were tall and 25 per cent were short. This was his amazing and revolutionary discovery. He found the same thing to be true of the other contrasting traits.

To explain his results, Mendel postulated the existence of discrete hereditary units which occured in pairs. He called them factors and divided them into two categories, dominant and recessive. He reasoned that the factor for tallness was dominant. If we designate the factor for tallness as T and the factor for shortness as t, then pure-bred tall plants contained TT; pure-bred short plants tt. The F_1 hybrids contained Tt. One fourth of the F_2 generation contained TT, one-half Tt, and one-fourth tt. It was like tossing two coins simultaneously. In enough tosses, one-fourth will yield two heads, one-half heads and tails, and one-fourth two tails.

Having read Chapter 27, the reader knows what Mendel could not know: his factors were the genes, the units of heredity in the chromosomes. It is amazing that Mendel arrived at his conclusions when he did. Ten years elapsed after the publication of his paper before it was demonstrated that the zygote was formed by the fusion of egg and sperm. Another 10 years went by before the process of meiosis was understood, and it was not until the 1920s that it was firmly established that the genes were located in the chromosomes.

We know today that the situation is not as simple as Mendel pictured it. Mendel was fortunate in choosing garden peas for his studies. Genes are not always completely dominant or recessive. An example is the familiar garden plant known as the four o'clock. If a plant with red flowers is crossed with one having white flowers, the F_1 generation will have pink flowers. But the F_2 generation will consist of one-fourth red-flowered plants, one-half pink-flowered plants, and one-fourth white-flowered plants. If we designate the gene for red by R and the one for white by r, the F_2 generation contains RR, Rr, and rr genes.

There are also complex situations in which a given trait is the result of the influence of more than one gene.

It is known from the work of Morgan and his colleagues that one pair of chromosomes is responsible for the determination of sex. The members of this pair, unlike other pairs, differ greatly in appearance. They have been called the X and Y chromosomes. If the zygote contains two X chromosomes, the organism will be female. If it contains one X and one Y it will be male.

VARIATION

It is evident from Chapter 27 that each pair of chromosomes in the cells of a multicellular organism consists of one chromosome from each of the parents. However, the distribution of these pairs in meiosis is the result of chance. The way in which the pair lands on the spindle determines whether the paternal chromosome or maternal chromosome migrates to one or the other pole of the spindle. Thus there is a wide variation in the make-up of the resulting sperm or egg cells. The human cell, it will be recalled, contains 23 pairs of chromosomes. Because of this random migration, more than 15 million different combinations of paternal and maternal chromosomes are possible in the sperm or egg.

This explains, for example, why the children of given parents, while usually resembling each parent in some respects, differ not only from each parent, but also from each other.

This random distribution of chromosomes in meiosis applies to all plants and animals and explains the wide differences in characteristics between members of the same species which the biologist refers to as variation.

MUTATIONS

At the turn of the century the Dutch naturalist Hugo De Vries found two evening primrose plants that were different from their parent stock. These differences were transmitted to succeeding generations. In further experiments with some 50,000 specimens of evening primrose he found several hundred such abrupt changes in traits that were subsequently inherited. He called these changes mutations. Other experimenters in the early decades of the twentieth century found that such mutations occurred with varying frequency in all bacteria, plants, and animals including man.

It is now understood that since the genes are the units of

heredity, such abrupt changes must be the result of changes in one or more genes. Such changes can be the result of errors in the sequence of nucleotides which occur during the process of replication. Cosmic rays, passing through an organism, can also cause mutations by changing the structure of a gene. A majority of mutations are lethal, but this is not true of all.

The ability of high-energy radiation to alter genes was first shown in 1927 by Herman J. Muller, who had been one of Morgan's graduate students. He was awarded the Nobel Prize for his experiments in which he used X-rays to alter the genes of the fruit fly, or Drosophila. Certain drugs are also known today which will cause mutations.

EVOLUTION

Biologists have long regarded evolution as a firmly established fact, not a theory. They believe that the microorganisms, plants, and animals inhabiting the earth today have all evolved from simpler forms. Once life had appeared on earth, it evolved into more and more complex forms. This means that all the organisms on earth today are related to each other and have descended from a common origin.

The operation of evolution is clearly indicated in the fossil record described in detail in Chapter 15. The oldest rock layers reveal no trace of life. Gradually the record of life begins to appear. Traces of primitive plants are found in rocks more than 2 billion years old. But traces of animal life are extremely rare until 600 million years ago, when suddenly they are found in great abundance. Life gradually left the oceans for land. Amphibians were followed by reptiles, they by mammals. Finally man appeared on the scene.

In many cases the fossil record is remarkably complete. This is particularly true of certain mammals of the Cenozoic era. It is possible to trace the present-day horse through a series of developments from a little four-toed animal, the eohippus, which flourished at the beginning of the era. The development of the elephant can be traced in similar fashion through a series of species of increasing size from an original one in the early part of the Cenozoic era who stood about 3.5 feet high.

A second piece of evidence for the descent of today's organisms from a common ancestor is the fact, as stated at the start of Chapter 26, that all living organisms, whether a typhoid bacillus, a tree, a man, or an elephant, are basically alike in their physical and chemical composition and in their metabolism.

Embryology offers a third piece of evidence for the fact of evolution. Every vertebrate begins as a zygote which grows by cell division into an embryo. In its development the embryo goes through a series of stages that resemble the corresponding embryonic stages of remote ancestral forms. Thus, at an early stage, the human embryo possesses both gill slits and a tail. At an early stage embryos of a fish, a chick, a rabbit, a calf, and a human being are so much alike that only an expert biologist can tell them apart. These similarities are evidence of a common ancestry. Embryonic differences, of course, become marked at later stages of development.

Vague notions of evolution are found in the writings of a number of the ancient Greek philosophers. Some authorities interpret certain statements of Aristotle to indicate that he believed in an ascending line of evolution from the simpler organisms to the more complex ones. But with the rise of Christianity the account of creation as given in Genesis was accepted and went unquestioned until the eighteenth century. It was held that each species represented an act of special creation.

The modern concept of evolution had its beginnings in the eighteenth century in the writings of the British physician, naturalist, and poet Erasmus Darwin, the grandfather of Charles Darwin, and the French naturalist Georges Louis de Buffon. Both of them believed that the environment was responsible for changes in organisms and that these changes were then transmitted to succeeding generations. The first detailed theory of evolution was presented by the French naturalist Jean Baptiste de Lamarck in 1801. Like his predecessors, Lamarck placed much emphasis on the effect of environment. The basis of Lamarck's theory was the idea of the inheritance of acquired characteristics. He believed that the characteristics of higher animals were influenced by the use or disuse of organs and that these acquired characteristics were passed on to succeeding generations. He sought in this fashion, for example, to account for the web feet of ducks and geese and the long neck of the giraffe. It is now agreed by biologists that acquired characteristics are not inherited.

CHARLES DARWIN

The greatest name in the history of the development of organic evolution is that of Charles Darwin. He assembled an overwhelming mass of evidence in support of the principle of evolution, and advanced a theory for it which is the basis of the modern concept of evolution.

Darwin was born in Shrewsbury, England, on February 12, 1809, and educated at Edinburgh and Cambridge Universities. In 1831, at the age of 22, he embarked on a five-year trip around the world as a naturalist on the British ship H.M.S. *Beagle*. It was his observations on this trip, particularly in South America and in the Galapagos Islands in the Pacific, that led him to his belief in the reality of evolution. He noted, for example, slight variations in the birds known as finches from island to island in the Galapagos group. However, they still showed strong similarities to each other and to the finches on the South American coast. He found it more reasonable to suppose that each species was the descendant of a common ancestor than to believe that each represented an act of special creation. The same thing was true of numerous other plants and animals as well. He was also impressed by the discovery in South America of the fossil of a large extinct mammal, now known as the glyptodon, which resembled the living armadillo. Again it seemed more reasonable to Darwin to explain the resemblance on the basis of evolution rather than special creation.

Two years after his return from the voyage of the *Beagle*, Darwin read "The Essay on Population" that Thomas Malthus had written in 1798. Malthus maintained that the population grew faster than the means of subsistence. Darwin saw that this observation applied to all of nature. Every plant produces more seeds than can ever mature into new plants. A sturgeon may lay 2 million eggs a year, a codfish 6 million. If all matured, the oceans would not be large enough to hold them. Consequently, all organisms face a severe competition for survival. Darwin called this "the struggle for existence."

Darwin saw that due to variation some individuals were better adapted to their environment than others. These survived and transmitted their characteristics to succeeding generations. Those least able to adapt to their environment died out and disappeared. Darwin regarded this as the main mechanism of evolution and called it "natural selection." He seems not to have made any clear distinction between heritable characteristics and non-heritable ones, taking variation for granted. Herbert Spencer termed this process of natural selection "the survival of the fittest."

Darwin spent 20 years after his return from the voyage of the *Beagle* assembling the evidence for his theory. In 1858, as he was preparing his work for publication, he received a manuscript from the Far East from another English naturalist, Alfred Russel Wallace. As the result of his studies in Indonesia, then known as the Malay Archipelago, Wallace, also inspired by Malthus' essay, had come to the same conclusion as Darwin. Wallace was completely unaware of Darwin's studies. In July 1858 Darwin, in the finest tradition of science, presented both his own conclusions and those of Wallace to the Linnaean Society in London. The following year his momumental book, *On the Origin of Species by Means of Natural Selection*, was published.

The publication of the book evoked a storm of controversy throughout the world. Many individuals everywhere who chose to accept literally the Biblical account of creation were furious. It was widely charged that Darwin was contending that man was descended from the monkey. The storm continued well into the twentieth century. In 1925 John T. Scopes, a high-school biology teacher in Tennessee, was arrested for breaking a state law which prohibited the teaching of evolution. The resulting "monkey trial," as it came to be called, saw William Jennings Bryan, statesman and three-time contender for the Presidency of the United States, leading the prosecution, while the famous Clarence Darrow marshaled eminent scientists for the defense. To students of the history of science, the widespread opposition to the idea of evolution was reminiscent of the furor in the seventeenth century which greeted Galileo's championship of the Copernican theory.

THE MODERN THEORY OF EVOLUTION

It was evident that variation, the shuffling of genes in meiosis, was sufficient to account for the rise of individuals best adapted to their environment to the extent made possible by existing genes. It could not, however, explain the appearance of new characteristics which sometimes were superior to any combination resulting from sexual reproduction.

In 1901 De Vries advanced an extension of Darwin's theory on the basis of his discovery of mutations. It was possible to explain the appearance of new characteristics as mutations. If these enabled an organism to make a superior adjustment to the environment, they could speed up the evolutionary process.

The modern view recognized environment as the directing force in evolution. This is quite different from the older notion of the inheritance of acquired characteristics. All the genes and their mutations present in a given population of a given species constitute the so-called gene pool. The environment selects those particular combinations which are best suited to insure the survival of the species by weeding out those less able to survive.

THE ORIGIN OF LIFE

We are left with an exceedingly important question. How did life originate in the first place? The ancient Babylonians and Egyptians took it for granted that certain forms of life arose spontaneously from inanimate substances or decayed organic matter. Frogs and toads were thought to develop from the mud on the bottom of rivers or ponds. Horsehairs falling into water turned into worms. Maggots developed from decaying meat. Bees sprang from the carcasses of dead animals. Such beliefs were the natural outcome of uncritical observations of nature. This notion of spontaneous generation, as it came to be called later, was held by the Greek philosophers including Aristotle. It continued to be believed until well into the nineteenth century.

It was first challenged by the Italian physician Francesco Redi in the seventeenth century. In 1668 he performed a series of simple but brilliant experiments. He placed meat in an uncovered jar and in another jar covered with gauze. As the meat decayed, maggots appeared in the uncovered jar but not the other. Redi proved that the maggots hatched from eggs laid by flies on the uncovered meat. He found eggs on the gauze which covered the other jar. The idea of spontaneous generation began to be questioned as biologists learned more about insects, worms, and reptiles, but with the perfection of the microscope and the discovery of microorganisms, spontaneous generation came to the fore once more. If a small quantity of hay is boiled and the liquid placed in a covered vessel, it is soon swarming with microorganisms. It was generally assumed that these organisms sprang from the so-called hay infusion, although a few biologists thought they were the descendants of organisms that had fallen in from the air.

A famous controversy arose in the eighteenth century. In 1745 the English naturalist John Needham published the result of experiments in which meat extracts and other infusions were sealed in jars and subjected to high heat. He found that all of them were soon swarming with microorganisms. He ascribed their appearance to the operation of a "vital force."

Needham's idea was challenged by the Italian scientist Lazzaro Spallanzani, who found that if various vegetable broths in sealed vessels were subjected to sufficient prolonged heating, they never developed microorganisms. He ascribed Needham's results to inadequate heating which failed to kill contaminating microorganisms already in his jars. Needham retorted that Spallanzani had killed the vital force by his prolonged heating.

The controversy raged for another century during which numerous experiments by various scientists yielded conflicting results. It was finally settled in 1862 by the great French chemist and biologist Louis Pasteur. He boiled a sugar solution in a flask with a long tubular neck bent into an S-curve like the neck of a swan. The end of the tube was open, as shown in Figure 43.

No microorganisms developed in the flask, although it was

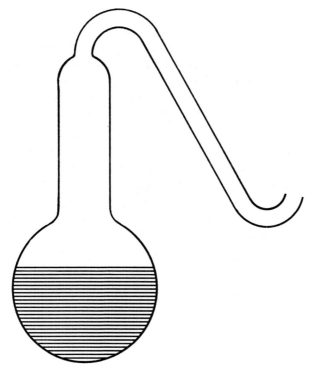

Figure 43. Pasteur's flask

open to the air. It was impossible for dust particles, water particles, bacteria, molds, and other microorganisms to get through the bent neck. But if the flask was tipped so that the liquid came in contract with the trapped organisms in the neck, the liquid was soon found to be swarming with bacteria. Pasteur had demolished a notion that had its beginning in ancient Babylon and Egypt. The French Academy of Sciences awarded him a special prize. But the downfall of the notion of spontaneous generation caused many biologists to return to the notion that there was some "life force" which distinguished living organisms from inanimate matter.

THE TWENTIETH-CENTURY VIEW

Early in the present century a number of biologists came to the view that life on earth had indeed evolved from inanimate matter, but not in any sudden spontaneous fashion. Their view was that life arose from inanimate chemical compounds by a slow process of evolution over a period of millions of years. In a sense this was a return to the notion of spontaneous generation, except that "spontaneous" is not a fitting description for a process that took millions of years. It might better be termed a theory of chemical evolution. It was given its first clear and logical presentation by the Russian biologist A. I. Oparin in 1924 in a book titled *The Origin of Life*. It was translated into English in 1936. His work is the foundation of present-day views.

It is now thought that living things first appeared in the ocean. However, the first organic molecules may have been formed in the earth's primitive atmosphere.

It is believed that the primordial atmosphere of the infant earth was quite different from today's atmosphere, consisting of hydrogen, water vapor, carbon dioxide, methane (a hydrocarbon), and ammonia (a compound of nitrogen and hydrogen). It is significant that these gases exist in the atmosphere of Jupiter and the other major planets and in the tails of comets. An important feature of the earth's primitive atmosphere was the absence of free oxygen.

It is assumed that the first organic molecules were formed in this atmosphere with energy supplied by the ultraviolet light of the sun, lightning, cosmic rays, the solar wind of subatomic particles, radiations from radioactive elements in the earth's crust, and volcanic heat. The possibility of such events has been attested by a great variety of laboratory experiments. In 1951 the American biochemist Melvin Calvin used a beam of alpha particles to bombard a mixture of gases like those thought to have formed the earth's primitive atmosphere. A number of simple organic molecules including lactic acid, formic acid, and malic acid were formed. Two years later another group of American biochemists under the leadership of Harold C. Urey achieved the synthesis of a number of amino acids by means of an electric discharge in a similar mixture of gases. In 1963 Calvin and his team synthesized adenine, one of the four bases in the nucleotides comprising DNA, by bombarding the mixture of gases with a stream of electrons. The following year Calvin and his team succeeded in creating even more complicated molecules, including adenosine phosphate and glucose phosphate.

In a spectacular experiment Sidney W. Fox of the University of Miami created primitive proteins which he called proteinoids. He placed dry, powdered amino acids on a block of lava which he heated for several hours in an electric furnace. The block was then removed and the molten amber mass which had formed on it was treated to a shower of water. Fox compares this shower to the primordial rains of the infant earth. As clouds of steam arose from the block the amber mass was dissolved, forming what he called a primordial soup. The electron microscope revealed millions of tiny globules of proteinoids in it. Fox thinks that proteinoids were formed on the infant earth when amino acids in the atmosphere landed on hot solidified lava flows in volcanic areas.

Organic molecules, however formed in the atmosphere, were washed into the ocean, where more such molecules formed in the surface waters, perhaps in tidal pools. These persisted for two reasons: There was no oxygen to oxidize them and there were no organisms to eat them. The primitive ocean thus became a "sterile broth."

Random collision of these organic molecules in the sea led to the formation of still more complex systems. In this fashion, over extremely long periods of time, monosaccharides became polysaccharides, fatty acids formed fats, amino acids were linked together to form proteins. Perhaps some of these proteins acted as enzymes, promoting and speeding up the formation of complex molecules.

James Bonner believes that the great turning point in this long process was the appearance of the first molecule of nucleic acid. Here was a molecule capable of duplicating itself by replication and subject to mutation by the impact of radiations.

The intermediate step between organic molecules and living things was the formation of intermolecular aggregations. Proteins were linked to carbohydrates, fats, or nucleic acids. These aggregations came about as the result of the polarization of molecules (see Chapter 19). It has long been known that molecules will form solid cyrstals. But they also can form liquid crystals when in solution. These crystals have definite architectural structure.

Oparin suggested that natural selection began to operate at this point. Certain colloidal mixtures were more successful than others in attracting organic molecules to themselves and bonding them into their structure. It may be that some such aggregation, containing a molecule of DNA, became the first living thing.

twenty-nine

BACTERIA AND VIRUSES

We have been scourged by invisible thongs, attacked from impenetrable ambuscades, and it is only today that the light of science is being let in upon the murderous dominion of our foes.

—JOHN TYNDALL

ANTONY VAN LEEUWENHOEK

THREE centuries ago a Dutch storekeeper went exploring and found a new world in a drop of stagnant water, a world teeming with life, whose denizens had helped and hurt man since the beginning of the race without his being aware until that moment of their existence. The explorer was Antony van Leeuwenhoek. He discovered the existence of microorganisms in a drop of water.

Leeuwenhoek was born in 1632 in Delft, Holland. At the age of 16 he became an apprentice in a drygoods store in Amsterdam, but at 21 he returned to Delft, opened his own drygoods store, and later became an assistant to the sheriff of Delft. It was a hobby that made Leeuwenhoek famous. Store-keepers used little magnifying glasses to count the threads in cloth. Leeuwenhoek was fascinated with the idea that a lens would reveal details too small to be seen with the unaided eye. The thrifty Dutchman set about grinding his own lenses and before long was making better ones than any in existence. With these he made small microscopes of a type known as the simple microscope because it contains only

407

a single lens. For 20 years he examined everything he could lay his hands on—the sting of a bee, the leg of an insect, the eye of a fly, whatever caught his fancy at the moment.

One day he told a fellow townsman, Regnier de Graaf, about his discoveries. De Graaf was a corresponding member of the Royal Society of England and he wrote a letter to the society, suggesting that they ask Leeuwenhoek to send his discoveries to them. Subsequently Leeuwenhoek sent the society a long letter. It was the first of many letters which he sent. He dispatched the last one when at the age of 91 he lay on his deathbed.

His greatest discovery came when some sudden fancy caused him to put a drop of stagnant water under his microscope. To his intense surprise, the drop of water was crowded with tiny living things. He called them "wretched beasties," in a letter to the Royal Society. Most of the creatures Leeunwenhoek saw were undoubtedly amebas and paramecia, but there probably were some bacteria present as well.

The members of the Royal Society were amazed by this communication from Leeuwenhoek. In fact, they refused to believe it and commissioned two of their members, Robert Hooke and Nehemiah Grew, to check his claims. On November 15, 1677, they reported that Leeuwenhoek was right. But almost two centuries were to slide by before the true significance of microorganisms was understood.

LOUIS PASTEUR

It was the eminent French scientist Louis Pasteur who taught the world the importance of bacteria. His discoveries laid the foundations for the science of bacteriology and changed the whole course of medicine and surgery.

Pasteur was born in Dôle, a small French city, on December 27, 1822, the son of the town's tanner. His father sent him to the famous École Normale Supérieure in Paris where he made a brilliant record. In 1854, at the age of 32, he was appointed professor of chemistry and dean of sciences at the University of Lille.

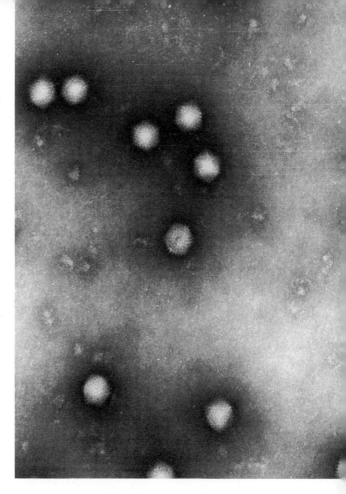

Adenoviruses, an example of spherical viruses, photographed with the electron microscope

The production of alcohol by the fermentation of the juice of sugar beets was a major industry in Lille. The juice, placed in large vats, soon began to froth and bubble as the sugar turned to alcohol. The process was regarded as a purely chemical one. While the appearance of yeast in the vats was known, chemists of the day regarded them as mere byproducts of spontaneous generation which played no role in the process of fermentation.

One day the father of one of Pasteur's students came to him to report that something had gone wrong with the big vats. Pasteur hurried to the factory, in all probability without the slightest idea of what he was going to do. He gathered up some of the liquid from the vats which were working well and some from the vats that were not functioning. Then he returned to his laboratory to examine the samples under his microscope.

In the liquid from the "healthy" vats Pasteur found tiny globules of yeast. The globules were growing, sprouting little buds, which separated and became globules in turn. Then he examined the liquid from the "sick" vats. There were no yeast globules to be seen here, but instead tiny rodlike things, which quivered and danced with incessant motion. He found that the beet juice which contained these microbes was producing lactic acid, the acid of sour milk, instead of alcohol. In both cases the production of alcohol or lactic acid increased as the population of yeast globules or microbes grew. Pasteur became convinced that these organisms were the cause, not the byproduct, of fermentation.

In 1857 he was called to Paris as assistant director of the École Normale. He continued his experiments in an attic laboratory. He concluded that yeasts and bacteria were in the air and that they fell into the liquids they fermented. To prove this, he drew samples of air through a tube with a cotton plug in it. Then he dissolved the plug in a mixture of alcohol and ether. Under the microscope he found bacteria in the solution.

Some intuition convinced Pasteur that microbes caused disease. No experiment had proved it yet, but he was positive of it. In a lecture at the Sorbonne he caused a ray of light to be directed across the darkened room. "Observe the thousands of dancing specks of dust in the path of this ray," he told his au-

Replica of Leeuwenhoek's microscope. Its single lens was set in the brass plate. Objects to be observed were put on the point of the screw.

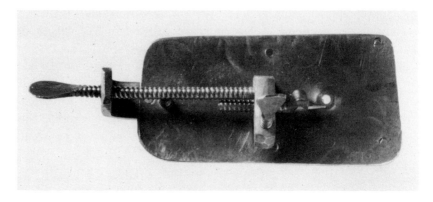

dience. "The air in this hall is filled with these specks of dust, these thousands of little nothings, that you should not despise always, for sometimes they carry disease and death, the typhus, the cholera, the yellow fever, and many other pestilences."

Eventually both Pasteur and the German physician and bacteriologist Robert Koch proved that certain diseases could be laid at the doors of specific microbes. Meanwhile a surgeon at the University of Glasgow, Joseph Lister, having read some of Pasteur's reports, revolutionized surgery by using carbolic acid to keep the bacteria of the atmosphere out of open wounds and surgical incisions. Modern surgery, with its emphasis on antiseptic methods, began with Lister's techniques. Pasteur, Koch, and Lister have been called the great triumvirate of bacteriology.

BACTERIA

The unicellular organisms known as bacteria are the most abundant form of life on earth. They thrive everywhere—in the air, in the soil, in food, in the bodies of plants and animals. Bacteriologists have described about 2,500 species. They are among the smallest and structurally simplest forms of life with dimensions ranging from as little as $1/250,000$ of an inch to about $1/5,000$. A drop of water can hold 50 million bacteria.

Bacteria can be divided into three types on the basis of their shapes. The majority are rod-shaped and are known as bacilli. They occur singly, in pairs, or joined end to end in filaments or strings. The next most numerous group are the sphere-shaped, known as cocci. They occur singly, in pairs, in groups of four, in strings, and in clumps. The strings are known as streptococci, the clumps as staphylococci. The least numerous group are shaped like spirals or corkscrews and are known as spirilla.

The bacterial cell possesses a cellular membrane surrounded by a rigid cell wall sometimes composed of cellulose. The wall is covered by a gelatinous coat, or slime layer, of varying thickness, composed chiefly of polysaccharides. When it is quite thick, it is known as a capsule. The most virulent disease-producing bacteria are encased in such capsules, which apparently protect them from the body's normal defenses.

The cytoplasm of the bacterial cell lacks an endoplasmic reticulum, mitochondria, and the Golgi complex. Ribosomes are free in the cytoplasm. There is a nuclear region, but no nuclear membrane. Chromatin bodies containing DNA occur in this region. It is thought by some biologists that these bodies constitute a single bacterial chromosome. Various bacilli and spirilla possess whiplike appendages known as flagella which enable them to move about in fluids.

The great majority of bacteria are either parasites, invading plants or animals and obtaining their nourishment from the tissues of their hosts, or saprophytes, living on dead organic matter. A small number possess a blue-gray pigment differing from the chlorophyll of plants, but enabling them to carry on photosynthesis. Certain other bacteria can obtain their energy from the oxidation of sulphur compounds, iron compounds, or nitrogen compounds. Most bacteria require oxygen, but a few are anaerobic, growing only in the absence of oxygen.

Bacteria reproduce by division or transverse binary fission. The cell divides into two equal parts. Under favorable conditions some bacteria will divide every 20 minutes. Fortunately, this rate cannot be maintained for long because of lack of space and food. If it could, one bacterium would generate a mass of bacteria weighing 2,000 tons in 24 hours.

While the layman thinks only of the harmful effects of bacteria, the fact is that their beneficial activities outweight their harmful ones. Were it not for bacteria, the surface of the earth would be buried beneath the remains of dead plants and animals. By encompassing the decomposition of these dead bodies, bacteria return the carbon dioxide and water to the atmosphere. The nitrogen-fixing bacteria, which have the ability of converting the nitrogen of the air into nitrates, grow in the soil or on the roots of certain plants, providing the nitrogen which plants require for the synthesis of proteins. Many agricultural processes, including the production of butter and cheese, depend upon the presence of bacteria.

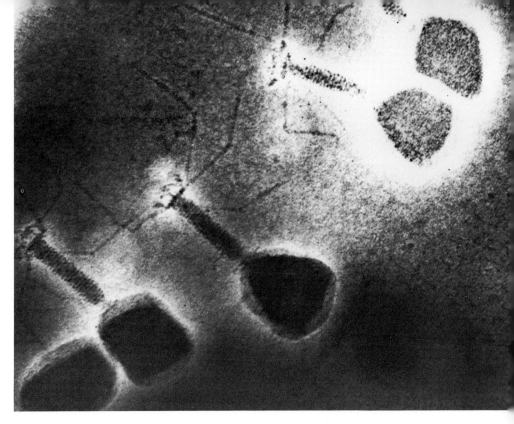

T4 virus, magnified 200,000 times by an electron microscope. Each T4 has six tail fibers.

VIRUSES

Before the end of the nineteenth century medical men were aware of contagious diseases for which no infecting bacteria could be found. Biologists were aware of similar situations with respect to plant and other animal diseases. In 1892 the Russian biologist Dimitri Iwanowsky found that if he squeezed some of the juice from tobacco plants suffering from tobacco mosaic disease and rubbed it on the leaves of healthy plants, they developed the disease. (The disease gets its name from the fact that it causes the leaves to develop a mottled appearance before they wither and die.) Iwanowsky found that if he passed the juice through an unglazed porcelain filter, it still had the ability to cause the disease. He regarded the juice as a contagious fluid. Six years later the Dutch biologist Martinus Beijerinck concluded that the juice contained an invisible agent small enough to get through the filter. He coined the name virus for this agent from the Latin word for "poison."

Almost 40 years elapsed before the existence of viruses was demonstrated. In 1935 Wendell Stanley, working at the Rockefeller Institute, ground up more than a ton of diseased tobacco leaves and extracted the juice from them. From this extract he obtained about a spoonful of needlelike crystals resembling the crystals of various chemical compounds. However, when they were suspended in water and rubbed on a healthy tobacco leaf, they produced the mosaic disease. Stanley received the Nobel Prize for this achievement.

The electron microscope has made it possible for biologists to see viruses and to determine some of their characteristics. In general viruses are either rod-shaped or spherical. They vary in size but are only a few millionths of an inch in length or diameter.

A virus consists of an outer shell, or coat, of protein surrounding a core of DNA or RNA. In some viruses both nucleic acids are present. Biologists are unable to agree as to whether the virus should be considered non-living or living. Because of its composition it has been called a "naked chromosome" and because of its behavior "a group of genes in search of a home." In isolation the virus is apparently lifeless, showing no ability to grow or reproduce and undergoing no metabolic activity. However, viruses possess the ability to invade the cells of bacteria, plants, and animals, doing no apparent harm in some cases and causing cell destruction and disease in others. About 300 disease-producing viruses are now known. Human diseases caused by virus include polio, smallpox, chicken pox, rabies, influenza, and the common cold. Viruses are highly selective in action. A given virus will invade only specific cells of a particular organism. Thus the polio virus invades only one kind of nerve cells in the brain and spinal cord.

Our knowledge of virus action comes largely from studies of the bacteriophages, known as phages for short, which invade certain bacteria. The amazing drama of the invasion can be recorded with the aid of the electron microscope. The bacteriophage is shaped much like a tadpole with a rounded or many-

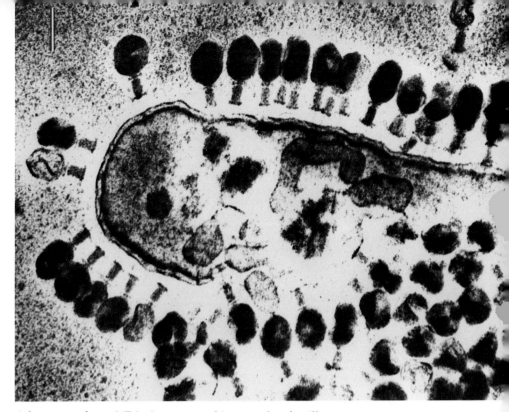

A large number of T4 viruses attacking a colon bacillus

sided head and a tail. A number of fibers extend from the tail.

Some of the most enlightening experiments have been conducted with phages known as the T_2 and T_4, which invade the common colon bacillus. When the phage collides with a bacillus, it attaches itself to the bacillus by its tail fibers. It secretes an enzyme which creates a hole in the wall of the bacillus. Next it inserts the tip of its tail into this hole. The tail then contracts, injecting the DNA content of the phage into the bacillus. Within 30 minutes the bacillus swells up like a balloon and then bursts, releasing 200 or more phages. It is apparent that the DNA of the phage has taken command of the machinery and contents of the bacillus, turning it into a factory for the manufacture of more phages. It is possible to think of the phage DNA as consisting of dominant genes which take command away from the bacterial genes.

thirty

PLANTS AND ANIMALS

> *And God said, Let the earth bring forth the living creature after his kind, cattle, and creeping thing, and beast of the earth after his kind: and it was so.*
>
> —GENESIS I: 24

To date, biologists have listed about 450,000 different kinds of plants and approximately a million different kinds of animals. About three-fourths of the animals, however, are insects. The list continues to grow as new plants and animals are discovered. The branch of biology concerned with the classification of living things is known as taxonomy. While attempts to classify plants and animals into groups go back to the time of Aristotle, the modern method of classification had its beginnings in the work of the Swedish physician and biologist Carolus Linnaeus in the 1730s. He called each kind of plant and animal a species, and grouped those species which resembled each other into genera (singular: genus). He gave each organism a double name derived from the Latin which denoted both the genus and the species. The system is still used today. Thus, present-day man is designated as Homo sapiens.

Biologists are not in agreement in the field of taxonomy, and a number of different systems of classification exist. One school divides all living things into two major groups, the plant kingdom and the animal kingdom, regarding bacteria as unicel-

Bison group

lular plants and protozoa as unicellular animals. Another school prefers to put all unicellular organisms into a separate kingdom known as the protista. Kingdoms are divided into phyla. These in turn are subdivided into classes, orders, families, genera, and finally species.

PLANTS

Leaving aside the technical ramifications of conflicting systems of classification, it will suffice for our purpose to note that plants can be divided into three major groups, the thallophytes, the bryophytes, and the tracheophytes.

The thallophytes are structurally the simplest plant forms. They lack roots, stems, and leaves. They can be divided into two principal subgroups, the algae and the fungi.

Algae are the most abundant plants in streams, rivers, ponds, lakes, and vast stretches of the world's oceans. Some 30,000 species of algae have been described, ranging from microscopic unicellular types to huge seaweeds like the giant kelp of the Pacific Ocean which reach lengths of 100 feet or more. A few species occur on land and can be found growing on wet soil or the bark of trees.

Practically all algae contain chlorophyll and carry on photosynthesis. They are the basic food supply for all aquatic life. Because many algae contain pigments in addition to chlorophyll, algae are known by their colors as blue-green, green, yellow-green, brown, golden-brown, and red algae.

The fungi range from microscopic unicellular yeasts to mushrooms and puffballs weighing more than a pound. About 75,000 species are known, including molds, mildews, blights, yeasts, rusts, smuts, morels, mushrooms, and puffballs. While many fungi do immense damage to growing plants, foodstuffs, and other products, other fungi, like some yeasts and the edible mushrooms, are highly useful. Fungi, like bacteria, help clear the earth of dead plants and animals.

Unlike the algae, the fungi lack chlorophyll and are either parasites on living organisms or saprophytes living on dead organic matter.

The bryophytes consist of the mosses and their less familiar relatives, the liverworts. They are small green plants possessing stems and leaves, but no true roots, only weak, hairlike appendages, known as rhyzoids, which anchor them poorly. They live mostly in wet places, in marshes, swamps, bogs, on damp rocks, in deep woods, and along streams or pools. While most abundant in the temperate and tropical zones of the earth, they are the dominant plants of the treeless tundras of the Arctic. About 25,000 species are known.

The tracheophytes are vascular plants whose stems have systems of tubes for the circulation of sap. About 300,000 species have been listed. The chief subgroups are the ferns and their relatives, known to botanists as the felicineae, the gymnosperms, and the angiosperms.

Ferns were the dominant plant life 300 million years ago in the great swamps and marshes of the Mississippian and Pennsylvanian periods. About 10,000 species are known today. The fern is far better equipped for the battle of life than moss. The roots penetrate the soil, where they can absorb water and mineral salts. The tall stems display the leaves to the atmosphere, where they can obtain an adequate supply of sunlight for the process of photosynthesis.

The gymnosperms and angiosperms are the seed plants. There are only some 700 species of gymnosperms today, but nearly

300,000 species of angiosperms are known, making them the most numerous species in the plant kingdom.

A typical seed is an embryo plant surrounded by a protective coat. When conditions of moisture and temperature are right, the seed coat softens and the young plant takes root in the soil and begins to grow.

The gymnosperms are the cone-bearing trees and shrubs. The chief ones are the conifers, which include such well-known trees as the pines, spruces, firs, cedars, cypresses, and sequoias. The seeds are formed on the upper sides of the scales of the cones.

The angiosperms are the flowering plants. They constitute most of the familiar trees, shrubs, and farm and garden plants. The flowers are the reproductive organs of the angiosperms. Sperm cells are formed in the pollen grains in the anthers, the

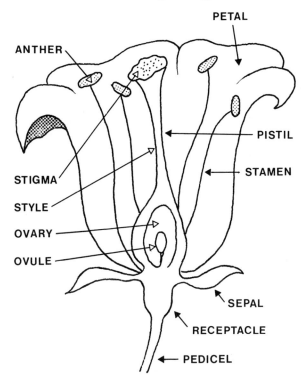

Figure 44. Structure of a flower

knoblike sacs at the tops of the stamens. The base of the pistil contains the egg cells, or ovules. This is shown in Figure 44.

A pollen grain, landing on the sticky top of the pistil, develops a pollen tube which penetrates the pistil, permitting the sperm cell to reach the ovule, forming a zygote which develops into an embryo plant. But growth stops shortly and a protective coat forms over the embryo, creating a seed.

PROTOZOA

Protozoa have long been regarded as unicellular animals. However, some biologists today prefer to classify them with the bacteria in the kingdom of the protista. If your schooling included a class in elementary biology, you have had the experience of watching various protozoa under the microscope. Estimates of the number of species of protozoa range widely from 15,000 to 100,000. They are found in the soil, in fresh water, in the ocean, and as parasites in plants and animals.

One ameba, found in stagnant water, is about 1/100 of an inch in diameter, but other species are smaller. The ameba looks like an animated blob of jelly. Its shape changes continuously, as it pushes out fingerlike extensions called pseudopodia. It moves slowly in any direction by extending these "fingers." Its food consists of bacteria, small algae, protozoa smaller than itself, and bits of organic matter. It approaches a food particle, surrounds it with pseudopodia, and engulfs it. Excess water, engulfed with the food, is concentrated in the so-called contractile vacuole and expelled by the sudden contraction of the vacuole. Details of the ameba are shown in Figure 45.

Other protozoa have definite shapes. One of the common ones found in stagnant water is the paramecium, which is shaped like a microscopic slipper.

ANIMALS

Biologists variously classify the animal kingdom into some two dozen phyla. However, the greater proportion of all animals oc-

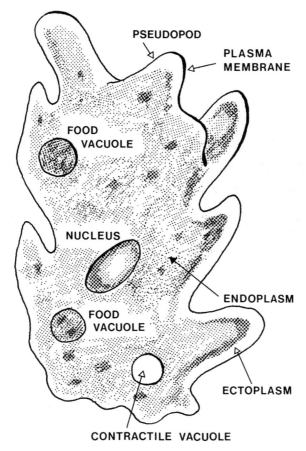

Figure 45. The ameba

cur in 10 major phyla, if we include the protozoa. While these show an ascending order of complexity of structure and function, the mistake must not be made of assuming that they represent a single line of evolution. The situation is far more complicated than that.

The simplest and most primitive of the multicellular animals are the sponges, known to biologists as the porifera. Aristotle though they were plants. About 15,000 species are known, ranging in size from a fraction of an inch to a diameter of several feet. They are mostly marine forms, living in shallow waters, firmly anchored to the ocean floor.

We come next to the coelenterates (10,000 species). They

Black bear group

are predominantly marine creatures, including the jellyfish, corals, and sea anemones. There are a few fresh-water forms like the tiny hydra. They are characterized by a primitive plan of organization consisting of radial symmetry centered on a single internal digestive cavity. All of the coelenterates possess special cells capable of shooting out stinging harpoonlike barbs.

Bilateral symmetry is found in the next two phyla, the flatworms (10,000 species), and the roundworms (10,000 species). The remaining phyla of the animal kingdom show a steady progression in complexity.

The mollusks (100,000 species) include the snails, slugs, clams, mussels, oysters, squids, octopuses, and nautiluses. We come next to the annelids (10,000 species), largely marine worms with segmented bodies, but including the familiar earthworm.

This brings us to the arthropods, the largest of the animal phyla. It includes about a million species of which about 800,000 are insects. All other animal phyla only total about 150,000 species. The arthropods have segmented bodies, jointed legs, and highly developed central nervous systems. They include the insects, spiders, millipedes, centipedes, shrimps, crayfish, crabs, and lobsters.

The echinoderms (6,000 species) show radial symmetry, five or more rays, or arms, radiating from a central body. They include the starfish, sea urchins, sea lilies, sea cucumbers, and sand dollars.

We come finally to the phylum which is most highly developed and which includes mankind. It consists of the chordates (50,000 species). The principal members of this phylum are the vertebrates, but taxonomists include in it a number of marine animals, relatively unknown except to biologists. These animals, which include the sea squirts, have at some time during their life cycles primitive backbones known as notochords.

The vertebrates include the fish, amphibians, reptiles, birds, and mammals. They all possess internal skeletons of bone and cartilage, including a vertebral column or backbone. Some of the fish, however, including the sharks, skates, and rays, have

Prairie dog village, eastern New Mexico

Wildebeests or gnus, Tanzania

Cape buffaloes, Uganda

completely cartilaginous skeletons. Fish are cold-blooded, possessing two-chambered hearts.

The smallest group of vertebrates are the amphibians including the salamanders, newts, frogs, and toads. The reptiles include the snakes, lizards, turtles, crocodiles, and alligators. Both amphibians and reptiles are cold-blooded, but possess three-chambered hearts. Birds and mammals are warm-blooded and have four-chambered hearts. Among the distinguishing characteristics of the mammals are the presence of hair, which serves as insulation, and mammary glands for nursing the young. The mammals include such well-known animals as the moles, shrews, and other insect-eaters, squirrels, mice, rats, and other rodents, cats, dogs, wolves, lions, and other meat-eaters, horses, cattle, elephants, whales, and dolphins. The most advanced mammals are the primates, including the lemurs, tarsiers, monkeys, apes, and man.

thirty-one

MAN

> *... the heir of all the ages, in the foremost files of time.*
> —TENNYSON

MODERN man, or Homo sapiens, made his appearance on earth about 200,000 years ago, but the precursors of man split off from the primate stock about 14 million years ago. Anthropologists are not yet in agreement as to the time. However, we must go back some 100 million years to the middle of the Mesozoic era, or Age of Reptiles, for the start of the story. The mighty dinosaurs ruled the earth. But tiny, archaic mammals had already made their appearance. The jawbone of one of these mammals was smaller than a single tooth of a big dinosaur. Sixty-three million years ago the Age of Reptiles came to an end. The Cenozoic era, or Age of Mammals, began.

The primate line is believed to have started 75 to 80 million years ago, before the end of the Mesozoic, with a tiny shrewlike creature that took to living in trees, a sort of life that developed agility, dependence on eyesight, and grasping paws. The present-day tree shrews of southern Asia and the Indies are thought to be the descendants of these first shrews. They are small insect-eating animals, with pointed faces and long tails.

Ring-tailed lemurs in the Washington National Zoo

LEMURS AND TARSIERS

About 60 million years ago the next primates arose. They were small primitive creatures whose nearest relatives today are the lemurs and the tarsiers. Today's lemurs are found on the island of Madagascar. Various species range in size from that of a mouse to that of a cat. They have foxlike faces, monkeylike bodies, and long tails. They have hands but make practically no use of them except to grasp tree branches.

The tarsier, found in the Philippines and Borneo, is a strange bug-eyed skinny-fingered creature that looks like a teddy bear with a long tail. He is about the size of a big rat. He leaps easily from bough to bough and perches on one in an upright position. The fossil record shows that the lemurs and tarsiers were numerous and widespread during the first half of the Age of Mammals. Then they dwindled in numbers.

Brown lemurs, Washington National Zoo

The evolution of these early primates was not the result of sudden, major mutations, but the end product of a whole series of minor mutations which were preserved by the process of natural selection. The accumulation of these moderate improvements gave them new traits of which the most important one was fairly efficient grasping paws. Others were nails in place of claws, five well-separated fingers or toes on each hand or foot, an ability to rotate the thumb and great toe, and good movable arms. Tree shrews, lemurs, and tarsiers are grouped together as the lower primates, or prosimians.

THE MONKEYS

About 40 million years ago the primitive monkeys, the first of the higher primates, arose. Not too much is known about how this happened. The earliest monkeys are fairly well developed and fossils of transitional types are missing.

The bug-eyed long-fingered tarsier

Two lines arose, apparently separate from the very start. One appeared in America and became the New World monkeys, including marmosets, spider monkeys, howler monkeys, cebus, or organ-grinder, monkeys, and others. Some of them are distinguished by the ability to swing by their tails. The other line appeared in Asia and Africa and became the Old World monkeys. Some are bigger than the New World monkeys. These include the macaques and baboons. The baboon, the largest and fiercest of the monkeys, lives on the ground and is a savage fighter.

THE ANTHROPOID APES

We come next to the group of primates known as the hominoids. These include the anthropoid apes and man. It got its start about 40 million years ago. Anthropologists do not known whether it arose from mutations among the Old World monkeys or whether it split off from a common ancestral type which preceded both lines. But at one time it contained a greater variety of creatures than today's hominoids. The hominoids differed from the monkeys in that they developed a tendency to hang from tree branches by their arms while feeding and to swing by their arms from branch to branch.

The present anthropoid apes indicate that all the hominoids did not follow the same line of development. The gibbon was the first to split off from the main ancestral line. It is a small ape and can show acrobatic proficiency that is denied its bigger cousins. It swings overhand from one limb of a tree to another with ease and elegance.

The next line to split off was the orang. It has about the bulk of a man. It also swings by its arms but in a more deliberate fashion than the more agile gibbon.

The main line split in two about 14 million years ago, perhaps earlier. One branch, which later split again, gave rise to the chimpanzee and the gorilla, the other became man. Both the chimp and the gorilla can swing about in trees by their arms, but they spend much of their time on the ground.

Anthropologists are not sure whether they have found the

Chimpanzee

BELOW: *Orang*

common ancestral form which gave rise to the chimpanzee and the gorilla on the one hand and man on the other. It may be one of a number of known fossils. If you wish, you may consider this uncertain form as the proverbial missing link.

The hominoids include both the anthropoid apes and man. The human line is known to anthropologists as the hominid line. The earliest known hominid may be Ramapithecus, first described and named by George E. Lewis of Yale University. It is known from upper and lower jaws found in India and Kenya. It is 13 million years old.

THE MAN-APES

A large number of fossils have been found of an early form of man which flourished from 5 million years ago to 1.5 million. These fossils, found in South Africa, are known to anthropologists as man-apes, but they are definitely hominids and among man's ancestors. The first skull of a man-ape was found by the South African anthropologist Raymond Dart in 1925. He christened it Australopithecus. He found it in a box of fossils that had been given him by the director of a limeworks northwest of Kimberly, where great bluffs of limestone were being blasted down. Subsequently a great many more fossils of man-apes were found by Dart and Robert Broom, a Scottish medical man who had given up medicine to become a museum curator in Pretoria. Additional specimens have been found in recent years by the East African anthropologists Louis S. B. and Mary Leakey and their son Richard, and also by French and American expeditions.

At first glance the skull of a man-ape looks like that of an anthropoid ape because of the small size of the brain and the large size of the jaw. However, careful examination, particularly of the teeth, shows that the creature was manlike rather than apelike. Skeletal remains confirm this view, showing clearly that the man-ape walked on the ground in an erect posture. He was at home on the ground and not in the trees. The man-apes ranged in height from about 4 feet to the height of present-day man.

The man-apes flourished in South Africa up to about a mil-

The agile gibbon

lion years ago. The Ice Age had begun in the northern hemisphere, but conditions were mild in the hills and valleys of South Africa. The caves in which the man-apes lived reveal heaps of animal bones. Among them are the skulls of baboons which had been split open by crushing blows. Dart believes that the man-apes used the long bones of animals as clubs with which to kill the baboons and split open their skulls. Apparently they regarded baboon brains as a special delicacy.

The next form of man made his appearance on earth about a million years ago. Known to anthropologists as Homo erectus, it included Java Man and Pekin Man.

The Ice Age began in the early part of the Pleistocene, perhaps 2.5 million years ago. In the latter part of the Pleistocene there were four periods of glaciation and three interglacial periods when the climate was warmer. The Fourth Glaciation was severe and Recent Time or Post-Glacial Time began at its close, 15,000 years ago. The first evidence of man is found in the Pleistocene Period. It consists of crudely chipped pebbles known to anthropologists as pebble tools. It is not too certain how old these are. They may go back 3 million years to the start of the Pleistocene Period. Better stone tools, known as hand axes, are found later in the period.

JAVA MAN

Java Man, or Pithecanthropus, whom you probably met in your high-school days, is the oldest example of Homo erectus known to anthropologists. He flourished from a million to 500,000 years ago during the First Interglacial Period or perhaps the Second Glacial Period. The first skull of Pithecanthropus was found in 1891 in a fossil-bearing bed in the bank of the Solo River in central Java by Eugene Dubois, a Dutch army surgeon. He had gone to the East Indies with the express purpose of hunting for the fossils of ancient man. The next year he found a human thighbone about 45 feet from where he had found the skull. It proved that Pithecanthropus walked erect. Pithecanthropus was about the same size as a modern man but he had a thick skull with little room for brains. His brain was about 850 cubic centimeters (modern man's brain averages 1,450 cubic centimeters). His jaw was large but the teeth were human. He had beetling bony brows from which his head sloped back with no suggestion of a forehead. Additional skulls of Java Man were found in the 1930s.

PEKIN MAN

The next oldest man known to anthropologists is Pekin Man, or Sinanthropus. William Howells of Harvard University calls him Java Man's brainy brother because his brain had a volume of 1,075 cubic centimeters. He flourished about 400,000 years ago. In general he resembled Java Man but his mouth and teeth were smaller and a little more like those of modern man.

The story of Pekin Man begins in 1900, when a fairly ancient human tooth showed up in a Peiping (now Peking) drugstore. This attracted the attention of anthropologists to the area where it had been found, the town of Choukoutien, about 27 miles southwest of Peking.

Pekin Man was discovered in a limestone cave in the area in 1926 by Davidson Black of the Peiping Union Medical College. By 1936 skulls and bones of more than 40 individuals had been recovered from various caves. In December 1941, at the request of the director of the Geological Survey of China, it was decided to move the fossils to the United States. North China had been occupied by the Japanese since 1937. The fossils disappeared while in transit to the port of Chinwangtao on the day of the Pearl Harbor attack and have never been seen since. Fortunately, casts of some of the skulls had been sent to the United States a few years earlier.

The Peking caves give ample evidence that Pekin Man was a cannibal, splitting skulls and bones to get at brains and marrow. The presence of charcoal in the caves proves that he had learned the use of fire and roasted his meat.

A number of other fossil finds appear to be as old as Pekin Man. These include several human jaws found at Ternifine in Algeria, and a skull found by Louis S. B. Leakey and his wife, Mary, at Olduvai Gorge in Tanganyika (now Tanzania). These have features very much like those of Java and Pekin Men.

While Europe, in contrast to Asia, yields a profusion of stone tools that make it possible to trace the cultural development of mankind from about the time of the First Interglacial, anthropologists have had almost no luck in finding the fossil remains of earliest men in Europe. The two exceptions are Heidelberg Man, represented by a jaw found in a sand pit near Heidelberg, Germany, in 1907, and a tooth and skull fragments found at Vertesszollos in Hungary. They are about as old as Pekin Man.

THE FIRST MODERNLIKE MAN

A number of skulls found in Europe indicate the transition from Homo erectus to Homo sapiens. The skull of a young woman was found in a gravel pit at Steinheim in West Germany. It shows a higher forehead resembling that of Neanderthal Man. A skull found in a gravel pit at Swanscombe in southern England is also more modern than Homo erectus. The skulls from both sites are about 250,000 years old. A number of skulls found in a cave at Fontechevade, France, date from the Third Interglacial and are even more modern.

Restoration of Neanderthal Man

Restoration of a Neanderthal family

NEANDERTHAL MAN

Most anthropologists today regard Neanderthal Man as the earliest form of Homo sapiens. He gets his name from the fact that the first skull was found in Neander Gorge near Düsseldorf, Germany, in 1856. Neanderthal Man appeared about 200,000 years ago. A few specimens found in southern Europe date from the Third Interglacial. But remains from the Fourth Glacial are found all over Europe, North Africa, Palestine, and Central Asia. Neanderthal Man persisted for about 100,000 years, gradually evolving into fully modern man as we now know him. Then he disappeared from the scene. Neanderthal Man possessed characteristics which set him off sharply from both Homo erectus and fully modern man. He had a low skull and a long face. Some specimens had prominent ridges over the eyes. These ridges were

The powerful gorilla

arches and not bars as with earlier types of men. The jaw was large and heavy. Among European Neanderthals the middle portion of the face protruded forward, throwing the large nose into unusual prominence. The brain was about as large as that of today's man. The bones were short and heavy. The joints were thick. He was probably 5 to 6 feet tall. His chief difference from later forms of Homo sapiens was his big front teeth.

TODAY'S MAN

Our kind of modern man first appeared in Europe about 35,000 years ago, having evolved from Neanderthal Man.

The first fully modern man is known as Cro-Magnon Man. The name comes from the site where the first fossil remains were found in 1856, the Cro-Magnon cave near the village of Les Eyzies in France. Cro-Magnon Man exhibited the refinement of the head which distinguishes modern man from his predecessors. Perhaps the most obvious characteristics are the smaller face, small front teeth, and a higher and shorter skull. The features are delicate and the nose has a bridge.

Like Neanderthal Man, the Cro-Magnons made a variety of tools, including spearpoints, knifeblades, and scrapers, by striking flakes from flint rocks. But they also made harpoons and bone needles. They were better hunters than the Neanderthals. They used the same caves that the Neanderthals did and their remains are found in higher layers in these caves. They were good artists and decorated the walls of the caves with amazingly realistic polychrome paintings of the animals they hunted.

thirty-two

MIND

The brain is the citadel of the senses.
—Pliny the Elder

MAN rules the earth today because he has the most complex and highly developed brain. Physically he is no match for the teeth and claws of the lion or the bear. The bird has better eyesight. The monkey is more agile, the deer swifter. The dog and many other animals have a better sense of smell. But no living creature's brain can match that of man. It has been said that the human brain is the greatest and most complex wonder of the known universe.

It is a curious fact that Aristotle was completely confused about the roles of the heart and the brain. Although Alcmaeon of Croton had recognized the main functions of the brain almost two centuries earlier, Aristotle taught that the heart was the seat of intelligence and that the brain was only a sort of sponge whose function was to cool the blood. His mistake gave rise to literary expressions still in use today—an understanding heart, a kind heart, and the like.

THE NERVOUS SYSTEM

A characteristic of all living things is the ability to react to the environment and to respond to stimuli. The ameba will surround and engulf an edible particle. It will move away from irritating stimuli. A great variety of reactions are shown by the ameba and other protozoa. In multicellular animals the recognition of stimuli and the response to them is controlled by the nervous system. It is possible to trace an increasing complexity in the nervous system from the simplest animals to man. As the organization of the nervous system becomes more complex, the level of behavior becomes higher.

The simplest nervous systems are found in jellyfish, sea anemones, and other coelenterates. These are merely networks of nerve cells extending through the body of the organism. The sea anemone has a small hollow stalklike body and a mouth surrounded by a ring of tentacles. If it is touched, the network of nerves causes the whole creature to respond to the stimulus. The tentacles are drawn in, the water is ejected from the central cavity, and the body is contracted to its smallest possible volume.

The beginnings of a centralized nervous system are found in some of the flatworms. The network of nerve cells is supplemented by two cords of nerves which run longitudinally through the body. A further development is found in the segmented worms. In the earthworm the two cords are fused into one marked by a succession of swellings which are clusters of nerve cells known as ganglia. Nerves run from the ganglia to the skin and organs of the earthworm. Further development is found in the mollusks. In the most advanced mollusks, the octopus and the squid, the ganglia in the head form a complex and competent brain. The greatest development of the brain is found in the vertebrates, reaching its climax in the human brain.

The vertebrate nervous system consists of two principal parts, the central nervous system and the peripheral nervous system. The brain and spinal cord make up the central nervous system. The brain is enclosed in the skull, the spinal cord in the vertebral column. The brain is an enlargement of nervous tissue at the top of the spinal cord. The peripheral nervous system consists of nerves which radiate from the brain and spinal cord to the skin, the sense organs, the internal organs, and the muscles.

Forty-three pairs of nerves emerge from the brain and

spinal cord to form the peripheral nervous system. Twelve pairs of cranial nerves arise from the undersurface of the brain, emerging through small openings in the skull. The 31 spinal nerves leave the spinal cord at intervals along its length.

The peripheral nervous system carries messages to and from the central nervous system, keeping it in touch with the external world and the body itself, and enabling the organism to react to stimuli.

In addition to the central and peripheral nervous systems, vertebrates possess a third system, known as the autonomic nervous system, which regulates certain vital functions of the body. Its action is involuntary and automatic. It is composed of two parts. One, called the sympathetic system, consists of two chains of ganglia, one on either side of the spinal column. Nerves originating in it run to the organs of the body and to the central nervous system. The other part, called the parasympathetic system, consists of nerves which emerge from the central nervous system and run to the vital organs. The parasympathetic system opposes the action of the sympathetic system, thus composing a system of checks and balances which regulates the vital functions of the body.

THE NEURONS

Like every tissue in the body, the nervous system is composed of cells. These are known as nerve cells, or neurons. There are 10 to 12 billion neurons in the nervous system of a human being. The neuron is unique in many ways. Each neuron consists of a cell body containing a nucleus and a cytoplasm, surrounded by a cell membrane. However, the cytoplasm with its enclosing membrane is extended into threadlike processes known as nerve fibers. A fiber may be less than 1/10,000 of an inch in length or as much as 3 feet. In man some fibers extend from the tips of the fingers or toes to the spinal column. Because of the variation in the size and shape of the cell bodies and in the number, size, structure, and arrangement of nerve fibers, there is an almost infinite variety of neurons. Many nerve fibers are enclosed in a whitish fatty covering known as the myelin sheath.

Nerves are bundles of nerve fibers enclosed in a wrapping of connective tissue. There are hundreds or even thousands of fibers in a nerve. The nerve can be likened to a telephone cable, which contains a large number of insulated wires.

The nerve impulse is now known to be an electrochemical wave which travels along the fiber. When one end of the fiber is stimulated, a change takes place in the membrane of the fiber, permitting positively charged sodium ions in the fluid around the fiber to get through the membrane. This gives the fiber a positive potential at the point of entry. But almost immediately potassium ions flow out through the membrane, restoring the original condition of the fiber. The process repeats itself along the length of the fiber so that a wave of electrochemical activity travels the length of the fiber. In some ways the nerve impulse is much like the spark traveling along a burning fuse.

Nerve fibers normally conduct impulses only in one direction. Two types of fibers are recognized. Those that conduct impulses toward the cell body are known as dendrites. Those that conduct impulses away from the cell body are called axons.

The neurons of the vertebrate nervous system can be divided into three general types: sensory, or afferent, neurons;

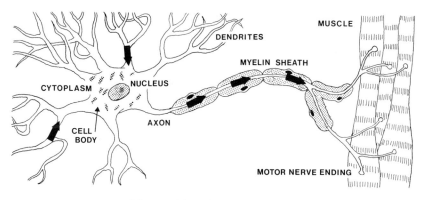

Figure 46. A motor neuron

motor, or efferent, neurons; and intermediate, or connector, neurons. A motor neuron is shown in Figure 46.

A sensory neuron has one very short fiber emerging from the cell body which almost immediately divides into two longer branches, a dendrite and an axon. Each ends in a proliferation of smaller branches. The endings of the dendrite terminate in a sense organ, or receptor. These range from such highly specialized structures as the eye or the ear to specialized cells. In some cases they are modified nerve fibers or even the actual nerve fibers themselves which end in the skin or deeper tissues of the body. There are millions of receptors in the body, each type responding to a particular type of stimulus. However, the nerve impulses which they initiate are essentially alike. The axon of the sensory neuron enters the spinal cord or brain, where its branches transfer the impulse to one or more intermediate neurons.

The intermediate neuron has many short fibers. The dendrites receive nerve impulses from the axons of other neurons. Its axons transmit impulses to the dendrites of other neurons.

The motor neuron has a number of short branching dendrites and one long axon which ends in a number of small branches that transfer the nerve impulse to the cells of muscles or organs.

Nerves are bundles of nerve fibers. Three types of nerves are known. Those consisting of fibers from afferent neurons are known as afferent nerves. Those consisting of fibers from efferent neurons are called efferent nerves. Nerves which include both types of fibers are known as mixed nerves.

Nerve impulses are transmitted from the axon endings of one neuron to the dendrites of another. The juncture of the two fibers is known as the synapse. It is important to realize that there is no actual contact between the two fibers. They are separated by an exceedingly minute gap. The neuron is a chemical factory as well as an electric wire. When the nerve impulse reaches an axon ending, it releases a chemical substance which crosses the synapse and initiates a new impulse in the next neuron. Some neurons, however, release chemical substances which inhibit further action. A variety of chemical substances are released by different types of neurons including noradrenalin and acetylcholine.

THE REFLEX ARC

At breakfast, let us assume, you are engrossed in your newspaper and while turning the page, accidentally touch the hot coffeepot. Immediately you do a number of things. You pull your hand away with a quick jerk. You feel some pain. You exclaim "Ouch!" or perhaps something more sulphurous. Neurologists have established that you do these things in exactly that order. You pull your hand away in about half the time it takes the sensation of pain to reach your brain. This is a wise provision of Mother Nature. At the moment the most important thing is to get your hand away from the hot coffeepot. The purpose of the pain is to teach you to keep your hand away from hot coffeepots in the future, and that, at the moment, is of secondary importance. The drawing away of your hand is a purely defensive mechanism which takes place automatically before you have become conscious of the incident. This is known as a reflex action and occurs in organisms whose nervous systems are so elementary that they do not possess a brain.

Innumerable reflex actions are constantly occurring in the body. They control the beating of the heart, breathing, the secretion of hormones and other body fluids, the behavior of the digestive system, and so on.

The basis of reflex action is the so-called reflex arc. This,

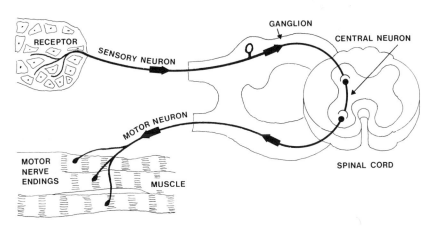

Figure 47. The reflex arc

in its simplest form, consists of three neurons: a sensory neuron, an intermediate neuron, and a motor neuron. In the case of the coffeepot incident a receptor in the skin sensitive to heat stimulates the sensory neuron to action. It transmits the impulse to an intermediate neuron in the spinal cord. This transfers the impulse to a motor neuron which in its turn stimulates the action of a muscle cell. However, the intermediate neuron also activates other intermediate neurons and eventually an impulse reaches the brain. The reader will understand that the coffeepot incident involves a great many reflex arcs and not merely one arc of three neurons. A simple reflex arc is shown in Figure 47.

Connections of the association neuron with others which carry the message to the brain are not shown in Figure 47.

THE BRAIN

The spinal cord of the average adult human being is a hollow tube about 18 inches long. A cross-section of the spinal cord shows that it consists of grayish and creamy-white material known to neurologists simply as gray matter and white matter. The gray matter consists chiefly of the bodies of neurons, the white matter of nerve fibers. The gray matter forms the inner portion of the spinal cord and in cross-section has an outline like that of a butterfly.

The brain is an enlargement at the top of the spinal cord. The human brain weights about 3 pounds. The development of the brain can be traced from the most primitive vertebrates to man. The brain of the most primitive fish consists of three portions known as the hindbrain, the midbrain, and the forebrain. It is possible to think of the human brain as consisting of three such divisions. The structure of the human brain is shown in Figure 48.

The hindbrain consists of the medulla oblongata, the cerebellum, and the pons.

The medulla oblongata is the lowermost part of the brain and is continuous with the spinal cord. A number of the most vital reflex centers are located in the medulla oblongata. These

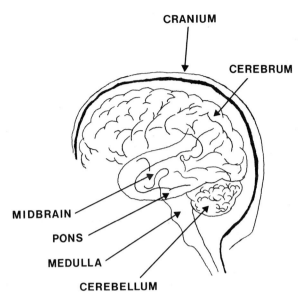

Figure 48. The human brain

control heartbeat, breathing, maintenance of blood pressure, and swallowing.

Immediately above the medulla is the pons, a name derived from the Latin for "bridge." Its neurons form a center which conducts nerve impulses to the other portions of the brain.

Behind the pons there is a large part of the brain which fills the back of the skull. It looks something like a smaller version of the cerebral hemispheres. This is the cerebellum. Three pairs of stalks connect it to the medulla, the pons, and the midbrain. It coordinates all the movements of the body, making them smooth, steady, and precise.

The midbrain is just above the pons and below the lower part of the cerebrum. It plays a role in coordinating muscular behavior. Certain visual and auditory reflexes lie within the midbrain. One of them controls the contraction of the pupil of the eye when it is exposed to a bright light. Both the pons and the midbrain serve largely as pathways between the spinal cord and other parts of the brain.

The forebrain consists of the thalamus, hypothalamus, and the cerebrum. The thalamus is essentially a relay center, receiving nerve impulses that have been brought to the brain from sensory areas of the body and transmitting them to the cerebrum, where they give rise to conscious sensations.

The hypothalamus is a tiny but extremely important portion of the brain which lies below the thalamus. It is the center which controls and organizes whole acts of behavior, coordinating the reflexes which originate in the spinal cord and medulla. The medulla, for example, controls breathing. But the hypothalamus alters the rate of breathing to meet special occasions. It controls the metabolism of the body and the activities of the internal organs. Such basic sensations as hunger, thirst, fear, and rage originate in the hypothalamus. It controls the body temperature, maintains the waking state, and produces sleep. It also controls the sexual drive.

The pituitary gland, about the size of the meat of a hazelnut, grows down on a little stalk from the base of the hypothalamus. It has been called the master gland of the body because its hormones control the action of the thyroid, adrenals, and other endocrine glands. But the pituitary, in its turn, is controlled by the hypothalamus.

Anatomists frequently divide the brain into two parts, the cerebrum and the rest of the brain, which they call the brain stem. The cerebrum is by far the largest part of the brain. It is at the top of the brain, overlapping the brain stem. It is divided by a deep groove into two parts known as the cerebral hemispheres. These are each divided by grooves, or fissures, into five lobes. Unlike the spinal cord, which has the white matter on the outside and the gray matter on the inside, the cerebral hemispheres are composed of gray matter, or cell bodies, with the white matter, or nerve fibers, underneath. These fibers connect the hemispheres to each other and to the rest of the brain. The gray matter is known as the cortex. It is arranged in folds, or convolutions, resembling the surface of the meat of a walnut.

There are about 2 billion neurons in the cerebral hemisphere. The dendrites of a given neuron may connect it to as many as 25,000 others. It is estimated that the total number of interconnections between neurons in the cerebral hemispheres is $10^{2,783,000}$ (this would be 1 followed by 2,783,000 zeroes, a number which defies the imagination).

MIND 449

The highest functions of the nervous system are located in the cortex of the cerebral hemispheres. These include intelligence, insight, personality, judgment, imagination, memory, the interpretation of the senses, and the voluntary control of the muscles.

It was already known to the ancient Greeks that injury to one side of the brain caused paralysis to the opposite side of the body. They concluded that control of the right side of the body resided in the left hemisphere and vice versa. By the start of the present century it was known from the experience of surgeons in dealing with brain tumors and head injuries that many functions are sharply localized in areas of the cortex. This knowledge was greatly expanded in succeeding years by surgical experience, particularly during the two World Wars, and by experiments in which areas of the cortex of both animals and humans were stimulated by mild electric currents. It is possible during a brain operation to do this to the conscious patient without causing pain or damage to the brain.

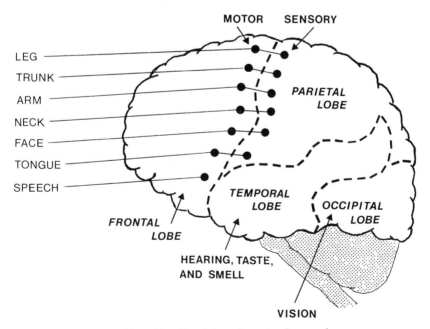

Figure 49. *Localized functions in the cerebrum*

The cortex is now known to contain sensory areas, motor areas, and association areas. Some of these are shown in Figure 49. More than one area is involved in many types of behavior, and some functions are thought to involve the entire cortex.

The sensory areas interpret the sensory impulses which reach the cortex. They include, among others, the visual area, the auditory area, and the olfactory area. These areas integrate and coordinate the impulses reaching them into meaningful concepts. Thus, for example, the impulses generated in the optical nerves are not interpreted as a mere pattern of light and shadow but as a significant entity—a man, for example, whom we recognize as an old friend. At a concert the sound waves striking the ear are heard as a Beethoven symphony or some other piece of music.

The motor areas control the voluntary movement of specific muscles, those of the fingers, the hands, the arms, and so on.

The association areas are believed to be the seats of various mental processes, including memory, intelligence, learning, imagination, and the emotions. It is important to realize, however, that we still know very little about these mental processes which translate the nerve impulses reaching the cortex into our picture of the world and our ability to make decisions for dealing with it.

BRAIN WAVES

By 1928 it was established that the cerebral cortex constantly undergoes small rhythmical changes in electric potential. These are known as brain waves because they can be recorded as waves on a moving strip of paper by an instrument known as the electroencephalograph. Electrodes placed on a person's scalp pick up differences in potential which are amplified and recorded as a wave. When the person is resting quietly with his eyes closed, a characteristic wave known as the alpha wave is recorded. However, it undergoes drastic changes as the result of mental, visual, or muscular activity, or certain types of dreams. Neurologists have found it possible to diagnose brain damage with the aid of the machine.

thirty-three

THE UNITY OF THE UNIVERSE

> *And God saw every thing that he had made, and, behold, it was very good.*
>
> GENESIS I: 31

FROM THE PRIMEVAL FIREBALL TO MAN

MODERN science has achieved its greatest triumph in establishing the unity of the universe. The universe is one. The same basic conservation laws govern the behavior of the elementary particles in the atom, the living organisms on the face of the earth, the stars in the Milky Way, and the billions of galaxies in the vast reaches of space. Evolution is now regarded as a cosmic phenomenon. It is a continuous story, from the explosion of the primeval fireball to the emergence of man as the dominant creature on earth. The story is not yet ended, for evolution in the heavens and on earth continues.

This concluding chapter will summarize briefly the course of cosmic evolution as it emerges from the facts and theories discussed in the preceding 32 chapters. Obviously, there are still many unanswered questions. The picture is by no means complete. There is still much work for scientists to do.

The most generally accepted view of the universe today is that of an oscillating universe, a universe without beginning or end, going endlessly through cycles of about 82 billion years each. The present cycle is believed to have started about 12 billion years ago, when the entire universe was concentrated in the primeval fireball, or superatom. It was about 4,000 miles in diameter. A cubic inch of it weighed 2 billion tons.

The fireball, for some unknown reason, exploded with a "big bang," and the universe began to expand. It was thought that the fireball was composed of neutrons. These disintegrated into protons and electrons which subsequently formed atoms of hydrogen, so that after a few million years the expanding universe was filled with a great fog of hydrogen gas. In time turbulence and gravitation caused local condensations in the fog and it broke up into immense clouds, the protogalaxies.

Just as the universal fog of hydrogen separated into protogalaxies, so each of these fragmented into protostars. The evolution of these protostars into stars is told in detail in Chapter 6 and need not be repeated here.

It is important, however, to keep in mind that the stars were the atomic stoves which "cooked" the chemical elements we now find in the universe. The atomic furnaces at the centers of the stars transformed hydrogen into helium, then helium into carbon, and in the giant stars carbon into the heavier elements.

There may be a trillion galaxies in the universe, each containing from 10 million to a trillion stars. But as inhabitants of the earth we are particularly interested in one of the 100 billion stars in one galaxy, our Milky Way. It is our own sun. Astronomers classify it as a yellow dwarf star. It is about 5 billion years old and is thought to have a life expectancy of at least another 5 or 6 billion years before it expands into a giant red star and burns our earth to a cinder.

The sun is the center of our solar system. You will want to ask whether our sun is the only star in our galaxy surrounded by planets. Astronomers cannot answer this question. The stars are too far away. But many astronomers think it unreasonable to suppose that there are not many systems of planets in the Milky Way, perhaps a million such systems.

As the sun formed from a protosun it left behind a disk of dust and gas. This fragmented into protoplanets which condensed into our present planets, including the earth. In the last few years astronomers have found a number of very young red stars which give evidence of being surrounded by great clouds of gas. These are thought to be going through the same process that resulted in the birth of our solar system. Apparently only protostars of a certain size result in the formation of planetary systems. We do not yet know why this is so.

One of the planets in our solar system is the earth. Geophysicists think it is 4.6 billion years old. This makes the earth almost as old as the sun. Only the four inner planets, the terrestrial planets, namely, Mercury, Venus, Earth, and Mars, and the asteroids have solid surfaces. If we think of the universe as a whole, there is only an infinitesimal amount of solid matter in it.

The 100 billion stars in our Milky Way galaxy are incandescent globes of gas. The galaxy contains additional clouds of gas, mixed with a little dust, equal to some billions more stars. Two gases, hydrogen and helium, account for 96 to 99 per cent of the universe. The atmosphere of our sun, a fairly typical star, is about 80 per cent hydrogen and almost 20 per cent helium. All the other chemical elements make up only about 0.2 per cent.

The protoplanets are thought to have formed before the protosun had become hot enough to shine. As it warmed up its heat and the atomic particles of the solar wind drove most of the hydrogen and helium and other light elements out of the nearer protoplanets which eventually formed the terrestrial planets. Five chemical elements account for 94.4 per cent of the earth by weight. They are iron, oxygen, silicon, magnesium, and nickel.

Geologists believe that the earth has an iron core, surrounded by the mantle, a layer of dense, heavy rock. The crust, which is relatively thin, rests on the mantle.

We have no way of knowing what our earth originally looked like. The oldest rocks now found on the surface of the earth are 3 billion years old. Rocks brought back from the moon by the Apollo astronauts show an age of 4.6 billion years. It is generally assumed today that the rocks of the continents, the waters of the oceans, and the gases of the atmosphere were squeezed out of the once molten mantle as it solidified. Two opposing sets of forces have shaped the surface of the earth in the course of the last 3 billion years: the forces of erosion, which wear away the land, and the forces of diastrophism, which alter the height of the land and create mountain ranges.

Traces of primitive plants are found in rocks more than 2 billion years old, but traces of animal life are extremely rare until 600 million years ago. We do not know how life originated on earth. The generally accepted theory is one of chemical evolution, a process that took millions of years to produce the first unicellular organisms. This is thought to have taken place when the earth still possessed its original primitive atmosphere, an atmosphere that contained hydrogen, water vapor, carbon dioxide, methane, and ammonia, but no free oxygen.

Organic molecules formed in this atmosphere with the aid of energy supplied by the ultraviolet light of the sun, lightning, cosmic rays, the solar wind of subatomic particles, radiations from radioactive elements in the earth's crust, and volcanic heat. Organic molecules originating in the atmosphere collected in the ocean, perhaps in tidal pools. Random collision of these molecules in the sea led to the formation of more complex molecules. The turning point in this long slow process came with the appearance of the first molecules of nucleic acid. Here was a molecule capable of duplicating itself by replication. Some colloidal mixture of molecules containing nucleic acid among other constituents eventually became the first living cell. The record of the evolution of life on earth is told in Chapter 15. The evolution of life reached its present climax in man.

Six chemical elements compose the bulk of any living organism. They are carbon, oxygen, hydrogen, nitrogen, sulphur, and phosphorus. These account for 97 per cent of your body. Some 25 elements make up the other 3 per cent. The organic compounds are all compounds of carbon. Thus carbon may be regarded as the foundation of life on earth. It is interesting to note, however, that water, one of the simplest chemical elements, composes from 60 to 90 per cent of all living organisms.

We can think of cosmic evolution as a process of growing complexity. Not long after the explosion of the primeval fireball, the universe was filled with a fog of hydrogen, the simplest of all atoms. With the formation of the galaxies, the stars by atomic transformations "cooked" the other chemical elements. In the earth we find crystals of simple inorganic compounds whose molecules contain only a few atoms. These crystals form the minerals which in their turn are combined into rocks. The organic compounds found in living organisms have molecules which contain in some cases hundreds of atoms, in others thousands, in still others millions. These organic compounds form the complex structure of the living cell. The most complex cells are the neurons, which form the nervous systems of animals. We find an ascending order of complexity in the nervous system from the simplest multicellular animals to man. Most complex of all is the cerebral cortex of the human brain where the fibers of one neuron may make connections with 25,000 other neurons. The total number of connections in the cortex is estimated to be 1 followed by 2,783,000 zeroes ($10^{2,783,000}$).

THE EVOLUTION OF SCIENCE

It has taken mankind some 5,000 years to arrive at the present unified view of cosmic evolution. Let us take a brief look at the history of science.

To ancient man, the universe was a chaos governed by caprice. In order to explain its phenomena, he found it necessary to people the heavens with a host of minor gods and goddesses, and the mountains and streams and trees with a varied throng of giants, nymphs, and spirits. The occurrence of an eclipse, the appearance of a comet, the gathering of the thunderstorm and the flash of lightning, the occurrence of sickness, particularly an epidemic, were interpreted as the activities of these ancient mythological personages.

Gradually, science revealed the order of the cosmos. The revelation began while the belief in the ancient mythologies still flourished. The Babylonian astronomer-priests regarded the heavenly bodies as luminous gods, but they contributed to the rise of science by their observation of the planets and their ability to predict lunar eclipses even though their astronomy was mixed up with astrology and magic.

Science took a giant step forward in the sixth century B.C.

with the rise of the Greek philosophers. They brushed aside the notion of luminous gods and sought to construct a rational picture of the heavens, the earth, the atom, and life. But mistaken notions about the structure of the solar system and Aristotle's theories of forces and motion plagued the advance of science for many centuries. During the Dark Ages, Arabian and Jewish scholars did much to keep Greek science and medicine alive. Greek learning returned to Europe after A.D. 1000 as the continent emerged from the Dark Ages.

The year 1543 is often cited as the great awakening in the world of science. In that year two books were published which upset ancient ideas and authority. Nicholas Copernicus, the Polish astronomer, published his *De revolutionibus orbium coelestium* (Concerning the Heavenly Spheres) in which he revived the theory that the sun was the center of the solar system. Andreas Vesalius, the Belgain anatomist, established the knowledge of human anatomy with the publication of *De humani corporis fabrica* (On the Anatomy of the Human Body.)

The seventeenth century was a time of great expansion of scientific knowledge. Its first decades saw Galileo turn his little telescope on the heavens while Kepler formulated the laws describing the motions of the planets. Its last decades witnessed the development of Newton's theory of gravitation. Let us recall a few of the important dates of the century. Galileo made his first little telescope in 1609. William Harvey of London demonstrated the circulation of the blood in 1628. Galileo completed his *Two New Sciences* in 1638, the foundation of the modern concept of energy. Robert Boyle published *The Sceptical Chymist* in 1661, thereby laying the foundations of modern chemistry. In 1667 the Royal Society verified the discovery of microorganisms which had been communicated to it in a letter by Antony Van Leeuwenhoek. And in 1687 Sir Isaac Newton published his monumental *Principia*.

The eighteenth century brought important advances in every field of science. In 1774 Joseph Priestley discovered oxygen, paving the way for the development of the true explanation of combustion by Anton Lavoisier, whose head was later cut off by order of a French revolutionary tribunal. In 1781 Sir William

Herschel, organist and amateur astronomer, discovered Uranus, the first time in history that the number of planets known to the ancients was augmented. In 1795 James Hutton published his *Theory of the Earth, with Proofs and Illustrations*, laying down the basic principle of modern geology that the past history of the earth must be explained in terms of the forces now at work.

The nineteenth century began auspiciously with the formulation of the modern atomic theory by James Dalton, the Quaker schoolmaster of Manchester, England, who in 1808 published his *New System of Chemical Philosophy*. In 1814 Joseph von Fraunhofer, the instrument-maker of Munich, discovered the dark lines in the solar spectrum. These were explained half a century later by Kirchhoff and Bunsen, who founded the sciences of spectroscopy and astrophysics. The middle of the century witnessed the discovery of Neptune; the beginning of Pasteur's work, which led to the development of modern bacteriology; and the publication of Darwin's *On the Origin of Species*. In 1873 James Clerk Maxwell published his electromagnetic theory of light.

In 1895, as the century was drawing to a close and physicists were congratulating themselves upon what they thought was their complete understanding of the universe, Roentgen announced the discovery of X-rays. In quick succession came the discoveries of the radioactivity of uranium, the existence of radium, and the electron.

The twentieth century has been truly the Golden Age of Science. Greater advances in science have been made in it than in the previous 5,000 years of history. It began with Einstein's theory of relativity and Rutherford's discovery of the nuclear structure of the atom. Physicists have explored the structure of the atomic nucleus. At the moment they are plagued by the existence of more elementary particles than they know what to do with. Astronomers have extended the limits of the universe, establishing the fact that there are billions of galaxies far beyond our own Milky Way. Perhaps the most spectacular astronomical advance was Hubble's discovery of the expansion of the universe. Biologists have a new understanding of the structure and functioning of the living cell. The most exciting developments in biology have been the discovery of the role of nucleic acids and the "cracking of the code," identifying the nature of the genes with the arrangement of the nucleotides in the strands of DNA.

THE FUTURE

It is reasonable to suppose that scientific progress will continue at an increasing rate. There are more scientists in the world today than at any time in the past. They are equipped for their researches with better apparatus than existed in the past—giant optical and radio telescopes, electron microscopes, computers, better instruments of every kind. Beyond a doubt many of the problems troubling scientists today will be solved as the twentieth century draws to a close and the twentieth-first century draws near.

Will the decades ahead see discoveries as revolutionary and as profound as relativity, the atomic nucleus, quantum theory, the expanding universe, or the role of DNA? This is a question no one can answer. One can only say that man's comprehension of the universe will grow greater and greater with progress in the Space Age, and his mastery of nature and of himself will increase. The ancient Psalmist, standing beneath the stars, exclaimed:

> When I consider thy heavens, the work of thy fingers, the moon and the stars, which thou hast ordained;
> What is man, that thou art mindful of him? and the son of man, that thou visitest him?

But the ancient Psalmist understood the greatness of man as well as the greatness of the universe, for he added:

> For thou hast made him a little lower than the angels, and hast crowned him with glory and honor.
> Thou madest him to have dominion over the works of thy hands; thou hast put all things under his feet. . . .

Science looks forward with confidence and courage to the day when man shall realize the best that is in him.

APPENDIX

✽✽✽✽✽✽✽✽✽✽✽✽✽✽✽✽✽✽✽✽✽

table one

THE PLANETS

Name	Average Diameter (in miles)	Average Distance from the Sun (in millions of miles)	Period of Revolution	Orbital Velocity (in miles per second)	Axial Rotation	Number of Moons
Mercury	3,010	36	88 days	24 to 36	59 days	0
Venus	7,600	67	225 days	22	247 days	0
Earth	7,913	93	365.25 days	18.5	23 hrs. 56 min. 4 sec.	1
Mars	4,200	141	687 days	15	24 hrs. 37 min. 22.5 sec.	2
Jupiter	86,800	483	11.86 years	8.1	9 hrs. 50 min.	12
Saturn	71,500	886	29.5 years	6	10 hrs. 40 min.	10
Uranus	29,400	1,783	84 years	4.2	10 hrs. 45 min.	5
Neptune	28,000	2,794	165 years	3.4	16 hrs.	2
Pluto	3,600	3,670	248 years	3	6.5 days	0

table two

THE FIRST-MAGNITUDE STARS

The 21 brightest stars are known as the first-magnitude stars. However, their apparent magnitudes, as determined with the photoelectric photometer, actually range through three magnitudes, designated as 1, 0, and −1. The value of the apparent magnitude decreases as the brightness of the star increases.

No.	Name	Constellation	Magnitude	Color
1	Sirius	Canis Major	−1.42	White
2	Canopus	Carina	−0.72	Yellow
3	Alpha Centauri	Centaurus	−0.27	Yellow
4	Arcturus	Boötes	−0.06	Orange
5	Vega	Lyra	+0.04	White
6	Capella	Auriga	0.05	Yellow
7	Rigel	Orion	0.14	Blue
8	Procyon	Canis Minor	0.38	Yellow
9	Achernar	Eridanus	0.51	Blue
10	Beta Centauri	Centaurus	0.63	Blue
11	Betelgeuse	Orion	0.70	Red
12	Altair	Aquila	0.77	White
13	Aldebaran	Taurus	0.86	Red
14	Alpha Crucis	Crux	0.90	Blue
15	Spica	Virgo	0.91	Blue
16	Antares	Scorpio	0.92	Red
17	Pollux	Gemini	1.16	Orange
18	Fomalhaut	Piscis Austrinus	1.19	White
19	Deneb	Cygnus	1.26	White
20	Beta Crucis	Crux	1.28	Blue
21	Regulus	Leo	1.36	Blue

table three

THE GEOLOGIC TIME SCALE

Era	Period	Beginning date (in millions of years ago)	Duration (in millions of years)
Cenozoic	Pleistocene	1	1
	Pliocene	13	12
	Miocene	25	12
	Oligocene	36	11
	Eocene	58	22
	Paleocene	63	5
Mesozoic	Cretaceous	135	72
	Jurassic	181	46
	Triassic	230	49
Paleozoic	Permian	280	50
	Pennsylvanian	310	30
	Mississippian	345	35
	Devonian	405	60
	Silurian	425	20
	Ordovician	500	75
	Cambrian	600	100

table four

THE CHEMICAL ELEMENTS

Atomic Number	Name	Chemical Symbol	Atomic Number	Name	Chemical Symbol
1	Hydrogen	H	28	Nickel	Ni
2	Helium	He	29	Copper	Cu
3	Lithium	Li	30	Zinc	Zn
4	Beryllium	Be	31	Gallium	Ga
5	Boron	B	32	Germanium	Ge
6	Carbon	C	33	Arsenic	As
7	Nitrogen	N	34	Selenium	Se
8	Oxygen	O	35	Bromine	Br
9	Fluorine	F	36	Krypton	Kr
10	Neon	Ne	37	Rubidium	Rb
11	Sodium	Na	38	Strontium	Sr
12	Magnesium	Mg	39	Yttrium	Y
13	Aluminum	Al	40	Zirconium	Zr
14	Silicon	Si	41	Columbium	Cb
15	Phosphorus	P	42	Molybdenum	Mo
16	Sulphur	S	43	Technetium	Tc
17	Chlorine	Cl	44	Ruthenium	Ru
18	Argon	A	45	Rhodium	Rh
19	Potassium	K	46	Palladium	Pd
20	Calcium	Ca	47	Silver	Ag
21	Scandium	Sc	48	Cadmium	Cd
22	Titanium	Ti	49	Indium	In
23	Vanadium	V	50	Tin	Sn
24	Chromium	Cr	51	Antimony	Sb
25	Manganese	Mn	52	Tellurium	Te
26	Iron	Fe	53	Iodine	I
27	Cobalt	Co	54	Xenon	Xe

Atomic Number	Name	Chemical Symbol	Atomic Number	Name	Chemical Symbol
55	Cesium	Cs	80	Mercury	Hg
56	Barium	Ba	81	Thallium	Tl
57	Lanthanum	La	82	Lead	Pb
58	Cerium	Ce	83	Bismuth	Bi
59	Praseodymium	Pr	84	Polonium	Po
60	Neodymium	Nd	85	Astatine	At
61	Promethium	Pm	86	Radon	Rn
62	Samarium	Sm	87	Francium	Fr
63	Europium	Eu	88	Radium	Ra
64	Gadolinium	Gd	89	Actinium	Ac
65	Terbium	Tb	90	Thorium	Th
66	Dysprosium	Dy	91	Protoactinium	Pa
67	Holmium	Ho	92	Uranium	U
68	Erbium	Er	93	Neptunium	Np
69	Thulium	Tm	94	Plutonium	Pu
70	Ytterbium	Yb	95	Americium	Am
71	Lutetium	Lu	96	Curium	Cm
72	Hafnium	Hf	97	Berkelium	Bk
73	Tantalum	Ta	98	Californium	Cf
74	Tungsten	W	99	Einsteinium	Es
75	Rhenium	Re	100	Fermium	Fm
76	Osmium	Os	101	Mendelevium	Mv
77	Iridium	Ir	102	Nobelium	No
78	Platinum	Pt	103	Lawrencium	Lw
79	Gold	Au			

BIBLIOGRAPHY

✽✽✽✽✽✽✽✽✽✽✽✽✽✽✽✽✽✽✽✽✽✽✽

BIBLIOGRAPHY

HISTORICAL

BUTTERFIELD, H., *The Origins of Modern Science* (New York: The Macmillan Co., 1957).

DAMPIER, SIR WILLIAM, *A History of Science* (New York: The Macmillan Co., 1949).

DINGLE, HERBERT, *The Scientific Adventure* (New York: Philosophical Library 1953).

DREYER, J. L. E., *A History of Astronomy* (New York: Dover Publications, Inc., 1953).

GLASSER, OTTO, *Dr. W. C. Röntgen* (Springfield, Ill.: Charles C Thomas, 1945).

KOESTLER, ARTHUR, *The Sleepwalkers* (New York: The Macmillan Co., 1959).

LODGE, SIR OLIVER, *Pioneers of Science* (London: Macmillan & Co., 1922).

LUCRETIUS, *On the Nature of Things*, H. A. J. Munro, trans. (London: G. Bell & Sons, 1864).

MUNITZ, MILTON K., editor, *Theories of the Universe* (Glencoe, Ill.: The Free Press, 1957).

NORDENSKIOLD, ERIC, *A History of Biology* (New York: Alfred A. Knopf, 1928).

PANNEKOEK, A., *A History of Astronomy* (New York: Interscience Publishers, 1961).

SARTON, GEORGE, *A History of Science* (Cambridge, Mass.: Harvard University Press, 1952).

SHAPLEY, HARLOW, and HOWARTH, HELEN, E., *A Source Book in Astronomy* (New York: McGraw-Hill Book Co., 1929).

TAYLOR, F. SHERWOOD, *A Short History of Science and Scientific Thought* (New York: W. W. Norton & Co., 1949).

TOULMIN, STEPHEN, and GOODFIELD, JUNE, *The Fabric of the Heavens* (New York: Harper & Row, 1961).

VALLERY-RADOT, PASTEUR, *Louis Pasteur* (New York: Alfred A. Knopf, 1958).

ASTRONOMY

BAKER, R. H., *Astronomy* (Princeton, N.J.: D. van Nostrand & Co., Inc., 1960).

BONDI, HERMANN, ET AL., *Rival Theories of Cosmology* (New York: Oxford University Press, 1960).

BOK, B. J. and BOK, P. F., *The Milky Way* (Cambridge, Mass.: Harvard University Press, 1957).

DIETZ, DAVID, *Stars and the Universe* (New York: Random House, 1968).

GAMOW, GEORGE, *The Creation of the Universe* (New York: The Viking Press, 1952).

HOYLE, FRED, *Frontiers of Astronomy* (New York: Harper & Brothers, 1955).

HOYLE, FRED, *Galaxies, Nuclei, and Quasars* (New York: Harper & Row, 1965).

MCLAUGHLIN, DEAN B., *Introduction to Astronomy* (Boston: Houghton Mifflin Co., 1961).

MENZEL, DONALD H., *Our Sun* (Cambridge, Mass.: Harvard University Press, 1957).

MENZEL, DONALD H., WHIPPLE, FRED L., and DEVAUCOULEURS, GERARD, *Survey of the Universe* (Englewood Cliffs, N.J.: Prentice-Hall, Inc., 1971).

PAGE, THORNTON, ET AL., *Stars and Galaxies* (Englewood Cliffs, N.J.: Prentice-Hall, Inc., 1962).

RUSSELL, H. N., DUGAN, R. S., STEWART, J. Q., *Astronomy* (Boston: Ginn & Co., 1926).

STRUVE, OTTO, LYNDS, BEVERLY, and PILLANS, HELEN, *Elementary Astronomy* (New York: Oxford University Press, 1959).

STRUVE, OTTO, and ZEBERGS, VELTA, *Astronomy of the 20th Century* (New York: The Macmillan Co., 1962).

VAN DE KAMP, PETER, *Basic Astronomy* (New York: Random House, 1952).

WHIPPLE, FRED L., *Earth, Moon, and Planets* (Cambridge, Mass.: Harvard University Press, 1965).

GEOLOGY

DE CAMP, L. S., and DE CAMP, C. C., *The Day of the Dinosaur* (New York: Doubleday & Co., Inc., 1968).
DUNBAR, CARL O., *The Earth* (Cleveland: World Publishing Co., 1966).
HOLMES, CHAUNCEY D., *Introduction to College Geology* (New York: The Macmillan Co., 1962).
MOORE, RUTH, *The Earth We Live On* (New York: Alfred A. Knopf, 1956).
OGBURN, JR., CHARLTON, *The Forging of Our Continent* (New York: American Heritage Publishing Co., 1968).
PUTNAM, WILLIAM C., *Geology* (New York: Oxford University Press, 1964).
VIORST, JUDITH, *The Changing Earth* (New York: Bantam Books, 1967).

PHYSICS

AMALDI, GINESTRA, *The Nature of Matter* (Chicago: University of Chicago Press, 1966).
EDDINGTON, SIR ARTHUR, *The Nature of the Physical World* (New York: Cambridge University Press, 1953).
EINSTEIN, ALBERT, and INFELD, LEOPOLD, *The Evolution of Physics* (New York: Simon & Schuster, 1938).
GAMOW, GEORGE, *Thirty Years that Shook Physics* (Garden City, N.Y.: Doubleday & Co., Inc., 1966).
GLASSTONE, SAMUEL, *Sourcebook on Atomic Energy* (Princeton, N.J.: D. Van Nostrand Co., 1950).
HOFFMAN, BANESH, *The Strange Story of the Quantum* (New York: Harper & Brothers, 1947).
LAPP, RALPH E. and ANDREWS, HOWARD L., *Nuclear Radiation Physics* (Englewood Cliffs, N.J.: Prentice-Hall, Inc., 1954).
MCCUE, J. J., *The World of Atoms* (New York: The Ronald Press Co., 1956).
PAULING, LINUS, *The Nature of the Chemical Bond* (Ithaca, N.Y.: Cornell University Press, 1960).
RICHTMYER, F. K., *Introduction to Modern Physics* (New York: McGraw-Hill Book Co., Inc., 1934).

ROTHMAN, MILTON A., *The Laws of Physics* (New York: Basic Books, Inc., 1963).

RELATIVITY

BORN, MAX, *Einstein's Theory of Relativity* (New York: Dover Publications, Inc., 1965).

EINSTEIN, ALBERT, *Relativity* (New York: Crown Publishers, Inc., 1931).

LANCZOS, CORNELIUS, *Albert Einstein and the Cosmic World Order* (New York: Interscience Publishers, 1965).

SCIAMA, D. W., *The Physical Foundations of General Relativity* (New York: E. P. Dutton & Co., Inc., 1969).

BIOLOGY

BAKER, J. J. W. and ALLEN, GARLAND E., *Matter, Energy, and Life* (Reading, Mass.: Addison-Wesley Publishing Co., 1965).

BEADLE, GEORGE and BEADLE, MURIEL, *The Language of Life* (New York: Doubleday & Co., Inc., 1966).

GOIN, COLEMAN J., and GOIN, OLIVE B., *Man and the Natural World* (New York: The Macmillan Co., 1970).

HOTTON III, NICHOLAS, *The Evidence of Evolution* (New York: American Heritage Publishing Co., 1968).

NASON, ALVIN, *Textbook of Modern Biology* (New York: John Wiley & Sons, 1965).

OTTO, JAMES H., and TOWLE, ALBERT, *Modern Biology* (New York: Holt, Rinehart & Winston, Inc., 1969).

SWANSON, CARL P., *The Cell* (Englewood Cliffs, N.J.: Prentice-Hall, Inc., 1960).

ANTHROPOLOGY

HOWELLS, WILLIAM, *Mankind in the Making* (Garden City, N.Y.: Doubleday & Co., Inc., 1967).

ROMER, ALFRED SHERWOOD, *Man and the Vertebrates* (Chicago: University of Chicago Press, 1941).

NEUROLOGY

CALDER, NIGEL, *The Mind of Man* (New York: The Viking Press, Inc., 1971).

NATHAN, PETER, *The Nervous System* (Philadelphia: J. B. Lippincott Co., 1969).

It is anticipated that unmanned spacecraft, particularly those put in orbit around the planets, will extend our knowledge of the solar system. Mariner 9, which went into orbit around Mars in November 1971, has revealed that many of the Martian craters are of volcanic origin. One crater, known as Nix Olympico, is the collapsed summit of a gigantic volcano larger than any on earth.

INDEX

✼✼✼✼✼✼✼✼✼✼✼✼✼✼✼✼✼✼✼✼✼

INDEX

Absolute magnitude, 89, 97
Acceleration, concept of, 292
Adams, John Couch, 55
Adams, Walter S., 141
Adenoviruses, 408
Afferent neurons (sensory neurons), 443-44
Africa, origins of, 179
Age
  of continental rocks, 174
  of earth, 155-56, 171, 473
  of Milky Way, 452
  of Niagara Falls, 197
  of ocean ridges, 176
  of oceans, 173
  of sun, 158-60
  of volcanoes, 218, 220
Airy, Sir George, 55
Alchemy, 252-53
Alcmaeon of Croton, 361, 440
Alcor (star), 93
Aldrin, Edwin E., Jr., 28, 32
Algae, 417-18
Algol (star), 95
Allfrey, Vincent G., 392
Alpha Centauri (star), 69, 109
Alpha particles, 269, 270, 326
Alpine glaciers (valley glaciers), 203
Alpine Valley (lunar valley), 27, 29
Alps (mountains)
  birth of, 242
  glaciers of, 203
  lunar, 23, 27
Altair (star), 109
Ammonia
  in atmosphere of primitive earth, 404
  in Jupiter atmosphere, 51
Amphibian fossils, 235

Amphibians, present-day, 425
Anatomy, early, 363-64
Anaximander, 251
Anaximenes, 251
Ancient Life, Age of (Paleozoic era), fossil record of, 230-36, 418
Anderson, Carl D., 334-35, 337
Andes (mountains)
  birth of, 242
  glaciers of, 203
Andromeda galaxy, 107, 122, 124-27
Andromedids (meteor showers), 67
Angiosperms, defined, 418-20
Angular momentum
  law of conservation of, 342
  of solar system, 158, 159
Animals, phyla of, 421-25
Antares (star), 91
Antarctica, origins of, 179
Anthropoid apes, evolution of man and, 430-32
Antineutrinos, 340
Antineutrons, 340
Antiparticles, 340
Antiprotons, 340
Apennines (mountains), lunar, 23, 27
Apes, anthropoid, 430-32
Apollo 11 (spacecraft), 18, 28, 32
Apparent magnitude, 87
Aquarids (meteor showers), 67
Arcturus (star), 90
Aristarchos, 6
Aristarchus Crater (lunar crater), 30
Aristotle
  animal life and, 425
  biological discoveries of, 361-62
  energy and, 290
  functions of the mind and, 440

geological notions of, 150
heredity and, 393, 398, 402
nature of matter and, 252-53, 255, 257
planetary motion and, 6
Arizona Desert, 196
Arp, Halton C., 134, 140-41
Arthropods, 422-23
Asclepius, 360-61
Asia, origins of, 179
Asteroids, 58-60
Aston, Francis W., 276-77
Astronomy, history of, 3-6
Atlantic Ocean, 175, 180, 181
Atlas of Peculiar Galaxies (Arp), 134, 140-41
Atmosphere
 of earth, 172, 183-85
 of primitive earth, 404
 as eroding force, 188
 of Jupiter, 51, 404
 of Mars, 47
 of Mercury, 41
 of Milky Way, 112, 114
 of Neptune, 55
 of Saturn, 53
 of sun, 75, 78-80, 160, 453
 of Uranus, 54
 of Venus, 41
Atomic energy, 315, 320, 324, 329
 stellar production of, 101
Atomic theory of matter, 254-57
Atoms
 Bohr theory of, 305-8, 312
 carbon, 282, 288
 metallic, 282
 primeval, 142
 structure of, 268-79
  cloud of electrons in, 272-74
  neutrons in, 321-22
  protons in, 274
  Rutherford atom, 270-72, 457
  separation of isotopes, 276-78
Aurora Borealis (Northern Lights), 83
Australia, origins of, 179
Australopithecus, 432
Avery, Oswald T., 381
Avicenna, 362
Avogadro, Amadeo, 256-57

Baade, Walter, 110, 125-27
Bacilli, defined, 411
Bacon, Francis, 255
Bacteria, characteristics of, 411-12
Bacteriology, 407-11

Bacteriophages, described, 414-15
Balmer, Johann J., 305
Balmer series, 305
Barkla, Charles G., 272
Basaltic rocks, lunar rocks compared with, 34
Beadle, George, 386
Beagle (ship), 399, 400
Becker, H., 320
Becquerel, Antoine Henri, 263-65, 268, 269
Becquerel rays, 264
Beijerinck, Martin, 413
Berkeley, George, 353
Bernoulli, Daniel, 283, 295
Berzelius, Jöns Jakob, 254, 280, 369
Bessel, Friedrich, 87
Beta rays, 269
Betelgeuse (star), 91
Biela's comet, 65, 67
Big Dipper (constellation), 93
Big-bang theory of formation of universe, 142-46
Binding forces
 of molecules, 281-83
 in nucleus, 323-27
Biology, 359-67
 Greek, 361-62
 medieval, 362-63
 modern, 366-67
Bird fossils of Cenozoic era, 243
Birth of stars, 100-4
Black, Davidson, 434
Bode, Johann Elert, 58-59
Bohr, Niels, 274, 305-8, 311, 313
Bohr electrons, 307
Bohr theory of atom, 305-8, 312
Bondi, Herman, 143
Bonner, James, 392, 405
Born, Max, 311, 312
Boscovich, Ruggiero, 255
Bose, Satyendra Nath, 334
Bose-Einstein statistics, 334
Bosons, defined, 334
Bothe, Walter, 320
Boyle, Robert, 253, 255, 456
Brachiopod fossils, 232
Bragg, Sir William, 288, 309-10
Bragg, W. L., 288
Brahe, Tycho, 7, 8, 61, 98
Brain
 defined, 446-50
 of Java Man, 434
 of Neanderthal Man, 438-39
 of Pekin Man, 434

Brain waves, 450
Bridges, C. B., 386
Brightness
  of Jupiter, 49, 66
  of Mars, 45
  of quasars, 138-39
  of Venus, 66
  See also Luminosity
Brock, A. van den, 272
Broglie, Louis de, 310-12
Broom, Robert, 432
Brown, Harrison, 160
Bryan, William Jennings, 400
Bryce Canyon National Park (Utah), 202
Bryophites, 417, 418
Bubble chamber, 319
Buffon, Georges Louis de, 156, 157, 398
Bulwark plains (walled plains), 27
Bunsen, Robert, 73, 457
Bunsen burner, 73, 74

Calcar, Stephen, 364
Calvin, Melvin, 404
Cambrian period of Paleozoic era, 233-34
Canals, Martian, 48-49
Cannizzaro, Stanislav, 257
Cape buffaloes, 424
Cape Clouds (Magellanic Clouds), 125, 127-28
Carbohydrates, chemical composition of, 370-71
Carbon
  atomic structure of, 282, 288, 370
  as foundation of life, 454
Carbon dioxide in atmosphere
  of earth, 183
  of earth in primitive stage, 404
  of Mars, 47
  of Venus, 42
Carboniferous period, 235-36, 244, 418
Carpathians (mountains), lunar, 23
Cassini's division, 53
Cassiopeia (constellation), 97, 106
Casualties
  from earthquakes, 215
  from volcanic explosions, 217, 220
Catastrophism, theory of, 152
Cathode rays, 261, 262
Caucasus (mountains), lunar, 27
Cells, 377-92
  of bacteria, 411-12
  chromosomes and, 380-83

control of genes in, 391-92
  cytoplasm of, 383-85
  first theory of, 366
  meiosis of, 390-91
  mitosis of, 388-90
  nucleus of, 379-80
  protein production in, 387-88
Cellular membrane, 379
Cenozoic era (Age of Mammals), fossil record of, 230, 240-47, 397-98
Central bulge of Milky Way, 108
Central nervous system, 441-42
Cephalopod fossils, 232, 238
Cepheid variables (stars), 96-97, 102, 123, 126, 127
Cerebellum, 446, 447
Cerebral cortex, 449-50, 455
Cerebrum, 447-49
Ceres (asteroid), 59, 60
Cerro Negro Volcano (Nicaragua), 224
Cetus (constellation), 96
Chamberlain, Thomas C., 157
Chadwick, James, 315, 320, 321
Chemical elements
  in composition of earth, 164, 453
  present in solar system, 160-61
Chemical energy, 294-95
Chemistry, *see* Organic chemistry
Chimpanzees, 430
Chondrules of meteorites, 68
Chordates, 423-25
Chromosomes, nature of, 380-83
Chromosphere of sun, 80
Clark, Alvin G., 94
Classical Cepheids (stars), 96-97
Clavius (lunar crater), 28, 29
Cloud layers of Venus, 42
Clouds
  of electrons, 272-74
  of Jupiter, 51
  of Mars, 47
  of Saturn, 52
Cluster-type Cepheids (stars), 96, 102, 126, 127
Coal Sack (nebula), 115
Cocci, defined, 411
Cockroft, J. D., 320
Cockroft-Walton experiment, 319-20
Coelenterates, 421-22
Combination, principle of, 305, 306
Comets, 60-65
Composite motion of spheres, 5-6
Compton, Arthur H., 308, 339

480 INDEX

Compton effect, 308-9
Connector neurons (intermediate neurons), 444
Conservation laws
  of energy, 296, 342
  of matter, 254
  of particles, 342
Continental drift, 179-81
Continental glaciers (icecaps), 203
Continental rocks, types of, 173-74
Continents, rocks composing, 173-75; see also Oceans
Continuous-creation theory (steady-state theory) of formation of universe, 143-44
Convection currents affecting ocean floors, 177
Copernicus, Nicholas, 6-8, 11, 149, 363, 456
Copernicus Crater (lunar crater), 17, 21, 30, 35
Core of earth, 168
Corona, solar, 80, 81, 83-84
Correns, Karl, 394
Cosmic phenomenon, evolution as, 451-55
Cosmic radio radiation, low-energy, 145
Cosmic rays, 338-40, 397
Cosmology, defined, 141
Counterglow (gegenschein), defined, 70
Covalent binding of molecules, 282
Cowan, Clyde L., 336, 340
Crab Nebula (nebula), 98, 105, 132
Craters
  lunar, 27-31
  Martian, 48
  meteoric, 69
Cretaceous period of Mesozoic era, 236-38
Crick, F. H. C., 381, 388
Crick-Watson model of DNA, 381, 382
Croly, George, 16
Cro-Magnon Man, 439
Crookes, Sir William, 261, 266
Crookes tube, 261, 262, 266
Crust of earth, 171, 172
Curie, Marie, 263-64, 268, 320
Curie, Pierre, 263-65, 268, 320
Currents affecting ocean floors, 177
Curved space of universe, 136-38
Cycle of oscillating universe, expected duration of, 145
Cycles of sunspots, 78, 84
Cyclotrons, 316-17, 319, 326

Cygnus (Northern Cross; constellation), 106, 132
Cytoplasm of cells, 383-85

Dalton, John, 255-57, 283, 457
da Vinci, Leonardo, 151, 331, 364
Dart, Raymond, 432, 433
Darrow, Clarence, 400
Darwin, Charles, 398-401, 457
Darwin, Erasmus, 398
Davisson, C. J., 312
Davy, Sir Humphry, 295
De humani corporis fabrica (On the Anatomy of the Human Body; Vesalius), 456
De nova stella (concerning the New Star; Brahe), 99
De rerum natura (Lucretius), 255
De revolutionibus orbium coelestium (Concerning the Revolution of Heavenly Spheres; Copernicus), 7, 363, 456
Death of stars, 100-4
Deep Sea Drilling Project, 181-83
Deimos (moon of Mars), 49
Delta (star), 96
Deltas of rivers, 193
Democritus, 251, 254
Dempster, Arthur J., 277
Density
  of earth, 164
  of moon, 24
  of rocks, composing earth's crust, 172
  of Saturn, 53
Deoxynucleic acid (DNA), 380-83, 386, 456
  Crick-Watson model of, 381, 382
  in evolution process, 404, 406
  in meiotic process, 391
  in mitotic process, 388-89
  viral, 414, 415
Depth
  of deepest mines, 165
  of Marianas trench, 177
  of Mid-Atlantic Ridge, 176
  of oceans, 175
Deslandres, Henri, 79
Descartes, René, 255
Devonian forests, 240
Devonian period of Paleozoic era, 235
Diameter
  of asteroids, 60
  of atoms, 278-79
  of earth, 44, 163
  of Jupiter, 49

of Mars, 45
of Mercury, 39
of Milky Way, 107
of Milky Way nucleus, 108
of moon, 24
  of lunar craters, 27-28, 31
of nebulae, 115
of primeval fireball, 142
of quasars, 139
of Saturn, 52
  of Saturn's rings, 53
of stars, 91
of sun, 71
of sunspot umbra, 77
of Uranus, 54
of Venus, 41
Diastrophism, 454
  forces of, 205-6; *see also* Earthquakes; Mountains; Volcanoes
Dicke, Robert H., 145
Dietz, Robert S., 177
Differential calculus, discovered, 12
Dinosaurs, 228, 229, 231, 232, 239-40, 246, 426
Dioscorides, 362
Dirac, P. A. M., 311, 312, 334-35
Distances
  from earth
    to moon, 12, 17-18
    to nearest star, 14
    to quasars, 138, 140-41
    to stars, 89
    to sun, 12, 44, 72
    to Venus, 41
  stellar, 87-90
  from sun
    to center of Milky Way, 107
    to earth, 12, 44, 72
    to Jupiter, 49
    to Mars, 145
    to Mercury, 39
    to Neptune, 55
    to Pluto, 12
    to Saturn, 52
    to Uranus, 54
    to Venus, 41
Diurnal librations, defined, 23
DNA, *see* Deoxynucleic acid
Doppler, Christian, 90
Doppler effect, 90, 113, 135, 136
Disk population of stars in Milky Way, 110, 118-19
Dog Star (Sirius), 89, 91, 94, 109
Dollfus, Audouin, 40
Donati's Comet, 60

Double stars (visual binaries), 93-95
Downard motions of solar prominences, 81
Draco (constellation), 44
Draconids (meteor showers), 67
Dryden, John, 71
Dubois, Eugene, 434
Dust in Milky Way, 112-16
Dutton, Clarence, 226

Earth, 42-45, 155-204
  age of, 155-56, 171, 473
  atmosphere of, 172, 183-85
  beginnings of, 155-62
  characteristics of, 42-45
  chemical composition of, 164, 453
  crust of
    isostasy and, 225-26
    rocks composing, 172
  distances from
    to moon, 12, 17-18
    to nearest star, 14
    to quasars, 138, 140-41
    to stars, 89
    to sun, 12, 44, 72
    to Venus, 41
  erosion of, 187-204
    by glaciers, 201-4
    by lakes, 197-99
    by rivers, 190-94
    underground water and caves, 199-201
    by waterfalls, 195-97
    by waves, 204
  evolution of, 453-54
  rotation of, moon phases and, 20
  structure of, 163-70
    core, 168
    mantle, 167-68
  surface of, 163
    faults, 180, 209-11, 215, 216
Earth craters, lunar craters compared with, 30
Earthlight, 23
Earthquakes, 210-15
  annual number of, 211
  in California, 215
  classification of, 213
  along ocean ridges, 176
Earthy chemical elements, defined, 160
East Pacific Rise, 176, 177, 180
Echinoderm fossils, 234
Echinoderms, 423
Echinoid fossils, 238

Eclipses
    of moon, 35-36
    of sun, 36
Eclipsing binaries, 95
Ecliptic, the, defined, 38
Efferent neurons (motor neurons), 443-44
Eightfold Way of elementary particle physics, 343-44
Einstein, Albert, 136, 137, 141
    biography of, 345-46
    elementary particle physics and, 334
    mass-energy equivalence principle, 319-20, 324, 325, 350-52
    quantum theory and, 302, 304, 311, 312
    theory of relativity of, *see* Relativity, Theory of
Electric charge, law of conservation of, 342
Electric fields, 298-99
Electromagnetic radiation, unit of, 333
Electromagnetic theory of light, 300, 457
Electron Microscopes, 325
Electrons
    Bohr, 307
    cloud of, 272-74
    discovery of, 265-67
Elementary particle physics, 331-44
    cosmic rays in, 338-40
    Eightfold Way of, 343-44
    four interactions among, 341-42
    mesons in, 336-38
    neutrinos in, 335-36
    positrons in, 334-35
Elliptical galaxies, 123-24
    spiral galaxies compared with, 130
Elsasser, W., 312
Embryology, 398
Emerson, Ralph Waldo, 227, 268
Empedocles, 252
Empty space, universe as mostly, 15
Energy, 290-301
    atomic, 101, 315, 320, 324, 329
    chemical, 294-95
    kinetic, 293-94
    law of conservation of, 342
    nature of heat and, 295-96
    Newton's laws of motion and, 291-93
    of photons, 333
    potential and kinetic, 293-94
    Principle of Equivalence of Mass and Energy, 319-20, 324, 334, 350-52
    quantum theory of, 302-5
    radiated by quasars, 139
    radiation as, 298-301
    solar, 72; *see also* Sun
    sound waves and, 297-98
    thermodynamics and, 296-97
Entropy (time's arrow), defined, 297
Enzymes, 375
Eocene epoch, 241-43
Epicenter, defined, 211
Epicurus, 255
Equatorial belts of Saturn, 52-53
Equatorial bulge of earth axis, 44
Equinoxes, precession of, 44
Eros (asteroid), 60
Erosion, 187-204, 453-54
    by atmosphere, 188
    by glaciers, 201-4
    by lakes, 197-99
    by rivers, 190-94
    by underground water, 199-201
    by waterfalls, 195-97
    by waves, 204
Essay Toward a Natural History of the Earth and Terrestrial Bodies, especially Minerals; as also of the Sea, Rivers, and Springs, with an account of the Universal Deluge and of the Effects that it had upon the Earth, An (Woodward), 151
"Essay on Population, The" (Malthus), 399
Europe, origins of, 179, 180
Euxodus, 5
Evolution
    as cosmic phenomenon, 451-55
    of galaxies, 129-31
    heredity and, 397-406
        Darwin and, 399-401
        origin of life and, 401-3
        20th-century view, 403-6
    of man, 426-39
        anthropoid apes and, 430-32
        Cro-Magnon Man, 439
        first modernlike, 435
        Heidelberg Man, 435
        lemurs and tarsiers in, 427-29
        man-apes in, 432-33
        monkeys and, 429-30
        Neanderthal Man, 436-39
        Pekin Man, 434-35
    of Milky Way, 116-20
    See also Fossil record

INDEX 483

Exosphere, defined, 185
Exploding stars (nova stars), 97-100

Fabricius, David, 96, 365
Faraday, Michael, 265, 298-300
Fats, chemical composition of, 372-73
Faults in earth's surface, 209-10
Fermi, Enrico, 328-30, 336
Fermi-Dirac statistics, 334
Fermions, defined, 334
Ferns, 418
Feynman, Richard P., 335
Firemaking
  by Cro-Magnon Man, 439
  by Pekin Man, 435
First Interglacial Period, 434, 435
First quarter of moon, defined, 19
Fission of uranium, 329-30, 457
Fish fossils, 234-35
Fitzgerald, George F., 265, 348
Flares, solar, 82-83
Flip (Floating Instrument Platform), 182
Floating Instrument Platform (Flip), 182
Flood-plains, defined, 193
Flowers, structure of, 419
Focus of earthquakes, 211
Forces
  binding
    in molecules, 281-83
    in nucleus, 323-27
  defined, 293
  Van der Waals, 285
  of weathering, 188-90
Forebrain, parts of, 447-48
Fossil record, 227-47
  of Cenozoic era, 230, 240-47, 397-98
  of Mesozoic era, 236-40, 426
  of Paleozoic era, 230-36, 418
Four element theory of matter, 252-54
Fourth Glaciation period, 433
Fowler, William A., 133, 139
Fox, Sidney W., 405
Fraunhofer, Joseph von, 73, 74, 457
Frederick the Great, 302
Fresnel, Jean, 299
Friedman, Alexander, 138
Frigoris, Mare (Sea of Cold), 26
Frisch, Otto R., 329
Full moon, defined, 19
Fungi, nature of, 418
Furrowing of cells, 389

Galaxies, 123-34
  evolution of, 129-31
  light-years separating, 15
  Local Group of, 125-27
  Maffei, 128-29
  radio, 131-34, 141, 144
  types of, 123-24
  See also Milky Way
Galen, 362-65
Galilean-Newtonian Relativity Principle, 346
Galileo (Galilei Galileo), 345, 361, 401, 456
  astronomical discoveries of, 7, 10-11, 16, 24, 26, 40, 51, 53, 87, 106-7
  atomic theory championed by, 255
  modern concepts of energy and, 290-92
  relativity of motion and, 345, 346
Gamma rays, 300
Gamow, George, 142, 316
Ganges River (India), 191
Ganymede (moon of Jupiter), 51
Gaseous chemical elements, defined, 160-61
Gases
  in Milky Way, 112-16, 453
  in molecular formation, 283
  kinetic theory of, 283
  monatomic, 281
  on moon surface, 35
Gassendi, Pierre, 255
Gastropod fossils, 232
Gay-Lussac, Joseph Louis, 256
Gegenschein (counterglow), defined, 70
Geiger, Hans, 271, 318
Geiger counter, 318
Geissler, Heinrich, 260
Geissler tubes, 260-61
Gell-Mann, Murray, 343-44
Generalized cells, 378
Genes, 386
  control of, 391-92
Genetic code, cracking, 458
Geologic Time Scale, 228, 230
Geology, 149-54
  Greek and medieval, 150-51
  modern, 152-53
  20th-century, 154
Geosyncline, described, 222
Germer, L. H., 312
Ghost craters of moon, 26
Gibbons, 432
Glaciated valleys, 199-200
Glaciers, 201-4

484   INDEX

Glaser, Donald A., 319
Globular clusters of stars in Milky Way, 111-12
Glomar Challenger (drill ship), 181, 182
Gold, Thomas, 143
Goldstein, Eugen, 261, 274
Golgi, Camillo, 385
Golgi complex of cytoplasm, 385
Gondwana (split-off from supercontinent), 179, 180
Good Friday Earthquake (Alaska; March 1964), 212, 213, 215
Goodspeed, A. W., 262
Gorillas, 438
Graaf, Regnier de, 408
Grand Canyon (Ariz.), 192, 194
Gravitation, Universal Law of, 7, 12, 64, 356
Gravitation and Inertia, Principle of the Equivalence of, 353-55
Gravitational red shifts of quasars, 141
Great Ice Age, *see* Pleistocene Epoch
Great Nebula in Orion (nebula), 113, 115, 123, 128
Great Red Spot of Jupiter, 50-51
Great Rift (nebula), 106, 115
Great Salt Lake (Utah), 198
Greek biology, 361-62
Greek medicine, 360-61
Greenhouse effect produced by carbon dioxide, 42
Grew, Nehemiah, 408
Gum Nebula (nebula), 105
Gutenberg, Beno, 167
Gymnosperms, 418-19

Hale, George Ellery, 79
Hahn, Otto, 329
Halley, Edmund, 64, 89
Halley's comet, 60, 61, 64, 67
Halo of Milky Way, 109
Hand axes of Pleistocene Period, 433
Harkins, William D., 320
Harmonic law (Kepler's third law of planetary motion), 10, 94, 161
Harvey, William, 364-65, 456
Hauksbee, Francis, 260
Heat waves (infrared rays), 300-1
Heat-death, defined, 297
Heezen, Bruce, 176
Heidelberg Man, 435
Heisenberg, Werner, 311-13, 321, 336
Helium
    in Jupiter atmosphere, 51
    in Milky Way, 112
    in Saturn atmosphere, 53
    in sun's atmosphere, 75, 160, 453
Helium nucleus, composition of, 323
Helmholtz, Hermann von, 265
Henderson, Thomas, 87
Heraclitus, 252, 280
Hercules (constellation), 90, 116
Heredity, 393-406
    evolution and, 397-406
        Darwin and, 399-401
        origin of life and, 401-3
        20th-century view, 403-6
    Mendel's discoveries on, 394-96
Herschel, Sir William, 12, 53-55, 59, 93, 107, 121, 456-57
Hertz, Heinrich, 300
Hertzian waves (radio waves), 300-1
Hertzsprung, Ejnar, 91
Hertzsprung-Russell Diagram (H-R Diagram), 91-93, 99, 100, 102, 103, 112
Hess, Harry, 177
Hess, Victor F., 338
Himalayas (mountains)
    birth of, 242
    glaciers of, 203
Hindbrain, parts of, 446-47
Hipparchos, 6
Hippocrates, 361
Hittorf, Johann Wilhelm, 261
Homer, 360
Homo erectus, 433, 435, 437
Homo sapiens, 435, 439
Hooke, Robert, 285, 299, 366, 408
Hooveria (asteroid), 60
Horace, 393
Hoyle, Fred, 133, 139, 143
H-R Diagram (Hertzsprung-Russell Diagram), 91-93, 99, 100, 102, 103, 112
Huang Ru-chih, 392
Hubble, Edwin P., 123, 126, 129, 130, 135, 138, 145, 457
Hubble's Law of Red Shifts, 135, 136
Human body, chemical composition of, 368-69
Humboldt, Alexander von, 66
Hutton, James, 149, 152-53, 457
Huygens, Christiaan, 53, 299
Hydrocarbons, molecular formation of, 288-89
Hydrogen, 454
    around comets, 63
    in atmosphere of primitive earth, 404

atomic structure of, 273
  in Jupiter's atmosphere, 51
  in Milky Way, 112, 114
  molecular structure of, 282
  in Saturn's atmosphere, 53
  in sun's atmosphere, 160, 453
  structure of, 273
Hydrogen burning, birth and death of stars and, 101, 102
Hydrogen nucleus, composition of, 323
Hypothalamus, 447-48

Icecaps (continental glaciers), 203
Icy chemical elements, defined, 160
Igneous rocks, 173-74
IGY (International Geophysical Year; 1957-1958), 154
Iliad (Homer), 360
Illustrations of the Huttonian Theory (Playfair), 153
Imbrium, Mare (Sea of Showers), 23, 26, 27, 29
India, origins of, 179
Inert gases (monatomic gases), 281
Inertia
  concept of, 292
  Principle of Equivalence of Gravitation and Inertia, 353-55
Infrared rays (heat waves), 300-1
Insect fossils, 235, 238
Interior of earth, 165-67
Intermediate neurons (connector neurons), 444
International Geophysical Year (IGY; 1957-1958), 154
Ionosphere, 184-85
Iron meteorites, 68
Irregular galaxies, 123, 124, 131
Isostasy, defined, 225-26
Isotopes, separation of, 276-78
Iwanowsky, Dimitri, 413

Jacob, François, 388
Jansky, Karl, 131
Java Man (Pithecanthropus), 433, 434
Jeans, Sir James, 158, 294, 303
Jean's paradox, 303
Jeffreys, Harold, 158
Joints of earth's surface, 209-10
Joliot, Frédéric, 321, 328
Joliot-Curie, Irène, 321, 328
Joly, John, 155
Joule, James Prescott, 295
Jupiter, 49-51, 53, 56
  atmosphere of, 51, 404

characteristics of, 49-51
chemical elements in, 160
family of comets of, 63-65
meteor showers and, 66
moons of, 11, 38, 51, 161
orbit of, 58-60
in protoplanet hypothesis, 161, 162
Jurassic period of Mesozoic era, 236-37, 243, 246

Kant, Immanuel, 121, 156
Keats, John, 37
Kelvin, Lord, 155
Kepler, Johannes, 7, 9-11, 58, 99, 456
Kepler Crater (lunar crater), 30
Kepler's Nova, 98-100
Kinetic energy, 293-94
Kirchhoff, Gustav, 73, 74, 457
Kilauea-iki Volcano, eruption of (1959), 219, 221
Koch, Robert, 411
Kolhörster, W., 338
Kornberg, Arthur, 381
Krakatoa explosion (1863), 220
Kuiper, Gerard P., 53, 161-62

Lakes, 197-99
Lamarck, Jean Baptiste de, 398-99
Langmuir, Irving, 272-74
Langmuir-Lewis Model, 306
Laplace, Pierre Simon de, 156, 157, 159
Large Magellanic Cloud, 128
Larmor, Joseph, 205
Last quarter of moon, defined, 19
Laue, Max von, 288
Laurentia (split-off from supercontinent), 179, 180
Lavoisier, Antoine, 254, 456
Lawrence, Ernest O., 316, 319
Layers of earth's atmosphere, 183-85
Leakey, Louis S. B., 432, 435
Leakey, Mary, 432, 435
Leavitt, Henrietta, 97
Lee, T. D., 343
Leeuwenhoek, Antony van, 407-8, 456
Lemaître, Abbé Georges, 138, 142
Lemurs, 427-29
Lenard, Philipp, 261-63
Length of moon shadow during eclipses, 36
Leonardo da Vinci, 151, 331, 364
Leonids (meteor showers), 67
Leucippus of Abdera, 254
Leverrier, Urbain Jean Joseph, 55
Lewis, George E., 432

Lewis, Gilbert N., 272-74
LGM's (Little Green Men; pulsars), 104-5
Librations, defined, 23
Life
    on Mars, 49
    origin of, 401-3, 454
    See also Evolution; Fossil record
Life expectancy of Milky Way, 452
Light
    composition of, discovered, 12, 299
    earthlight, 23
    electromagnetic theory of, 300, 457
    quantum theory of, 304-5
    speed of, 300
    ultraviolet, 141, 300-1
    zodiacal, 70
Light-years
    defined, 14
    separating galaxies, 15
Lindblad, Bertil, 116
Liquids in molecular formation, 284-87
Lisbon earthquake (Portugal; 1755), 213, 215
Lister, Joseph, 411
Lithium, structure of, 273
Lithium nucleus, composition of, 323
Little Green Men (LGM's; pulsars), 104-5
Liverwort, characteristics of, 418
Living organisms
    attributes of, 368
    chemical elements composing, 454
Local Group of galaxies, 125-27, 136, 142
Lodge, Sir Oliver, 265, 352
Long-chain molecules, defined, 287
Long-period comets, 63
Long waves (L-waves), 165
Loop Nebula (nebula), 128
Lorentz, Hendrik A., 265-66, 274, 300, 333, 348-50
Lorentz transformations, 348
Lorentz-Fitzgerald contraction, 300, 348, 349
Lowell, Percival, 48, 56
Low-energy cosmic radio radiation, 145
Lucretius, 255
Luminosity
    of novae, 99-100
    of stars, 89
        in Milky Way, 109
        size and, 92-93
Luna 13 (Russian spacecraft), 33
Lunar craters, 26-30,

Lunar Orbiter 1 (U.S. spacecraft), 24, 35
Lunar Orbiter 2 (U.S. spacecraft), 21
Lunar Orbiter 3 (U.S. spacecraft), 31, 35
Lunar rocks, color of, 34
Lunar soil, 34
L-waves (long waves), 165
Lyell, Sir Charles, 153, 241
Lynds, C. R., 133
Lyot, Bernard, 80
Lyra (constellation), 91, 96, 116
Lyrids (meteor showers), 67

McCarty, Maclyn, 381
McCrea, W. H., 137-40
Mach, Ernst, 349, 353
McLeod, Collin, 381
McMillan, Edwin M., 329
Maffei, Paolo, 128
Maffei galaxies, 128-29
Magellan, Ferdinand, 128
Magellanic Clouds (Cape Clouds; galaxy), 125, 127-28
Magnetic field
    defined, 298
    effects of, on ocean ridges, 177-78
Magnetic storms, solar flares causing, 83
Magnitude
    absolute, 89, 97
    apparent, 87
    Richter, 213
    of stars, 87
Maimon, Moses ben, 363
Major planets, 37; *see also* Jupiter; Neptune; Saturn; Uranus
Malpighi, Marcello, 365
Malthus, Thomas, 399
Mammals, 425
Mammals, Age of (Cenozoic era), fossil record of, 230, 240-47, 397-98
Mammoth Cave (Ky.), 200
Mammoth fossils, 245-47
Man, evolution of, 426-39
    anthropoid apes and, 430-32
    Cro-Magnon Man, 439
    first modernlike, 435
    Heidelberg Man, 435
    lemurs and tarsiers in, 427-29
    man-apes and, 432-33
    monkeys and, 429-30
    Neanderthal Man, 436-39
    Pekin Man, 434-35

INDEX

Mantle of earth, 167-68
Maria of moon, 26
Marianas trench, 177
Mariner 2 (U.S. spacecraft), 41, 42
Mariner 4 (U.S. spacecraft), 48
Mariner 6 (U.S. spacecraft), 49
Mars, 37, 44-49, 53, 56
   characteristics of, 45-49
   distance of, to sun, 145
   moons of, 38, 49
   orbit of, 58-59
   in protoplanet hypothesis, 161
   surface of, 453
Marsden, E., 271
Marshak, Robert E., 337
Mass
   of quasars, 139
   rest-mass of particles, 333
   of stars, 91
Mass and Energy, Principle of Equivalence of, 319-20, 324, 335, 350-52
Mass spectograph, 277-78
Masson, Orme, 320
Mastodon fossils, 234, 245-47
Matter, 251-58
   atomic theory of, 254-57
   four element theory of, 252-54
   law of conservation of, 254
   periodic table of elements and, 257-58
Mauna Loa (Hawaii), 178
Maxwell, James Clerk, 157, 300, 457
Mechanics
   fundamental law of, 293
   wave, 310-13
Medicine, Greek, 360-61
Medieval biology, 362-63
Medulla oblongata, 446, 448
Meiosis of cells, 390-91
Meitner, Lisa, 329
Menard, Henry, 176
Mendel, Gregor, 394-96
Mendeleef, Dmitri Ivanovich, 257-58, 273
Mercury, 37-41, 51, 53
   atmosphere of, 41
   characteristics of, 38-40
   Halley's comet and, 64
   irregular motions of, 356
   in protoplanet hypothesis, 161
   surface of, 453
   Venus compared with, 41, 42
Mesons, defined, 336-38
Mesozoic era (Age of Reptiles), fossil record of, 230, 236-40, 426

Messier, Charles, 121
Metallic atoms, outer shells of, 282
Metallic binding of molecules, 283
Metamorphic rocks, 173-74
Meteor Crater (Ariz.), 30, 69
Meteor showers, 66-67
Meteorics (Aristotle), 150
Meteorite craters, 69
Meteorites, 29, 67-70
Meteors, 65-66
Methane
   in Jupiter's atmosphere, 51
   in Saturn's atmosphere, 53
Michelangelo, 364
Michelson, Albert A., 347
Michelson-Morley experiment, 346-48
Micrometeorites, 70
Microscopes
   electron, 325
   of Leeuwenhoek, 410
Mid-Atlantic Ridge, 175-77, 180
Midbrain, parts of, 447
Milky Way (galaxy), 106-20
   age of, 452
   evolution of, 116-20
   gas and dust in, 112-16
   globular clusters in, 111-12
   stellar population of, 109, 453
      disk population, 110, 118-19
      globular clusters, 111-12
   structural features of, 107-8
Millikan, Robert A., 267, 308, 339
Mills Cross (radio telescope), 132
Milton, John, 106
Mineur, Henri, 127
Minkowski, Herman, 352
Miocene epoch, 236, 241-45
Mira (star), 96, 97
Mirsky, Alfred E., 392
Mississippi River, 191, 193
Mississippian period of Paleozoic era, 235, 418
Mitochondria of cytoplasm, 384
Mitosis of cells, 388-90
Mizar (star), 93, 94
Model of universe, 13-15
Mohorovicic, Andrija, 166
Mohorovicic discontinuity, 166
Molas Lake (Colo.), 201
Molecules
   formation of, 280-89
      binding forces in, 281-83
      gases and, 283
      liquids in, 284-87
      solids in, 288-89

origin of life and collision of, 454
size of, 278-79
Mollusks, 422
Momentum
   angular
      law of conservation of, 342
      of solar system, 158, 159
   concept of, 292-93
Monatomic gases (inert gases), 281
Monkeys, 429-30
Monod, Jacques, 388
Moon, the, 16-36
   craters of, 26-30
   diameter and density of, 24
   distance of, from earth, 12, 17-18
   eclipses of, 35-36
   lunar spacecrafts and, 33-35
   maria of, 26
   motions of, 23
   mountains of, 27
   phases of, 19-23
   rays of, 30-33
   surface of, 17, 24-26, 35
Moon landings, first, 34
Moons
   of Jupiter, 11, 38, 51, 161
   of Mars, 38, 49
   of Neptune, 55
   number of, in solar system, 156
   of Saturn, 53
   of Uranus, 54
Morgan, Thomas Hunt, 386, 395
Morley, Edward W., 347
Moscovium, Mare (Sea of Moscow), 35
Moseley, Henry G. J., 272
Mosses, 418
Motion
   Newton's laws of, 291-93
   relativity of, 346
   of stars, 82-90
   of sun, 15
   See also Planetary motion
Motor neurons (efferent neurons), 443-44
Moulton, Forest R., 157
Mountain formation
   in Cenozoic era, 242
   in Mesozoic era, 238
   in Paleozoic era, 235
Mountain ranges of the moon, 27
Mountains, 222-25
Mu mesons, 337
Muller, Herman J., 386, 397

Murray, Sir John, 190
Mutations, 396-97

Natural selection, defined, 400
Nature (magazine), 143
Neanderthal Man, 436-39
Nebulae, 114-16
Nebular hypothesis of earth's formation, 156-57
Neddermeyer, Seth, 337
Needham, John, 402
Ne'eman, Yuval, 343
Neptune, 55-57
   characteristics of, 55
   chemical elements on, 161
   discovered, 12, 457
   Halley's comet and, 64
   in protoplanet hypothesis, 161, 162
Neptunists, views held by, 152, 153
Nerve impulses, 444
Nerves, 444
Nervous system, 440-50
   brain in
      described, 446-50
      of Neanderthal Man, 438-39
      of Pekin Man, 434
   neurons in, 442-44
   reflex arc and, 445-46
Neurons
   complexity of, 455
   in nervous system, 442-44
Neutrinos, 335-36
Neutron bombardment, 328
Neutrons, discovered, 320-21
Nevados Huascarán Mountain explosion (Peru; 1970), 217
New moon, defined, 19
New System of Chemical Philosophy (Dalton), 255, 457
Newton, Sir Isaac, 7, 304, 345
   atomic theory championed by, 255
   biography of, 11-12
   discovers nature of light, 12, 299
   law of universal gravitation of, 7, 12, 354, 456
   laws of motion of, 291-93, 353
   relativity of motion and, 346
   spectroscope and, 72, 73
Niagara Falls, 195-97
Nicholas of Autrucia, 255
Nicholson, Seth B., 39, 51
Nirenberg, Marshall W., 388
Nitrogen, 454
   in earth's atmosphere, 183

INDEX 489

in Mars' atmosphere, 47
in Venus' atmosphere, 41
Noah's flood in geology, 151-52
Nobel Prizes
  Becquerel, 265
  P. and M. Curie, 265
  Einstein, 304
  Millikan, Compton, C.T.R. Wilson, 308
  Pauling, 374
  Roentgen, 263
  Stanley, 414
  Wilkins, Crick, Watson, Kornberg, 381
Nonmetallic atoms, outer shells of, 282
North America, 179, 180, 237, 241, 242
North Star (Polaris), 44, 97
Northern Cross (Cygnus; constellation), 106, 132
Northern Lights (Aurora Borealis), 83
Novae (exploding stars), 97-100
Nucleonic charge, law of conservation of, 342
Nucleus, 314-30
  of cells, 379-80
  of comets, 62
  of galaxies
    of Milky Way, 108
    quasars and, 139-40
  structure of
    binding force in, 323-27
    Cockroft-Walton experiment, 319-20
    fission and, 329-30
    neutron discovered, 320-21
    radiation detectors and, 318-19
    Rutherford's atom-smashing experiment, 270-72, 274, 275, 305-6, 314-15
    transformations in, 327-28

Ocean floors, 175-77
Ocean trenches, 177
Oceans, 172-81
  continental drift and, 179-81
  depth of, 175
  ridges of, 175-77
  surface of, 172-73
  volcanic islands in, 178
Ochoa, Severo, 388
Oil-drop experiment, 267
"Old Moon in the new moon's arms," 23
Oligocene epoch, 241, 243

Omar (Arab scientist), 150
On the Origin of Species by Means of Natural Selection (Darwin), 400, 457
Oort, Jan Hendrick, 63, 116
Oparin, A. I., 404
Orang, 431
Orbits
  of comets, 67
  of double stars, 93-94
  of earth, 44
  of Halley's comet, 64
  of Jupiter, 58-60
  of Mars, 46, 58-59
  of Mercury, 39
  of moon, 17
  of Pluto, 56-57
  of Uranus, 54
  of Venus, 40
Ordovician period, 234
Organic chemistry, 368-76
  carbohydrate composition, 370-71
  defined, 369
  enzyme composition, 375
  photosynthesis and, 371-72
  protein composition, 373-75
  respiration and, 376
Organic molecules, origin of life and collision of, 454
Origin of Life, The (Oparin), 404
Orion (constellation), 91, 109, 115
Orion arm of Milky Way, 109
Orionids (meteor showers), 67
Oscillating universe, 451-52
  described, 143
  duration of cycles of, 145
Ovid, 106
Oxygen
  discovered, 456
  molecular structure of, 282
Ozonosphere, 184

Pacific Ocean, 176-79, 181, 215
Page, Thornton, 131
Paleocene epoch, 241-42
Paleozoic era (Age of Ancient Life), fossil record of, 230-36, 418
Pallas (asteroid), 60
Pangaea (supercontinent), 179, 180
Parallax
  of moon, measuring, 19
  stellar, 87-89
Paricutín Volcano (Mexico), birth of, 218, 220, 223
Parity, repeal of, in particles, 342-43

Parkfield-Cholane earthquake (Calif.; 1966), 216
Parsec, defined, 89
Pasteur, Louis, 402-3, 408-11, 457
Pasteur's flask, 402-3
Pauli, Wolfgang, 307, 336
Pauli's exclusion principle, 334
Pauling, Linus, 374
Peebles, P. J. E., 145
Pekin Man (Sinanthropus), 434-35
Pelée, Mt., eruption of (1902), 220
Pennsylvanian period of Paleozoic era, 235, 236, 418
Penzias, Arno A., 145
Peroidic table of elements, 257-58
Peripheral nervous system, 441-42
Permian period, 235-36, 244
Perseids (meteor showers), 67
Perseus (constellation), 95
Pettit, Edison, 39
Philosophiae naturalis principia mathematica (The Mathematical Principles of Natural Philosophy; Newton), 12, 346, 354, 356
Phobos (moon of Mars), 49
Phosphorus, 454
Photoelectric theory, 304
Photons, energy of, 333
Photosynthesis, 371-72
Phyla of animals, 421-25
Physics, *see* Elementary particle physics
Physiology, beginnings of, 365
Pi mesons, 337
Piazzi, Guiseppe, 59
Pickering, Edward C., 94
Pickering, William H., 56
Pitchblende, 264
Pithecanthropus (Java Man), 433, 434
Plages of chromosphere, 80
Planck, Max, 303-4
Planck's constant, 303, 308-10, 313, 333
Planetary motion
  composite, 5-6
  Kepler's laws of, 11, 456
    third law, 10, 94, 161
  Mercury, 356
  around sun, 38
  theories of, 3-7
Planetary nebulae, 115-16
Planetesimal hypothesis of earth formation, 157-58
Planets, *see specific planets; for example*: Earth; Mercury; Venus
Plants, 416-20
Plato, 255, 361

Playfair, John, 153
Pleistocene epoch (Great Ice Age), 237, 241, 245-47
  first evidence of man in, 433
  First Interglacial Period during, 434, 435
  Third Interglacial Period during, 435, 437
Pliny the Elder, 362, 440
Pliocene epoch, 241
Plücker, Julius, 261
Pluto
  characteristics of, 37, 56-57
  discovered, 38
  distance of, from sun, 12
Poincaré, Henri, 349
Polar cap of Mars, 46-47
Polaris (North Star; Thuban), 97
Polido, Donisio, 218-20
Pons of brain, 446, 447
Pope, Alexander, 377
Population I stars, 110, 118-27
Population II stars, 110, 118-19, 126-27
Pores of sunspots, 78
Porifera, 421
Positive ray tubes, 274-75
Positrons, 334-35
Post-Glacial Time (Recent Time), 433
Potassium, structure of, 273
Potential energy, 293-94
Powell, Cecil F., 337
Power, defined, 293; *see also* Energy
Prairie dogs, 423
Priestley, Joseph, 456
Primary waves (P-waves), 165, 166
Primates, 425-32
  apes, 430-32
  lemurs and tarsiers, 427-29
  monkeys, 429-30
  See also Man
Primeval fireball (primeval atom), 142
Primitive earth, atmosphere of, 404
Procyan (star), 109
Prominences, solar, 81-83
Proper motion of stars, 89
Proteins
  chemical composition of, 373-75
  production of, 387-88
  synthesis, 387
Protons, 274-76
Protoplanet hypothesis of earth's formation, 161-62
Protostars, defined, 115
Protozoa, 420
Proust, Joseph, 256

INDEX

Prout, William, 257
Ptolemaic system, 6, 8
Ptolemy, Claudius, 6
Pulsars, 104-5
Purkinge, Jan, 366
P-waves (primary waves), 165, 166
Pythagoras, 252

Quantum theory, 302-13
  Bohr theory of the atom and, 305-8
  Einstein and, 303-4
  Planck and, 303-4
  wave mechanics and, 310-13
  waves vs. particles and, 308-10
Quarks, defined, 344
Quasars (quasi-stellar radio sources), 138-41
Quiescent prominences, 81-82

Radial motion of stars, 89-90
Radiation
  cosmic radio, 145
  detectors of, 318-19
  electromagnetic, 333
  as form of energy, 298-301
Radio communications, solar flares affecting, 82-83
Radio galaxies, 131-34, 141, 144
Radio noises from Jupiter, 51
Radio radiation, cosmic, 145
Radioactivity, 263-65
Radiolara, 184
Radium, discovered, 264
Rain, as eroding force, 188, 190
Ranger (U.S. spacecraft), 27, 33
Ranger 7 (U.S. spacecraft), 25
Raphael, 364
Rate of expension of universe, 145
Rays
  alpha, 269
  beta, 269
  cathode, 261, 262
  cosmic, 338-40, 397
  gamma, 269, 300
  infrared, 300-1
  of moon, 30-33
  x-rays, 259-63, 300, 457
Recent Time (Post-Glacial Time), 433
Red shifts
  Hubble's Law of, 135, 136
  of quasars, 138, 140-41
Redi, Francesco, 402
Reflex arc of nervous system, 445-46
Reines, Frederick, 336, 340

Relativity, Theory of, 141, 302, 304, 310, 319, 345-56, 457
  equivalence of gravitation and inertia and, 353-55
  equivalence of mass and energy and, 319-20, 324, 335, 350-52
  experimental confirmation of, 355-56
  General Theory of, 141, 353-55
  Michelson-Morley experiment and, 346-48
  Special Theory of, 348-52
Replication of DNA, 389-90
Reptiles, Age of (Mesozoic era), fossil record of, 230, 236-40, 426
Respiration, function of, 375-76
Rest-mass of particles, defined, 333
Retreat of the Sea, The (Omar), 150
Ribonucleic acid (RNA), 380, 383
Riccioli, Giovanni Battista, 93
Richards, Theodore W., 276
Richter, Charles F., 213
Richter magnitude, defined, 213
Ridges, ocean, 175-77
Riemann, Bernard, 136, 354
Rigel (star), 109
Rills of moon's maria, 26
Ring-mountain craters, 28
Ring Nebula (nebula), 116
Rings of Saturn, 52, 53, 157
Ritz, W., 305
Rivers, functions of, 190-94
RNA (ribonucleic acid), 380, 383
m-RNA, 387, 388
t-RNA, 387, 388
Robertson, Howard P. 138, 143
Rocks
  basaltic, 34
  composing earth's crust, 172
  continental, 173-74
  erosion of, 188
  formation of, 206-9
  granitic, 172-74
  igneous, 172-74
  lunar, 34
  metamorphic, 172-74
  sedimentary, 173-74, 207-8
Rocky Mountains, 222-25, 238
Roentgen, Wilhelm Conrad, 259-60, 262, 263, 457
RR Lyrae (cluster-type Cepheids), 102, 126, 127
Rumford, Count, 295
Russel, Henry Norris, 91
Rutherford, Ernest Lord, 155, 268-72, 280, 457

atom-smashing experiment of, 270-72, 274, 275, 305-6, 314-15
neutron discovery and, 320
Rutherford atom, 270-72, 457
Rydberg, Johannes, 305
Rydberg's constant, 305
Ryle, Martin, 105

Sagittarius (constellation), 106
St. John, Charles E., 141
San Andreas fault (Calif.), 180, 209, 211, 215, 216
San Francisco earthquake (1906), 214, 215
Sandage, Alan, 133, 138, 141, 145
Sarpi, Fra Paolo, 290
Saturn, 66
  characteristics of, 52-53
  chemical elements in, 160
  in protoplanet hypothesis, 161, 162
  rings of, 53
Sceptical Chymist, The (Boyle), 253, 456
Schiaparelli, Giovanni, 48
Schleiden, Matthias Jakob, 366, 368
Schmidt, Maarten, 138
Schrödinger, Erwin, 311, 312
Schwann, Theodor, 366
Schwarzchild, Karl, 141
Schwarzchild singularity (black hole), 141
Science, evolution of, 455-58
Scopes, John T., 400
Scorpio (constellation), 91
Seaborg, Glenn T., 329
Searchlight and shutter effect of quasars, 140
Second Glacial Period, 434
Secondary waves (S-waves), 165, 166
Sedimentary rocks (stratified rocks), 173-74, 207-8
Sediments
  of rivers, 193
  rock formation from, 206-8
Seismic waves in earth's interior, 166-67
Seneca, 259
Sensory neurons (afferent neurons), 443-44
Sequence of galaxies, 130
Serenitatus, Mare (Sea of Serenity), 26
Seven zones of stars, 104
Shakespeare, William, 58, 314
Shapley, Harlow, 107, 110, 111, 130
Short-period comets, 63

Showers
  meteor, 67
  of stars, 117
Silurian period, 234-35
Sinanthropus (Pekin Man), 434-35
Sinkholes, defined, 200
Sirius (Dog Star), 89, 91, 94, 109
Sitter, Wilhelm de, 137
Size
  of atoms, 278-79
  of galaxies, 123
  of Great Red Spot, 50
  of Mauna Loa, 178
  of meteorites, 68
  of micrometeorites, 70
  of molecules, 278-79
  of Neptune, 55
  of Pluto, 57
  of Saturn, 52
  of stars, luminosity and, 92-93
  of Venus, 41
Skulls
  of man-apes, 432
  of Pithecanthropus, 434
Slipher, Vesto Melvin, 134-35
Small Magellanic Cloud (galaxy), 97
Smith, Adam, 143
Smith, William "Strata Smith," 153
Soap, molecular structure of, 287
Soddy, Frederick, 269, 276
Sodium, structure of, 273
Sodium chloride, molecular formation of, 281-82, 288
Soil, lunar, 34
Solar apex, defined, 90
Solar disk, 159, 160
Solar energy, intensity of, 72
Solar flares, 82-83
Solar system, 13-14, 37-38, 156-62
Solids in molecular formation, 288-89
Sommerfield, Arnold, 306
Sound waves, energy and, 297-98
South America, origins of, 179
Space Age, dawn of, 33
Spacecrafts, 33-35
Space-time continuum, 352-53
Spallanzani, Lazaro, 402
Special Theory of Relativity, 348-50, 352
Spectroscope of sun, 72-75
Speed
  of light, 300
  of stars in motion, 90
Spencer, Herbert, 400
Spicules of chromosphere, 80

Spiral arms of Milky Way, 108-9
Spiral galaxies, 123-24, 128, 130
Spiral track of electrons, 332
Spirilla, 411
Spontaneous generation, 401-3, 404
Stanley, Wendell, 414
Stars, 86-105
  birth and death of, 100-4
  Cepheids, 96-97, 102, 123, 126, 127
  distance between and motions of, 87-90
  double, 93-97
  dwarf, 92
  exploding, 97-100
  Hertzsprung-Russell diagram of, 91-93, 99, 100, 102, 103, 112
  giant, 92
  in Milky Way, 109, 453
    disk population, 110, 118-19
    globular clusters of, 111-12
  pulsars and, 104-5
  supergiant, 93
  variable, 96-97
Steady-state theory (continuous-creation theory) of formation of universe, 143-44
Stellar parallax, determining, 87-89
Stevenson, E. C., 337
Stone tools, European, 435
Stoney, G. Johnstone, 265
Stony meteorites, 68
Stony-iron meteorites, 68
Strassman, Fritz, 329
Strata Identified by Organized Fossils (Smith), 153
Stratified rocks (sedimentary rocks), 173-74, 207-8
Stratosphere, 183-84
Street, J. C., 337
Streptococci, 411
Sturtevant, A. H., 386
Struve, Friedrich, 87
Sulphur, 454
Sun (star), 71-85
  age of, 158-60
  atmosphere of, 75, 78-80, 453
  distance from
    to earth, 12, 44, 72
    to center of Milky Way, 107
    to Jupiter, 49
    to Mars, 145
    to Mercury, 39
    to Neptune, 55
    to Pluto, 12
    to Saturn, 52
    to Uranus, 54
    to Venus, 41
  eclipses of, 36
  evolution of, 452-53
  motion of, 15
  motion of stars and, 90
  prominences of, 80-83
  role of, in photosynthesis, 371-72
  spectrum of, 72-75
  spots on, 77-78, 84
  surface of, 75-76
  winds of, 84-85
Sunpots, 77-78, 84
Superexplosions, radio galaxies and, 131-34
Supernovae, 99, 117, 139-40
Surface
  of earth, 163
    faults of, 180, 209-11, 215, 216
  lunar, 17, 24-26, 35
  of Mars, 453
  solar, 75-76
  of Venus, 453
Surveyor 1 (U.S. spacecraft), 33
Surveyor 3 (U.S. spacecraft), 33-34
Survival of the fittest, 400
S-waves (secondary waves), 165, 166
Swedenborg, Emanuel, 156, 157
Synodic month, defined, 21

Tails of comets, 63
Tarsiers, 427-29
Tatum, Edward, 386
Taurids (meteor showers), 67
Taylor, Frank B., 197
Telescopes
  invented, 10
  giant, 2, 3, 138
  radio, 132
Temperature
  of earth, 164
  of exploding primeval fireball, 142
  of Jupiter, 51
  of lunar surface, 35
  of Mars, 47-48
  of Mercury, 39
  of Neptune, 55
  of Saturn, 52
  of solar corona, 84
  of stars, 91
  of stratosphere, 184
  of sun, 72
  of troposphere, 183
  of Venus, 41-42

INDEX

Temperature changes as eroding forces, 188
Tennyson, Alfred Lord, 155, 157, 426
Terrestrial planets, *see* Earth; Mars; Mercury; Venus
T4 viruses, 413, 415
Thalamus, 447-48
Thales of Miletus, 251
Thallophytes, 417
Theophrastus, 362
Theory of the Earth, with Proofs and Illustrations (Hutton), 152, 153, 457
Thermodynamics, first and second law of, 296-97
Third Interglacial Period, 435, 437
Thomson, G. P., 312
Thomson, Sir Joseph John, 266-67, 271, 274, 276-77, 312
Thuban (star), 44
Tidal theory of earth formation, 158
Titan (moon of Saturn), 53
Titian, 364
Titius, Johann, 58, 59
Titius-Bode series, 59
Tokyo earthquake (Japan; 1923), 215
Tolman, Richard C., 143
Tombaugh, Clyde W., 48, 56
Tools
  of Neanderthal Man, 439
  of Pleistocene period, 433
  stone, 435
Tracheophytes, 417, 418
Tranquilitatis, Mare (Sea of Tranquility), 26, 34
Transit of planets, defined, 40
Trapezium (star), 115
Treatise on Electricity and Magnetism (Maxwell), 300
Triassic period of Mesozoic era, 236-37
Trifid Nebula (nebula), 118
Trilobite fossils, 231-32
Tropical belt of Jupiter, 49
Troposphere, 183
Trumpler, Robert, 110
Tschermak, Carl, 394
T Tauri stars, 115
Two meson theory, 337
Two New Sciences (Galileo), 456
Tycho Crater (lunar crater), 30
Tycho's Nova (star), 99-100
Tyndall, John, 407

Ultraviolet light, 141, 300-1
Umbra of sunspots, 77

Underground caves, 199-201
Universe, 121-46
  big-bang theory of, 142-46
  curved space of, 136-38
  expanding, 134-36
  galaxies in, 123-34
    evolution of, 129-31
    light-years separating, 15
    Local Group of, 125-27
    Maffei, 128-29
    radio, 131-34, 141, 144
    types of, 123-24
    See also Milky Way
  model of, 13-15
  origin and nature of, 141-46
  oscillating, 145
  scale of, 3-13
  steady-state, 143
  unity of, 451-58
  See also specific stars and planets
Unmanned spacecraft, 33-35
Uranium, fission of, 329-30, 457
Uranus, 53-57, 156
  characteristics of, 53-55
  chemical elements of, 161
  discovered, 12, 57, 457
  in protoplanet hypothesis, 161, 162
Urey, Harold C., 404
Ursa Minor (constellation), 44
U-shaped glaciated valleys, 199
Ussher, James, 155

Valley glaciers (Alpine glaciers), 203
Van Allen, James A., 185
Van Allen belt, 185
Van der Waals forces, 285
Vanguard 1 (U.S. spacecraft), 163
Variable stars, 96-97
Vaucouleurs, Gerard de, 123
Vega (star), 91
Veneg-Meinez, F. A., 168
Venus, 37, 53, 66, 156
  atmosphere of, 41-42
  characteristics of, 40-42
  Mars and, 45
  in protoplanet hypothesis, 161
  surface of, 453
Vertebrates, 423-25
Vesalius, Andreas, 363-65, 456
Vesta (asteroid), 60
Vesuvius, Mount (Italy), 30
  eruption of (A.D. 79), 22
Viruses, characteristics of, 413-15
Visible light waves, 300
Visual binaries (double stars), 93-95

Volcanic islands, 178
Volcanic sand, 170
Volcanoes, 215-22
  age of, 218, 220
  cones of, 218
  location of active, 215-16
Volume
  of earth, 163
  of sun, 71
Vries, Hugo De, 394, 396
V-shaped glaciated valleys, 200
Vulcanists, views held by, 152-53

Waals, Johannes van der, 285
Wallace, Alfred Russel, 400
Walled plains (bulwark plains), 27
Walton, E. T. S., 320
Water
  molecular formation of, 286-87
  underground, 199-201
Water vapor in earth's atmosphere, 183
Waterfalls, 195-97
Watson, James D., 381
Watson, William, 260
Wavelengths, 301
Waves
  brain, 450
  as eroding factor, 204
  heat, 300-1
  long, 165
  mechanics of, 310-13
  particles vs., 308-10
  primary and secondary, 165, 166
  seismic, 166-67
  sound, 297-98
Weathering, forces of, 188-90
Wegener, Alfred L., 179
Weight
  of brain, 446

  of primeval fireball, 142
  of protons, 275
Weizsäcker, Carl F. von, 159
Werner, Abraham Gottlob, 152, 153
Whipple, Fred L., 48, 62, 160, 162
Wildebeests, 424
Wilkins, M. H. F., 381
Wilkinson, David T., 145
Wilson, Charles T. R., 308-9, 318
Wilson, Robert W., 145
Wilson cloud chamber, 318-19, 321, 335
Winds
  as eroding force, 188
  solar, 84-85
Wöhler, Friedrich, 369
Wolf 359 (star), 109
Woodward, John, 151
Work, defined, 293
Wright, Thomas, 107, 121
Wu Chien-shiung, 343

X chromosomes, 395-96
X-rays, 259-63, 300, 457

Y chromosomes, 396-97
Yang, C. N., 343
Yangtze River (China), 191
Yentna Glacier (Alaska), 188
Yellowstone National Park, 191, 195, 197, 200
Yokohama earthquake (Japan; 1923), 215
Yosemite Falls (Calif.), 197
Young, Thomas, 299
Yukawa, Hideki, 337

Zach, Franz von, 59
Zeeman, Peter, 78, 300
Zeeman effect, 78
Zodiacal light, 70